STEM CELLS

From Mechanisms to Technologies

STEM CELLS

From Mechanisms to Technologies

Michal K Stachowiak
Emmanuel S Tzanakakis
State University of New York at Buffalo, USA
editors

World Scientific

NEW JERSEY • LONDON • SINGAPORE • BEIJING • SHANGHAI • HONG KONG • TAIPEI • CHENNAI

Published by

World Scientific Publishing Co. Pte. Ltd.
5 Toh Tuck Link, Singapore 596224
USA office: 27 Warren Street, Suite 401-402, Hackensack, NJ 07601
UK office: 57 Shelton Street, Covent Garden, London WC2H 9HE

British Library Cataloguing-in-Publication Data
A catalogue record for this book is available from the British Library.

STEM CELLS
From Mechanisms to Technologies

ISBN-13 978-981-4317-70-2
ISBN-10 981-4317-70-5

Typeset by Stallion Press
Email: enquiries@stallionpress.com

Printed in Singapore by Mainland Press Pte Ltd.

To Ewa, Karina and Rafal and to the memory of my parents and brother.
Michal K. Stachowiak
To my parents and to Persefoni, Stylianos and Rhea.
Emmanuel S. Tzanakakis

Preface

Stem cells have an astounding capability to self-renew or differentiate under a plethora of seemingly chaotic external and internal inductive signals which may lead to contradicting fate decisions. The current popular view relies on the identification of isolated signals or combinations based on the premise that by varying the signaling milieu, one can effectively control the development of stem cells and enable their medical applications.

The phenomenology of stem cell fate decisions for self-renewal, differentiation, as well as the reprogramming of committed cells into pluripotent cells, results from a complex interplay of a multitude of regulatory elements on the molecular level. One way to formally describe these complex interactions is the application of network models which represent the relevant molecular components and map their interactions (Chapters 1 and 2). Studying the dynamic properties of such networks at different levels will allow for a better systemic understanding of the regulatory principles underlying stem cell fate decisions.

In this book, a perspective is presented from a systems level of the mechanisms regulating stem cell development. Information resulting from the application of systems biology concepts can be incorporated in developing technologies for the realization of stem cell-based applications. The recurrent network motifs of information processing could account for and integrate the "chaotic" signals and thus underwrite the biological "fits" of stem cells. We believe that the network perspective, which has been under-represented in the discussion of stem cell biology and applications, is beginning to attract attention from the research community.

One of the apparent merits of network motifs is prediction. The discovery of the pluripotency gene network suggested that over-expression of the few "pluripotent" genes linked via positive feedback loops may suffice to generate the pluripotent state in differentiated somatic cells. Testing this prediction has led to the development of a new successful technology for the generation of induced pluripotent stem cells (iPSCs) (Chapter 3). The generation of human iPSCs using viral and non-viral approaches, characterization of their pluripotent capabilities and development of culture conditions for their stable propagation are addressed in Chapter 5.

Feed-forward loops are frequently found motifs in biological networks (Chapter 1). Their functions may vary from pulse generators and gating systems to response delaying circuits. The novel integrative nuclear FGFR1 signaling (INFS) mechanism discovered in the recent years constitutes a nexus of a feed-forward-and-gate module which effectively regulates developmental gene activities (Chapter 4). Direct activation of this module via over-expression of the INFS genes should suffice to convert the proliferating stem/progenitor-like cells into their differentiated progeny, thus bypassing the daunting need to identify proper combinations of the many external and internal factors needed for such a transition. This has been shown to work not only *in vitro* in cultured cells (Chapter 4), but recently also *in vivo* whereby the transfer of the INFS genes reinstates an active neuronogenic development in stem cells in the mature brain (Chapter 10).

An important feature of developmental differentiation is the "permanent" — often irreversible — status of the achieved state. Such a state can result from positive feedback regulation whereby the expression of the central genes in the module is activated by their products (Chapter 1). These positive feedback loops may lead to a bi-stable "low" or a "high" signal as discussed in Chapters 1 and 4. Once initiated by external factors, the positive feedback loop that underlies the high activated state remains switched on even if the original activators are no longer present.

We have started untangling the effects of critical signals on stem cell biology only recently. A deeper understanding of these effects will be important for the development of technologies, enabling the testing of diverse developmental networks already implicated by systems biology as well as the development of new network models. Chapter 6 discusses the significance of physicochemical parameters such as oxygen tension, pH and biomechanical forces for the *ex vivo* expansion and differentiation of hematopoietic stem cells. Chapter 7 focuses on recent advances and challenges associated with the scalable culture of embryonic and induced pluripotent stem cells. Enabling bioreactor technologies and associated bioprocesses will also be critical for the realization of the clinical potential of stem cells. Given the important role of the stem cell milieu in guiding their commitment along specific lineages, intense research efforts have concentrated on appropriately designed biomaterials for the engineering of stem cell-based tissue equivalents. The engineering of bioactive scaffolds for vascular tissue constructs utilizing stem cells is presented in Chapter 8. Special microfluidic platforms can be used to further identify the critical signals controlling developmental networks in stem cells. Aspects of the stem cell niche can be recreated in microfluidic devices to obtain a better understanding of the interactions among stem cells and with their environment as described in Chapter 9.

Besides embryonic stem cells and iPSCs, alternative sources of stem/progenitor cells for regenerative medicine are also explored. One such promising source of stem cells is the human umbilical cord blood. Its multi-potent fetal stem-like cells can differentiate into diverse cell lineages including neural cells. Progress in elucidating the mechanisms, control networks and technologies for their neural differentiation and application in preclinical and experimental medical treatment are discussed in Chapter 10.

An understanding of the mechanisms that control the biology of adult stem cells could potentially be used to recruit endogenously produced stem cells for regenerative medicine. Nanotechnology has provided new therapeutic tools such as non-viral vectors for DNA delivery. DNA nanoplex targeting of developmental networks in brain stem/progenitor cells *in vivo* can reinstate the generation of new neurons by an adult brain and thus open a new horizon for the treatment of neurological disorders (Chapter 11).

As discussed in Chapter 12, an important development has been the identification of cancer stem cells, which may aid in improving disease diagnosis. We are just beginning to unravel the mechanisms underlying an aberrant control of the developmental networks in those cells. Successful corrective targeting the deregulated pathways should enable future advances in cancer treatments.

In summary, this book has two interconnected themes: (1) systems approach towards understanding the biology of stem cells and targeting network mechanisms for effective manipulation of stem cell biology and commitment, and (2) the development of technologies through the application of engineering and biological tools for advancing our ability to control aspects of stem cell biology critical for the realization of the clinical potential of stem/progenitor cells.

Michal K. Stachowiak and Emmanuel S. Tzanakakis

Contents

Foreword

Stem cells have traditionally served a fundamental role in tissue engineering and regenerative medicine approaches to address clinical challenges. As the understanding of stem cell biology continues to expand, the leverage of stem cells in tissue engineering and regenerative medicine presents ever-increasing potential to make a positive and lasting impact on patient care. However, a fundamental understanding of emerging stem cell technologies and methods is necessary to enable the fullest realization of the potential of stem cells in tissue engineering and regenerative medicine to advance clinical care. Michal Stachowiak and Emmanuel Tzanakakis clearly identified the critical need for a comprehensive overview of the cutting-edge, multidisciplinary technologies available to enable practical application of stem cells in medicine, which they masterfully address through this book.

The book comprises twelve chapters contributed from leading international authorities in stem cell biology and biotechnology, which Stachowiak and Tzanakakis skillfully weave into an insightful play on two interrelated themes. One such theme involves the application of systematic, computational methods to elucidate the complexities of stem cell biology and to enable the rational identification of targets that may be harnessed for the manipulation of stem cells. The second theme focuses on the leverage of biological sciences and engineering principles toward the development of technologies and methods to enable the control of critical aspects of stem cell biology and facilitate the feasibility of their use in the clinic. Although each theme involves different approaches to ultimately direct stem cells for clinical application, the book imparts the reader with a strong appreciation of the power of computational methods and biological tools, as well as the tremendous synergies between them.

Chapters 1 and 2 review the application of network models to describe the complex milieu of molecular regulators of stem cell fate and to map the interactions between the regulators. The application of such network models will allow for continued advances in the understanding of the complex mechanisms that influence stem cell fate and flexibility. Chapter 3 describes the application of global functional profiling of embryonic stem cells by comparative transcriptomics to identify potential molecular mechanisms involved in the pluripotency of these cells, which may translate to other types of stem and progenitor cells. Chapter 4 involves discussion of a novel signaling mechanism that may be manipulated to directly activate a feed-forward-and-gate module to provide a focused mechanism for differentiation of stem and progenitor cells. Chapter 5 provides an overview of human induced pluripotent stem cells, including methods for their generation, characterization and stable propagation. Chapter 6 focuses on methods and issues associated with the *ex vivo* expansion of hematopoietic stem and progenitor cells, while Chapter 7 outlines technologies and challenges associated with the expansion and processing of stem cells for regenerative medicine applications. The importance of the interactions between stem cells and scaffolds is highlighted in Chapter 8, which describes the engineering of biologically active scaffolds to support stem cell populations for the generation of vascular tissue constructs. Chapter 9 also underscores the significance of stem cell-material interactions in presenting an overview of technologies to leverage stem cells and biofunctionalized surfaces in bio-testing and tissue engineering applications. The discussion in Chapter 10 shifts to human umbilical cord blood-derived stem and progenitor cells, with a particular focus on the application of these cells in experimental therapies for neurological disorders. Chapter 11 illustrates the potential of nanotechnology-based tools to direct stem cell behavior through the example of DNA nanoplexes for the directed neurogenic differentiation of brain stem cells. Chapter 12 discusses the potential implications of recent advances in the understanding of cancer stem cell biology on the development of targeted therapies for cancer.

These twelve chapters complement each other in providing a multi-disciplinary, comprehensive overview of recent advances in stem cell technology that will pave the way for large-scale translation of stem cell-based therapies to the clinic. The material is highly accessible and presents tremendous value to readers from a wide range of backgrounds. The book stands as an outstanding resource for anyone interested in stem cells and their potential for clinical application.

Antonios G. Mikos, Ph.D.
Louis Calder Professor of Bioengineering and
Chemical and Biomolecular Engineering
Rice University

"I believe in the dreams of the future more than the history of the past"
— Thomas Jefferson

Part I

System Biology Lessons for Stem Cells

Computational Modeling of Biochemical Processes and Cell Differentiation

1

Max Floettmann, Till Scharp, and Edda Klipp

We introduce various modeling approaches that can be used to understand structure, dynamics and function of regulatory cellular networks. These techniques comprise network representation with graphs, Boolean networks, piecewise linear differential equations, ordinary differential equations and stochastic modeling. Mathematical modeling has been successfully applied to elucidate different biological processes and properties. Here, we demonstrate how computational techniques can be exploited in the analysis of stem cell differentiation and cellular reprogramming.

1.1. Introduction

Computational approaches and mathematical modeling have become increasingly important for biological sciences and also find more and more applications in stem cell research and regenerative biology. There are several developments supporting this trend: (i) the insight becomes increasingly accepted that complex biological processes cannot be understood solely by intuition or by studying individual genes, proteins or even large groups of genes or proteins at one moment. Biological objects are more commonly understood as systems with properties arising from the interplay of many components in space and time; (ii) biology has, to some extent, become an information-driven science due to the progress in genomics, characterized by rapidly developing sequencing and expression analysis technologies, and due to advances in proteomics and functional studies. This is accompanied by the establishment of large databases, bioinformatic algorithms, and data analysis techniques, (iii) computer power still increases exponentially, computational algorithms are massively developed and laboratories all over the world are quickly connected and exchange information via the internet; all this makes computational approaches faster and more interactive.

In this chapter we give a short and comprehensive introduction to well-established mathematical modeling approaches, and show the advantages and the beauty of building coherent mechanistic models from biological data. We present practical problems from the field of stem cell biology as an example to apply the basic concepts and different approaches of computational dynamical modeling. Since gene regulatory networks and signaling networks are considered important in this context, we will focus on those here. For a deeper understanding of the presented concepts of mathematical modeling, the interested reader should refer to more general textbooks on systems biology (Klipp, 2005).

Many stem cell researchers are still reluctant in adopting and taking advantage of computational modeling techniques. This is partly due to uncertainties about the system and a lack of sufficient data for dynamical models of regulatory processes contributing to stem cell development.

Traditionally, modeling often dealt with explaining principles such as the role of negative feedback in regulation of gene expression or was restricted to areas of biology that were better understood and more comprehensively described, such as metabolism, or to model organisms such as yeast or bacteria where experimental data is abundant compared to mammalian systems. In recent years, more data have become available due to the widespread use of high-throughput techniques (that have accumulated a wealth of various types of information). The field has reached a state where it can benefit greatly from systems biology approaches that help to understand the "unbelievably" complex system is stem cell development and differentiation. This change becomes apparent in the growing number of publications of stem cell biology models (Chavez *et al.*, 2009; Chickarmane and Peterson, 2008; Chickarmane *et al.*, 2006; Hanna *et al.*, 2009; Kalmar *et al.*, 2009). Although most of these models are still small and coarse grained, and sometimes restricted to structural aspects, they nevertheless provide us with an insight into important mechanisms involved in the regulation process (MacArthur *et al.*, 2009; MacArthur *et al.*, 2008).

The advantage of modeling is not always clear on first sight, but there are a number of ways in which it can improve biological research. The most apparent merit of mathematical models is prediction. This can mean proposing a mechanism inside the system that explains its behavior or predicting the system's behavior under altered conditions. These predictions can help to speed up research and facilitate efficient experimental design, but they only make sense if they can be tested experimentally.

Modeling creates a dialog between the experimentalist and the theoretician. Ideas and concepts must be explained. This way, modeling forces researchers to formulate verbal hypotheses in a precise, specific, and conceptually rigorous manner. Building models draws attention to gaps in current knowledge, possibly enforcing new experiments and may highlight quantitative aspects of the problem under study. Modeling can also result in improved experimental design; on the other hand, it may allow for rapid, inexpensive testing of ideas, hypothesis and possible interventions.

Although there is already an enormous amount of data available, the level of complexity limits the choice of modeling approaches that can currently be used in stem cell biology in practice. There is always a trade-off between the complexity of a model and the amount and type of data that is needed to show that the output of a model is coherent with reality to a certain degree. It makes no sense to build a detailed model of each and every reaction in a system if we do not have experimental data that can possibly falsify our model. In an optimal scenario, model building and data retrieval go hand in hand. That is why we will put the focus on qualitative modeling techniques such as Boolean models that can be constructed with limited knowledge about the system of interest. As they are the most common (Klipp *et al.*, 2007) and also one of the most precise ways to model a system, we will also introduce quantitative approaches in the form of ordinary differential equation systems.

Different scenarios in biological research with respect to data availability and scientific questions favor different modeling frameworks, but sometimes it is hard to decide how to model a system and what questions the model can answer. In the following sections, we will introduce different modeling techniques and discuss the applicability to a specific situation and what has to be taken into account in the process.

1.1.1. *Building a network structure*

Most examples of mathematical modeling in stem cells deal with gene regulation in the differentiation process of stem cells or the reprogramming of somatic cells to induced pluripotent stem cells (iPSCs). These studies utilize a number of different modeling approaches, from big structural models (Kim *et al.*, 2008) to small "quantitative" models (Chickarmane and Peterson, 2008). The majority of these models use a top down approach, meaning they create a model by the analysis of large scale studies of high-throughput data. The first step in these studies is the generation of a structural network, e.g. a protein interaction network, a gene regulatory network comprising transcription factors and target genes, or a combination of both. Networks have two types of components: a list of elements or nodes, i.e. anything from small molecules to proteins, and a set of interactions between these nodes.

A large number of biological methods are available to generate node lists. With gene regulatory networks (GRNs), the list of nodes often consists of differentially expressed genes that have been identified in high-throughput expression profiles taken during differentiation or reprogramming. Common approaches to find physical interactions between these parts are, e.g. mass spectrometry for protein–protein interactions and chromatin immunoprecipitation (ChIP) for transcription factor binding. Several studies have used these techniques to establish gene regulatory and protein interaction networks for stem cells in the last few years (Boyer *et al.*, 2005; Chew *et al.*, 2005; Zhou *et al.*, 2007). All resulting networks are centered on the core embryonic stem (ES) cell transcription factors. Most parts of these networks are relatively well established, although some studies show a poor overlap in the presented networks, thus causing some uncertainty about the structure of these networks (Kopp *et al.*, 2008; Pan *et al.*, 2006).

In addition to these transcription factor networks, there are other regulatory layers that play a role in the complex regulation of stem cell development. There have also been studies on the role of various miRNAs (Singh *et al.*, 2008; Xu *et al.*, 2009), epigenetic regulation (Arney and Fisher, 2004; Bernstein *et al.*, 2006; Loh *et al.*, 2007) and signaling pathways, and on their interplay with these networks (Guo and Wang, 2009; Valdimarsdottir and Mummery, 2005). Despite some obstacles and uncertainties, together these studies pave the ground for a systematic understanding of the control of stem cell fate. The results of these studies are a rich source of data that was already used by following computational studies that refined the results and combined them with other sources and databases (Chavez *et al.*, 2009). This fast progress gives modelers a structural foundation for various types of dynamical models that we will present in the following.

1.1.2. *The core network in regulation of pluripotency*

Throughout this chapter, we will use the gene regulatory network that governs differentiation in mammalian cells as an example. The core of the network consists of the genes POU5F1 (or OCT4), SOX2 and NANOG. These genes all encode highly connected transcription factors and are therefore considered as hubs. They are the best studied players in this network and are increasingly becoming the focus of research since it was found that the over-expression of OCT4 and SOX2 — together with KLF4 and MYC — can be used to reprogram somatic cells to iPSCs (Takahashi and Yamanaka, 2006). Although there is a lot of research focusing on this network, there are still many connections that are poorly understood or insufficiently proven (Niwa, 2007).

The network we chose as an example is described below by a toy model. We tried to stick as close as possible to the known facts, presenting all the necessary motifs and effects, while at the same time simplifying the network to make it useful as an example. Here, we proceed as follows: The three genes, OCT4, SOX2, and NANOG, are tightly regulated by one another and also show auto-regulation. Since the network that governs their interactions is well known, it forms the core of our example. First, we set up a static network of gene regulation based on information from literature, as shown in Figure 1.1. Second, as an introduction to dynamic models, we will further simplify this network to end up with only two nodes involved in a positive feedback loop, as presented in Figure 1.2. This is justified by the fact that two transcription factors, Oct4 and Sox2, have the same inputs and they only have an effect on NANOG when both are present. This feature of the network enables us to combine them to one node that we call OS. Third, since there are a

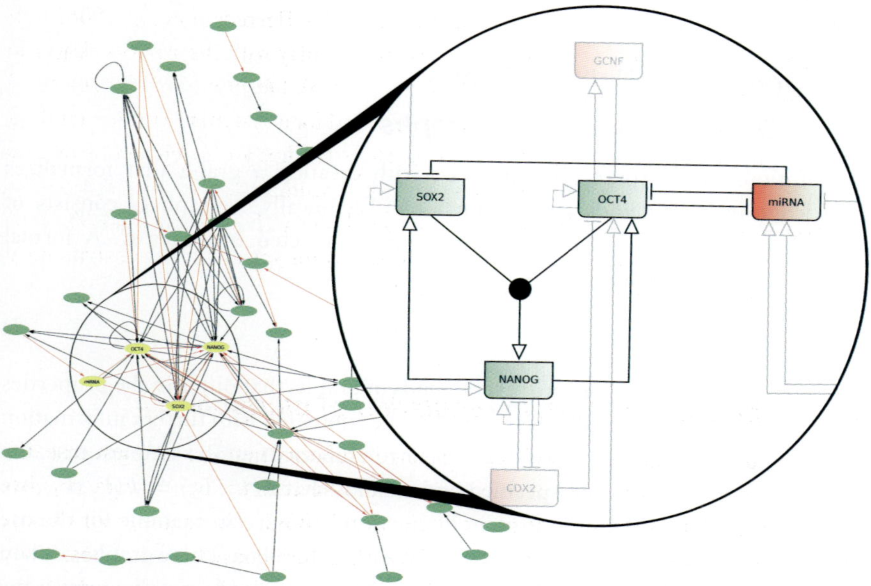

Figure 1.1. A close look at the complex gene regulatory network we used as an example in Chapter 1. Faded components indicate that we leave out many known connections to isolate the main part. The included species have a very high connectivity compared to the rest of the network. NANOG, OCT4 and SOX2 share 353 target genes altogether and each one of them regulates a large number of genes exclusively or in combination with other factors (Boyer *et al.*, 2005).

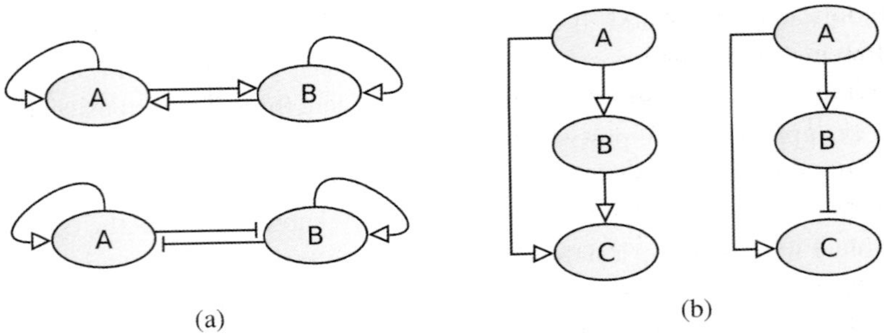

Figure 1.2. Important network motifs in developmental networks. (a) Two types of two-node positive feedback loops with additional positive auto-regulation. These motifs exhibit a distinct bi-stability that is often needed in regulatory networks. (b–c) Feed-forward loops that are often found in biological networks in general. They can have various functions such as response delay or pulse generators; (b) positive feed-forward, (c) negative feed-forward.

large number of different factors also influencing the genes of the core network, we will pick one of them to extend our example at a later modeling stage.

1.2. Graphs and Network Motifs

1.2.1. *General properties of graphs*

Descriptions of networks typically start with creating a graph that formalizes knowledge on compounds and interactions. Technically, a graph G consists of a number of nodes or vertices V, which are connected by edges E. A formal mathematical description is:

$$G = (V, E), \tag{1}$$

where $E = E(V_i, V_j)$. There are several kinds of graphs that differ in the properties of their edges. In general, edges can or cannot be annotated with further information about their interaction, e.g. classifications or references. If, in the simplest case, the direction of the edges is not specified and it holds that $E(V_i, V_j) = E(V_j, V_i)$, we speak of an undirected graph. Protein interaction maps are an example for the use of undirected graphs in stem cell research (Wang *et al.*, 2006). These graphs contain information about the association of nodes in the graph, but do not distinguish the contribution of both nodes to the interaction. This is different in directed graphs, which are a good formal representation of GRNs where we discriminate between the gene coding for a transcription factor that acts as the regulator and the regulated genes. Here, the edges have a direction, meaning that $E(V_i, V_j) \neq E(V_j, V_i)$. The edges can also be labeled to express whether they represent an activation or inhibition. This can be written as $(V_i, V_j, \{+, -\})$. As mentioned above, further annotation can be added.

Graphs are very powerful tools for exploring the topological properties of a network. They can be classified according to the pattern of node degrees expressing their connectivity. The degree of a node is essentially the number of edges it is connected to. Global measures for connectivity in a graph are the average degree and the distribution of degrees. Classical cases are random graphs with a Poisson distribution, scale-free networks with a distribution of node degrees following a power law (Barabasi and Oltvai, 2004), and hierarchical networks (Newman, 2003).

Typical GRNs have very different graph properties compared to random graphs. They often show a significantly higher number of so-called hubs, i.e. genes that are highly connected, while the majority of genes only have a low connectivity. The big graph shown in Figure 1.1 with its inhomogeneous distribution of connectivity is a good example of a scale-free graph. They have a very short average path length and are highly robust against random failure of single nodes (Albert *et al.*, 2000), both of which are desirable properties for a biological network. Short paths between nodes enable efficient communication in the network. Robustness against random failure of genes is an advantage when there are mutations in a few genes, which can happen at surprisingly high rates. However, scale-free networks are vulnerable when there are attacks aimed at their hubs, which causes the network to degenerate quickly. Nevertheless, scale freeness must be evolutionary favorable for GRNs, as mutations add a constant pressure by deleting connections and creating new ones in a random fashion, however these networks still maintain their scale-free condition.

1.2.2. *Network motifs*

On the local level, graphs can show specific connectivity patterns, which occur significantly more often in biological networks than random networks. These accumulated patterns are called network motifs (Milo *et al.*, 2002). Motifs are assigned to different functions in networks, both for structural organization and for the dynamics of the network. We will discuss some examples of network motifs and their properties in the following.

An important network motif in development and differentiation is positive auto-regulation, i.e. denoting direct or indirect activation of a gene's expression by its own product. Positive auto-regulation can lead to bistable behavior, i.e. the gene either reaches a state of high or low expression depending on the initial conditions (Becskei *et al.*, 2001). If the gene is switched on by another activator, it is expressed even if that activator is no longer present. Another motif that shows a similar behavior is the two-node positive feedback loop (Figure 1.2a). Such a feedback loop consists of nodes that are coupled either by mutual activation or mutual inhibition, both of which lead to a bistable system that can be used to make a permanent decision in the network. In our example, we have the mutual activation of OS and NANOG that holds the cell in a pluripotent state. As mentioned before, it has been shown that NANOG, SOX2 and OCT4 show positive auto-regulation, as well as forming positive feedback loops. Such a combination is not unusual for transcription factors in developmental networks because it enhances the bi-stability and leads to irreversible decisions (Demongeot *et al.*, 2000).

Negative auto-regulation, meaning a gene product suppresses its own expression, has two important functions that are useful for biological circuits. First, it enables very short response times (as explained below), and second they make a network robust to input variations because it exhibits high dampening for high input and weak dampening for low input (Rosenfeld *et al.*, 2002).

The response time of a signal transmitting function can be measured as the time in which it reaches its half maximal concentration. It depends on the slope of the function and the maximum concentration. The response time of a negative auto-regulated gene to a signal can be very short, because even if it has a very strong activator that enables a fast expression, it can have a low state that has to be reached to transmit the signal. When the activator is present, the gene is quickly expressed and its concentration rises until it reaches the point where it represses its production and reaches a limit. The time course data of such a motif almost looks like a step function, whereas without the repression, the limit would be reached at a much higher concentration and the response would be slower.

The analysis of these motifs makes it possible to classify structures and to draw conclusions about their dynamic behavior. However in practice, precise predictions can be very difficult, since such motifs do not usually appear isolated, but as overlapping structures in huge intertwined networks. This problem can sometimes be solved by applying time scale separation, i.e. distinguishing between slow and fast variables. Then one considers either only the fast variable while assuming the slow variable to be constant or only the slow variable, with the assumption that the fast variable quickly reaches a quasi-steady state.

1.3. Dynamic Modeling Frameworks

The core network for stemness has been determined in a number of studies, among them Boyer *et al.*, 2005; Ivanova *et al.*, 2006; Loh *et al.*, 2006; Rodda *et al.*, 2005; Zhou *et al.*, 2007. Although not very large, it is a complex system since its graph is highly interconnected. In order to understand its function, we may consider the network not just as a static graph, but also as the structural basis for a dynamical system. Such a system can have an infinite number of states and it may become hard to predict the system's behavior. Dynamical systems follow certain rules that can be characterized and formalized in the modeling frameworks, which we are going to present and hopefully increase our understanding of their behavior.

It is important to clarify the difference between a dynamical model and interaction graphs. Dynamical models, as the name suggests, change their state over time. This is the single feature all the various modeling approaches have in common. In this context, a model is usually based on the description of species and their interaction as a graph $G = (V, E)$. Vector V comprises the species, which in our case are mostly genes, proteins or other biological compounds or drugs. Vector E characterizes the relations between the species, i.e whether there is a relation and what the possible quantitative parameters are. The entirety of these species, their relations, and the quantitative parameters characterizing the relations make up the model. Knowing these quantities, we can predict the system's states (in deterministic models) or their probability (in stochastic models) in the future.

Following an allegory by Conrad Waddington (1957), it might be helpful to imagine the state of a dynamical system as a ball rolling around in a hilly landscape. The direction of the ball is determined by the slope of the surface at its current position. It will roll around and eventually find a valley to stop in. When perturbed, it will roll back to its place or roll in a different place to rest depending on its speed and on the landscape. This image was also proposed as a picture for cell fate by Waddington over 50 years ago, but is also helpful to understand dynamical systems in general. Before Waddington, Max Delbrück (1949) already suggested that different cell types could correspond to different stable states of a high-dimensional system. Experimental evidence for this claim has just recently been found by Huang *et al.* (2005).

The next step in modeling the system is to develop the network structure into a dynamical model that can be simulated and analyzed. To construct this type of model, more data is needed to be able to understand how the system behaves under different conditions over time. This data has to provide sufficient information as to how the system can be wired with regard to timing, which enables the modeler to create a model that has the same structure as the system and exhibits a similar behavior.

After setting up a base for our model in the form of an interaction graph, we now move on to dynamic systems. We will start with coarse-grained logical models and refine these to end up with a detailed continuous ODE model. This method should not be mistaken as a way of judging one approach to be superior to the other.

Logical and quantitative modeling deal with different problems and each framework has its own strengths and weaknesses.

1.3.1. *Boolean networks*

Known data in most cases suggest that it is a reasonable approximation to assume that a gene's product is either present or absent, and that there are no relevant expression levels in between. This assumption is a discretization of the network's states. One of the simplest and most common forms of discrete models are Boolean networks, in which each species is represented by a variable that can take only the values *on* = expressed (1) and *off* = not expressed (0). Also time is discretized to yield a series of time points at which gene expression states are updated, as explained in the following.

Combinations of transcription factors govern the expression of genes regulated by these transcription factors. This relation was already expressed in Figure 1.1. There was already a relation that defined whether the factor was an activator or an inhibitor, but there were no exact rules about how the inputs act on the targets when combined. To implement our system as a Boolean network we need some sort of rule set that specifies how these combinations work. These rules must take into account the current state of the system into account to evaluate its state in the next point in time.

Boolean networks are named after the algebraic system of logic as defined by George Boole in the 19th century, which is used here to compute the dynamics of the model. This system defines operations on binary values, either false and true, or 0 and 1. Depending on the number of inputs per gene, k, i.e. how many transcription factors regulate one gene, one may define 2^{2^k} different output rules. These output rules are summarized in truth tables (Tables 1.1 and 1.2). Practical applications frequently use only the three basic operators AND ($\wedge$), OR ($\vee$) and NOT ($\neg$) for computations. Combinations of these three operators allow for the presentation of the full set of rules. Boolean modeling of biological networks was introduced by

Table 1.1. Boolean rules for nodes with one input.

Input	Output			
X	0	X	not X	1
0	0	0	1	1
1	0	1	0	1
Rule	0	1	2	3

Table 1.2. Boolean rules for nodes with two inputs. All Boolean rules for nodes with two inputs, some of the operators generating the output in question in the upper row and the output rules' decimal number in the lower row.

Input		Output															
X	Y	0	AND		X		Y	XOR	OR	NOR		not Y		not X		NAND	1
0	0	0	0	0	0	0	0	0	0	1	1	1	1	1	1	1	1
0	1	0	0	0	0	1	1	1	1	0	0	0	0	1	1	1	1
1	0	0	0	1	1	0	0	1	1	0	0	1	1	0	0	1	1
1	1	0	1	0	1	0	1	0	1	0	1	0	1	0	1	0	1
Rule		0	1	2	3	4	5	6	7	8	9	10	11	12	13	14	15

Glass and Kauffmann (1973) and has since been refined and extended in a number of ways (Thieffry and Thomas, 1998).

We can use Boolean algebra with this simple set of rules to define a model:

$$x_i(t+1) = b_i(x(t)), \quad 1 \leq i \leq n, \tag{2}$$

where x is a vector of variables, x_i is the state of the network of n genes, and b is the vector of Boolean functions or rules $b : \{0,1\}^n \rightarrow \{0,1\}$ that map the *k inputs* at time t to the *output* at time $t+1$. In each step, all the variables are updated synchronously according to the set of Boolean rules, which leads to a sequence of states following each other.

If we take the most central components of our example model and implement it as a Boolean network, we end up with the following rules:

$$\begin{aligned} OS(t+1) &= Nanog(t) \\ Nanog(t+1) &= OS(t) \end{aligned} \tag{3}$$

This extremely simple model already shows some interesting behavior. In a Boolean network with n species, there are 2^n possible states. This property makes the analysis of this small network fairly straightforward. We can write down all four possible states and their successor states in a table (Table 1.3).

The example network has four states out of which two have themselves as a successor. When both genes are *on*, they activate each other, and as a result stay activated; and when none of the genes are active, there is no activation at all and both genes stay *off*. States like these, in which the system cannot escape from a single state, are called point attractors or cyclic attractors of length 1. The other two states, where only one of the species is *on* at a time, form a cycle of length 2 of states that the system cannot leave. In each time step, one of the variables is switched *on* while the other is switched *off*. These cycles of states are called *dynamic attractors*.

Because there is only a finite number of states, every sequence of states must reach an attractor in a finite number of steps. All states that lie on a path to an attractor, but are not attractors themselves, are called transient states. The set of all transient states leading to a specific attractor a in the following steps, is called the *basin of attraction* of a.

The sequence of states of Boolean networks (state space) can be represented as a transition graph, which often is a more convenient representation than a truth table. A transition graph — not to be confounded with the original gene regulation graph — has 2^n nodes and a variable number of edges. Its nodes represent the states of the gene regulation network and the edges represent the transition from state t to state $t + 1$. Because the network is synchronous and deterministic, each of the nodes has exactly 1 outgoing edge.

Figure 1.3a shows the wiring diagram and the transition graph of the example model from Eq. (3). The dynamic attractor, in which the system is constantly

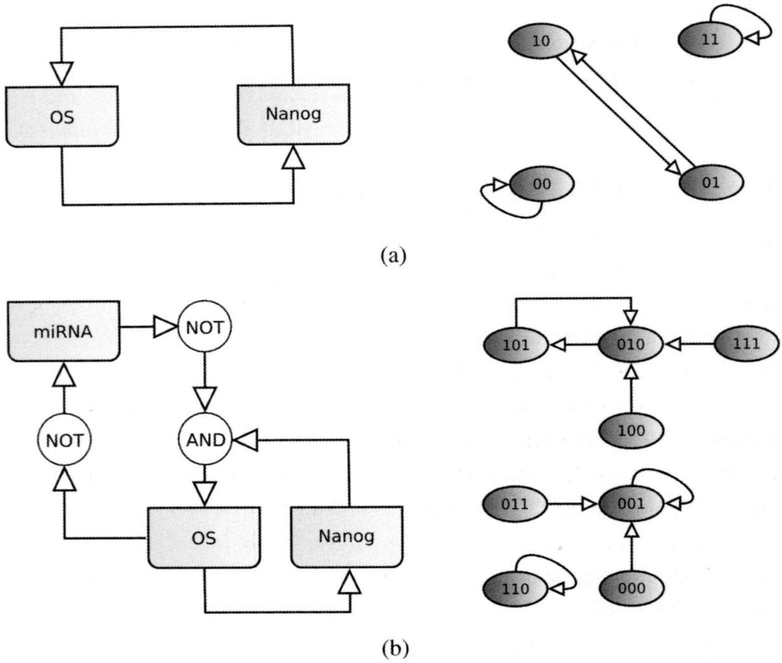

Figure 1.3. (a) The network from Eq. (3) as a wiring diagram (left) and its transition graph in the state space (right). The transition graph clearly shows three distinct attractors. (b) A more complex form of the network that uses Boolean operators and has transient states in its transition graph.

Table 1.3. Truth table of the example model.

Nanog(t)	OS(t)	Nanog(t + 1)	OS(t)
0	0	0	0
0	1	1	0
1	0	0	1
1	1	1	1

switching between one and the other gene, is not a biologically observed state in the network. Obviously oscillating developmental genes could not fulfill their biological role as a major switch between the stem cell state and the beginning of differentiation. Here, we use this argument to falsify the initial version of our model. To get to a better compliance of our model with the experiments, one may change the network structure as illustrated in Figure 1.3b. Here, we introduced an additional node, miRNA, yielding three basins of attraction. However, it is not always necessary to make changes to the structure of the model. We will now use a different way of simulation that introduces new possibilities for state transitions, which is enough to fix the problem in this case.

1.3.1.1. *Asynchronous Boolean networks*

The previously discussed form of Boolean networks makes the assumption that time is divided into discrete steps and state changes in all the species happen simultaneously. Networks that operate this way are called synchronous Boolean networks. They are deterministic and generate the same simulation results each time they are simulated. Opposed to this, we will now introduce stochastic simulation in the form of asynchronous Boolean networks. These are similar to the synchronous case, except that the simulation mechanism is changed to make the behavior of the network more realistic. In synchronous networks, each update rule is used in every time step and the result of the next time step is exactly one predetermined state. In asynchronous networks the updating of variables occurs in a random order. There is only one change in the state vector in each iteration and the order of variables to be changed is not fixed.

The outcome of the simulation is completely different if we use this method. In the synchronous case we only had one outgoing edge for each node in the transition graph, but now there can be up to n outgoing edges from each node, which offers new possibilities for transitions between states. In our example, it completely changes the properties of the transition graph. If we are in the state where *Nanog* is active and

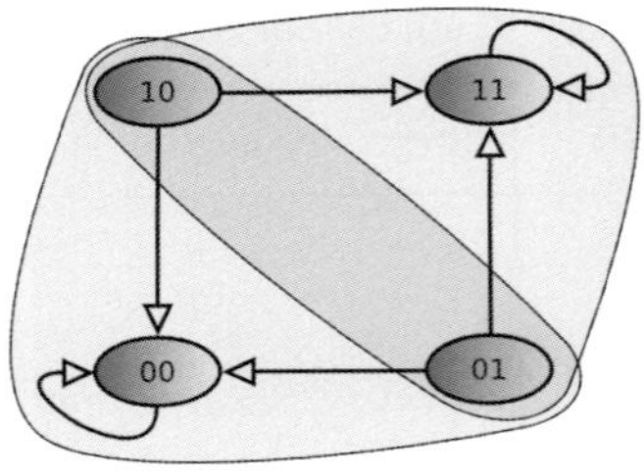

Figure 1.4. The transition graph of Eq. (3) is simulated as an asynchronous model. The basins of attraction are drawn as overlapping sets.

OS is inactive, we randomly choose one of the variables and switch it. This makes a transition either to the ON- or the OFF-point attractor. The transition between the two states where only one of the genes is ON is no longer possible as there can be only one change at a time. The model loses the dynamic attractor and only the two point attractors are preserved. The states that made up the dynamic attractor in the synchronous case have now become transient states. Because of the introduced stochasticity, the basins of both attractors are now overlapping and it is not clear in which attractor the simulation will end up as shown in Figure 1.4.

In the transition graph, it becomes clear that we now have a bistable structure with a clear separation of the attractors. If one of the genes is switched *off*, the system becomes unstable and can switch to the other attractor. It will stay in this state, even if the signal is switched *off* again. If we introduce an external signal that can switch *off* OS, we force this effect. This is a well-known property of dynamical systems and is called hysteresis.

Another variation of the Boolean approach is the ranked asynchronous Boolean network simulation. In this variant there are different groups of variables that are updated randomly inside each group, but the given groups have a distinct order. This can be useful to explicitly model different timescales in asynchronous networks, for example when the activation of a signal molecule activates gene expression (Albert *et al.*, 2008).

There have been several more additions and generalizations to the Boolean network formalism to make it more suitable to biology and to give it a wider space of applications. One of these approaches are generalized logical networks that extend logical modeling from binary state values to natural numbers and from Boolean functions to more general logical functions to govern the activation of genes. In this formalism, each variable has k thresholds that discretize its states to a finite set of natural numbers. Now each activation or inhibition of gene A on gene B has a value that determines at which level gene A influences gene B in which direction. Like in asynchronous Boolean networks, the model is also updated one variable at a time. The advantage of these models is that they still have a discrete state space that can be drawn as a transition graph and it is possible to find the logical steady states, but the interactions between the genes can be modeled in a much more detailed way than in Boolean networks. There are several variations of these types of networks which cannot be discussed here in detail, but logical networks might be useful to model the differentiation in our proposed network in more detail. A reason for applying logical networks might be that it has been proposed that there could be different levels of expression of Oct4 responsible for differentiation of stem cells to different types of cells (Chickarmane and Peterson, 2008), which could not be covered by Boolean models. Logical networks have already proven useful in studies of differentiation and development (Davidson *et al.*, 2002; Sanchez and Thieffry, 2003).

1.3.2. *Piecewise linear differential equations*

This takes us on the way to quantitative continuous modeling strategies, which we will now take one step further. Until now we only discussed logical modeling strategies that work well to describe the abstract properties of a system such as state space and attractors, but cannot give any quantitative information about the involved molecules. This is different in piecewise linear differential equations (Glass, 1975). These models are still similar to logical and Boolean models, because they use logical functions for the evolution of the system, but they also have linear degradation and production factors which influence the continuous level of activity of the different genes and they can provide quantitative information. The rules for these models are differential equations that define the rate of change of the variable at a given state. For example, the rate of *Nanog* activity change would look like this:

$$\frac{\mathrm{d}Nanog}{\mathrm{d}t} = k_1 \cdot s^{+}(OS, \theta) - k_{OS} \cdot Nanog, \tag{4}$$

where k_1 is the production rate constant and k_{OS} is the degradation rate constant. $s^{+}(OS)$ is a step function that is 1 if the concentration of OS is above a threshold θ and 0 if it is below the threshold. Thus if OS is sufficiently expressed, NANOG will be expressed with rate k_1. Step functions can be used to model repressors using $s^{-}(OS) = 1 - s^{+}(OS)$. They can also be combined arithmetically to express the regulation of a gene by different transcription factors.

1.3.3. *Ordinary differential equation models*

Ordinary differential equation (ODE) models are by far the most commonly used framework for biological models. There have been a number of studies using ODE models to describe certain aspects or parts of more complex regulatory systems in stem cells (Kalmar *et al.*, 2009). There are also many signaling pathways involved in the regulation of differentiation and stem cell development that can be modeled in greater detail than the actual gene regulation.

Ordinary differential equations are a way to model system dynamics in a much more detailed and — more importantly — in a quantitative way. The system is described by a number of differential equations that govern the state of each variable over time. There is again one equation per variable, which describes the change of the species depending on the current conditions.

An equation can consist of different terms that describe the different processes the species is involved in, e.g. synthesis, transport, activation, inactivation, degradation or dilution. In gene regulation, this is mostly restricted to synthesis and degradation of mRNA and protein. The equation for NANOG in the above example [Eq. (4)] could look like the following:

$$\frac{dNanog}{dt} = \frac{V_{OS} \cdot OS^{nos}}{(K_m + OS^{nos})} - k_{OS} \cdot Nanog. \tag{5}$$

This equation can be split into two terms. The first part is the gene regulation function that governs the transcription of NANOG which is dependent on OS. The second part of the equation is a linear term that describes the degradation of NANOG. So we simply subtract the degradation from the production of a protein and we end up with the rate of change. Let us discuss the terms separately before analyzing the entire system.

To explain the activation function, we have to think about how transcriptional regulation works. The activation happens when a transcription factor binds to a promoter. In many cases the transcription factor consists of several subunits that each can bind to inducers. The probability of binding becomes higher as the concentration of the transcription factor increases while the promoter concentration naturally remains constant. At very low concentrations, activation is very unlikely because there has to be a certain number of transcription factors to bind to all the inducers before the gene is activated. This is expressed by the factor K_m which marks the value of half maximal activation, i.e. when $OS = K_m$, the gene has reached 50% of its maximal activation by this factor. This maximum activation is given by the parameter V_{os}. It is reached when the concentration of transcription factor is so high that the promoter is constantly bound by it. In this case, the production can no longer be increased by this transcription factor. The last parameter n is called the Hill coefficient, and governs the steepness of the slope of the function. Simply put, it determines how many molecules have to bind to the promoter at a time to reach full activation. The function as a whole has a monotonic hyperbolic ($n = 1$)

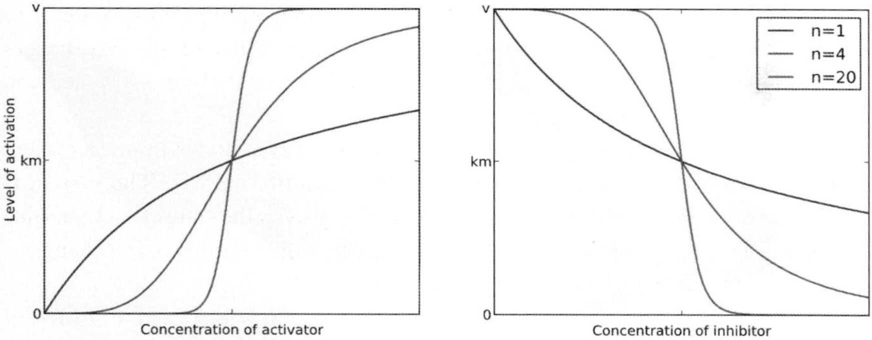

Figure 1.5. Plots of Hill functions with parameters $v = 1$ and $k_m = 0.5$. The Hill coefficient is set to different values to show the different behaviors of the function with changing coefficient.

or sigmoid ($n > 1$) shape as depicted in Figure 1.5. In gene regulation, the Hill coefficient is usually between 1 and 4. If it is any higher, the function becomes very similar to a step function like they are used in logical modeling.

The degradation term is a simple linear degradation that is assumed for the gene product, as long as there is no specific information about its degradation.

The differential equation for the change of *OS* is a bit more complicated than the simple activation of *NANOG*. It has two species regulating its production. Such regulatory functions are sometimes called multi- or two-dimensional input functions because the activity of the gene depends on both inputs. As seen in Figure 1.5, the gene has one activator and one repressor. Both are incorporated in the equation, and the activation can look the same as in Eq. (5). The effect of the repressor is represented by the term

$$\frac{V}{\left(1 + \left(\frac{miRNA}{K_{m2}}\right)^{n_m}\right)}. \tag{6}$$

This function has similar properties as the activating Hill function but is flipped around the $x = K_m$-line for $n > 1$. The decreasing hyperbolic or sigmoid function is dependent on similar parameters as the activating Hill function. Again there is a maximum activation, but it is now reached when there is no repressor present. Half maximum is reached when $miRNA = K_m$; the Hill coefficient n_m is again responsible for the steepness.

Now the two regulating functions have to be combined to a two-dimensional input function (Alon, 2007). There is no general rule as to how activators and repressors interact, and there are many different possible scenarios. One possibility

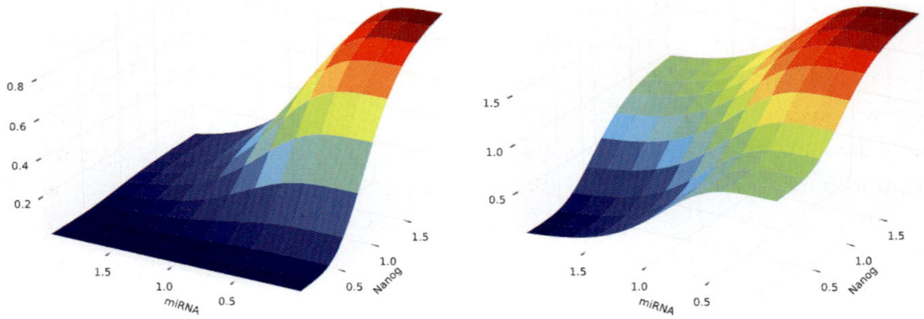

(a) (b)

Figure 1.6. Plots of different versions of two-dimensional input functions. (a) Plot of Eq. (8) which shows a complete deactivation if the repressor is present and a complete activation if only the activator is present. (b) An additive input function where the inhibitor can only partly silence the gene and the maximum is the sum of two activations.

is to sum up the activation and the inhibition terms. This would lead to a situation where the gene expression is at maximum when there is no repressor and a high concentration of activator (Figure 1.6b).

In this case, the repressor is not able to silence gene expression below the level of the activator because the terms are additive. The expression would also be at level V even if there was no activator and no repressor present, because without the repressor there is always an expression strength V. Additive coupling is comparable to using the OR operator in the Boolean formalism.

$$OS(t+1) = NANOG(t) \vee -miRNA(t). \tag{7}$$

There are some examples for this kind of coupling (Kalir and Alon, 2004), but in the Boolean case we used the AND operator for the function which we can translate to multiplicative coupling in the continuous scenario:

$$\frac{dOS}{dt} = \frac{V_{\max} \cdot NANOG^{n_n}}{(K_{m1} + NANOG^{n_n}) \cdot \left(1 + \left(\frac{miRNA}{K_{m2}}\right)^{n_m}\right)} - k_{OS} \cdot OS. \tag{8}$$

There is just one parameter $V_{\max}$ left for the maximum production and the repressor can inhibit the production completely if it is present in high amounts (Figure 1.6a). The type of coupling used in a model can have an enormous impact on the simulation results and a decision should be made based on known data about the transcription factors involved. It is also possible to describe the system in more detail with this approach and we can include the precise influence of transcription and translation by adding different species and reactions for these levels. But in most cases, even the current level of detail will pose enough problems to solve.

The next step that can be taken with these systems is to calculate qualitative properties of the system such as steady states. Steady states are an analog to attractors in the Boolean systems. A steady state is a state where all of the differential equations are equal to zero, meaning there is no more change in the system or the system is in equilibrium. As for attractors in the logical framework, there are different types of steady states. The set of all possible conditions of all variables is called the state space of the system. The state space has the dimension of the number of variables and is continuous. The set of ODEs describes a trajectory from an initial point towards the attractive state. If we stay within the previously mentioned picture of a hilly landscape, a ball starting somewhere rolls down to the next reachable valley. Some steady states are stable and all trajectories starting in their proximity run towards them like water in a sink, while others are like small plateaus on a hill and balls will roll away once perturbed (trajectories move away after slight perturbation). One major difference to logical models is that the number and properties of attractors can depend on the parameter values. In our example model, there are two stable steady states as in the previous logical models, but it is possible to create a third steady state that is unstable by choosing the parameters in a certain way. This change in the number and property of steady states is called bifurcation.

Given that the equations only give us the rate of change, we need a state to start in, which is defined as initial concentration. In the equations above, we also introduced a number of parameters that have yet to be specified and that strongly influence the system. Thus, we also need a set of parameters to start with. This enables us to calculate the time evolution of the system from the initial concentrations. A set of equations, parameters and initial concentrations is called an initial value problem (IVP). This can be numerically solved by following the rate equations from the initial concentration in tiny steps and walking along the concentration changes like following the transition graph before. There are very sophisticated algorithms for solving these types of equation systems available that give very accurate results.

Until now we have derived the equations, which is only one of the three parts we need for a numerical simulation of our model. To get the other two we need further knowledge about the system. This is the point where we have to face the fact that we are working in biology and we do not have enough data to characterize our system completely. For our small model, the initial conditions can be derived from the literature at least in a semi quantitative way by comparing RNA levels of the genes in the stem cell state as it was done by Chickarmane and Peterson (Chickarmane and Peterson, 2008). By selecting parameters based on literature and experience, we

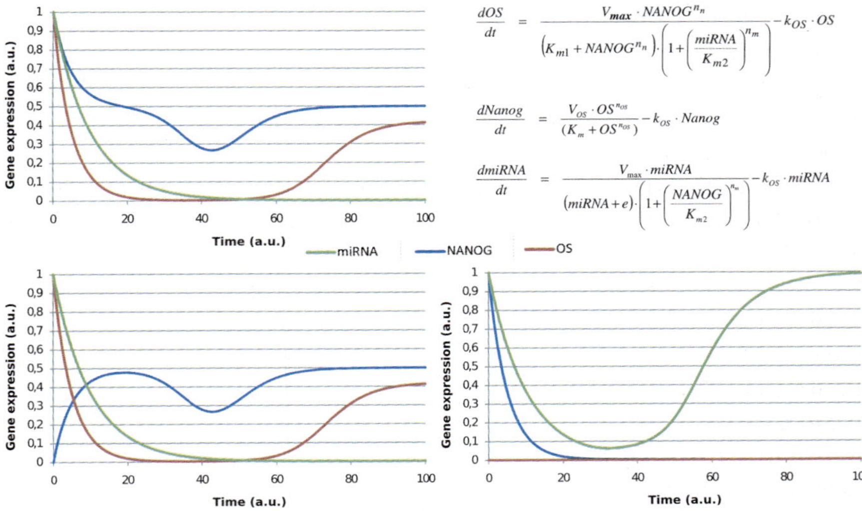

$$\frac{dOS}{dt} = \frac{V_{max} \cdot NANOG^{n_n}}{\left(K_{m1} + NANOG^{n_n}\right) \cdot \left(1 + \left(\frac{miRNA}{K_{m2}}\right)^{n_m}\right)} - k_{OS} \cdot OS$$

$$\frac{dNanog}{dt} = \frac{V_{OS} \cdot OS^{n_{os}}}{(K_m + OS^{n_{os}})} - k_{OS} \cdot Nanog$$

$$\frac{dmiRNA}{dt} = \frac{V_{\max} \cdot miRNA}{(miRNA + e) \cdot \left(1 + \left(\frac{NANOG}{K_{m2}}\right)^{n_m}\right)} - k_{OS} \cdot miRNA$$

Figure 1.7. Time courses of the ODE system shown in the upper right for 3 different sets of initial values. One can see that for the chosen parameter set, the simultaneous activation of OS and NANOG (shown in the upper right) is sufficient to turn off the miRNA which is also the case for sole activation of OS when the miRNA is activated (lower left). These states thus represent pluripotency related steady states. However, only activating NANOG in a state where the miRNA is active (shown in the lower right) will lead to a differentiation related steady state, where OS and NANOG are switched off and the miRNA is turned on.

can already simulate the model and observe its dynamical behavior for this special set of parameter values. An example for model simulation is given in Figure 1.7.

However, the parameters of our system are not known for gene regulatory networks and can only be found indirectly by computational methods or tuned by hand as it was done in some models of the core network (Kalmar *et al.*, 2009). We will now shortly describe how parameters can be inferred from data using computational optimization techniques. This concept is called parameter estimation or model fitting.

There are various methods for further analyses of ODE models. Sensitivity analysis, for example, can provide information as to the extent that the behavior of the system changes upon parameter changes.

1.3.4. *Parameter estimation in ODE models*

After having established an ODE system to mathematically describe the biological system, one can start to overlay data from biological experiments with the outcomes of this theoretical model. This process is called parameter estimation or model fitting and represents one of the central challenges in systems biology approaches. Usually, this step consists of the progressive adjustment of the model's parameters to experimentally measured time courses of observables, i.e. the minimization of an error function between both data sets.

However, the issue with experimental biological data resides in their noisiness and the limited number of measured time points. Experiments are often expensive, time consuming or simply difficult to carry out from a technical point of view. Moreover, since the theoretical model often gives a reduced or simplified picture of the biological process in question, it may be impossible to measure certain parameters that do not have a direct biological meaning. For the sum of these reasons, parameter estimation is a suitable and necessary tool to validate a model by quantitative results and to make predictions (Lillacci and Khammash, 2010).

Several efforts have been put into the application of optimization techniques to mathematical models of biological systems, such as linear and nonlinear least-square fitting (Mendes and Kell, 1998), simulated annealing (Kirkpatrick *et al.*, 1983), genetic algorithms (Srinivas and Patnaik, 1996), and evolutionary computation (Ashyraliyev *et al.*, 2008; Moles *et al.*, 2003). A detailed description of these methods would lead too far in here. Still, in principle all of these methods work in the same way; by heuristically searching the parameter space in order to minimize a certain cost function that reflects the error between the experimental data set and the theoretical predictions.

As mentioned above, the noise of biological data which is due to a combination of measurement inaccuracy and intrinsic noise, e.g. in case of low copy numbers of the species in question, constitutes a major problem in the data analysis. Therefore, methods have been developed to account for these stochastical processes. They include Bayesian methods such as Markov chain Monte Carlo simulations, ensemble methods (Brown and Sethna, 2003; Battogtokh *et al.*, 2002) and sequential Monte Carlo methods (Sisson *et al.*, 2007) as well as Maximum likelihood (Bortz and Nelson, 2006) and more recently control theory state observers, e.g. Luenberger-like, Kalman filter based and high gain observers.

In the rare case where the estimation problem has a unique solution, the result of parameter estimation is a set of parameters that works to produce the time course that fits the experimental data best. Once a set of parameters is available that can reproduce a significant amount of the experimental data, one can start to develop model-based predictions about the system from overlaying and comparing the model structure with the biological process in question. However, more often than not, one will be confronted with the problem of indetermination, especially for large models with many parameters. In that situation, the algorithm finds more than one set of parameters that fit the experimental data; in this case we can only define wide ranges for every parameter.

After having dealt with the problem of parameter estimation for deterministic ODE models, we will now focus on the introduction of stochastic processes into the mathematical models, its opportunities and the associated problems and restrictions.

1.4. Stochasticity

There are many examples in recent works on stem cell differentiation and reprogramming that favor a stochastic view of the regulation events (Yamanaka, 2009) and some use stochastic modeling in this context. For example, Hanna *et al.* (2009) discuss the dependence of reprogramming efficiency on cell division rate of murine fibroblasts and explain their findings with a simple stochastic model of reprogramming and cell division.

The previous models were all based on two rather strong assumptions that are needed to justify these approaches. The first one is to assume that all the included reactions take place in a "well-stirred" environment, meaning that the substances are equally distributed in the medium. The second assumption is that the concentration changes of the species are continuous throughout the experiment. Both of these assumptions are not completely justified in the case of gene regulation due to a low number of molecules in each cell. This makes purely deterministic models unsuitable for application in cases with low numbers of molecules.

As we mentioned before in the context of asynchronous Boolean models, biological systems do not work like deterministic models. Gene expression is an inherently stochastic process. On top of that there are a number of external influences such as signal molecules or changes in the environment that are stochastic as well. There are two main approaches to simulate stochasticity in a much more detailed way than asynchronous Boolean models did. The first one is the simulation of every single reaction as a stochastic process and a combination of these in a Markov chain. There are different algorithms for the simulation of these models; the first of which was introduced by Gillespie (1977). These models use continuous time but discrete numbers of molecules and calculate probabilities based on these numbers to randomly choose the next reaction. Simulation of these models is computationally very expensive and can only be realized for small models with very low particle numbers.

A compromise between this extremely detailed approach and deterministic ODE models (Gillespie, 2000) is the use of Langevin equations to introduce stochasticity in ODE models. These equations are very similar to the ones we have seen before, but in addition have a Gaussian noise term $\gamma(t)$ added to them.

$$\frac{dNANOG}{dt} = \frac{V_{OS} \cdot OS^{n_{OS}}}{(K_m + OS^{n_{OS}})} - k_{OS} \cdot NANOG + \gamma(t). \tag{9}$$

This noise term gives the system new capabilities of changing its state. Given a sufficiently strong noise, our bistable example model could randomly switch between its stable steady states. A very similar model has been used by Kalmar *et al.* (2009) to explain the different levels of Nanog expression they observed in a population of murine ES cells. By combining the model and their experimental results, they came to the prediction that the regulatory system has two steady states; one of which is stable while the other is unstable. The system is excited by the transcriptional noise and leaves the stable steady state for a short time. It runs towards the unstable state, which makes the cell amenable to differentiate, because this state shows a much lower Nanog expression. In a whole cell population, this results in a situation where there are always some cells that are in the low-Nanog state and therefore are able to differentiate.

1.5. Planning a Model

Before starting to model a process, one should always try to pose a question and then build a model that can somehow help to answer it or to ask a more precise question. There are two decisions that have to be made in advance. The first one is the choice of the formal framework, and the second one is the level of detail that is necessary to answer the question. Both decisions depend on each other, and the choice of the framework somehow limits the scale and level of abstraction of the model. For example if one decides to model a gene network with ordinary differential equations, we can build a detailed model with separate equations for transcription and translation, but the size of the model is limited to a few genes, because the model would become too complex. If we use Boolean modeling, the situation is turned around. The described network can be bigger, but it would probably not be wise to include different rules for translation and transcription in this case.

To choose the right dynamic modeling framework for a system is a very complex task. There can be no definite answer as to which technique is the best for a special project. The choice depends on many factors some of which are shown in Figure 1.8. Most often, the limiting factors for a model are the knowledge about the structure of the network and experimental data about its behavior. Compared to model systems such as yeast or *E. coli*, there is still a small amount of data available for gene regulation in stem cells. In many cases, it will be necessary to use strong simplifications and resort to logical modeling as shown in the examples presented here.

Sometimes, it also makes sense to model different parts of the system in different levels of detail and to merge the results to yield a more precise image of the system as a whole. There are also some examples of models of one and the same

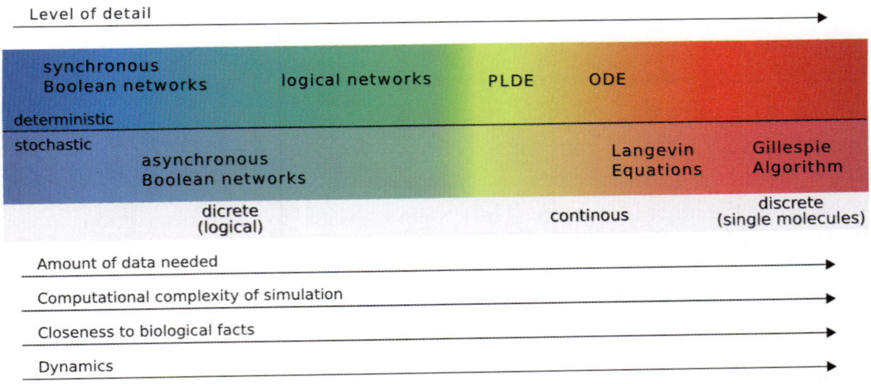

Figure 1.8. Comparison of different modeling approaches with respect to represented complexity. The different frameworks are positioned according to the maximum level of detail possible, with the maximum level of detail increasing from left to right.

process that have different levels of complexity and are useful to answer different questions.

As knowledge about the system is limited, it may also be necessary to build different alternatives of a model, analyze them independently and select the version that fits the data best.

Acknowledgements

We gratefully acknowledge funding from the German Ministry for Education and Research (Medical Systems Biology grant "Drug-iPS").

Bibliography

Albert, I., Thakar, J., Li, S., Zhang, R. and Albert, R., 2008. Boolean network simulations for life scientists, *Source Code for Biology and Medicine*. 3:16 http://www.scfbm.org/content/13/11/16.

Albert, R., Jeong, H. and Barabasi, A. L. 2000. Error and attack tolerance of complex networks, *Nature* 406:378–382.

Alon, U. 2007. Network motifs: Theory and experimental approaches, *Nature Reviews* 8:450–461.

Arney, K. L. and Fisher, A. G. 2004. Epigenetic aspects of differentiation, *J Cell Sci* 117: 4355–4363.

Ashyraliyev, M., Jaeger, J. and Blom, J. G. 2008. Parameter estimation and determinability analysis applied to Drosophila gap gene circuits, *BMC Syst Biol* 2:83.

Barabasi, A. L. and Oltvai, Z. N. 2004. Network biology: Understanding the cell's functional organization, *Nature Reviews* 5:101–113.

Battogtokh, D., Asch, D. K., Case, M. E., Arnold, J. and Schuttler, H. B. 2002. An ensemble method for identifying regulatory circuits with special reference to the qa gene cluster of Neurospora crassa, *Proc Natl Acad Sci USA* 99:16904–16909.

Becskei, A., Seraphin, B. and Serrano, L. 2001. Positive feedback in eukaryotic gene networks: Cell differentiation by graded to binary response conversion, *EMBO J* 20: 2528–2535.

Bernstein, B. E., Mikkelsen, T. S., Xie, X., Kamal, M., Huebert, D. J., Cuff, J., Fry, B., Meissner, A., Wernig, M., Plath, K., Jaenisch, R., Wagschal, A., Feil, R., Schreiber, S. L. and Lander, E. S. 2006. A bivalent chromatin structure marks key developmental genes in embryonic stem cells, *Cell* 125:315–326.

Bortz, D. M. and Nelson, P. W. 2006. Model selection and mixed-effects modeling of HIV infection dynamics, *Bull Math Biol* 68:2005–2025.

Boyer, L. A., Lee, T. I., Cole, M. F., Johnstone, S. E., Levine, S. S., Zucker, J. P., Guenther, M. G., Kumar, R. M., Murray, H. L., Jenner, R. G., Gifford, D. K., Melton, D. A., Jaenisch, R. and Young, R. A. 2005. Core transcriptional regulatory circuitry in human embryonic stem cells, *Cell* 122:947–956.

Brown, K. S. and Sethna, J. P. 2003. Statistical mechanical approaches to models with many poorly known parameters, *Phys Rev E Stat Nonlin Soft Matter Phys* 68:021904.

Chavez, L., Bais, A. S., Vingron, M., Lehrach, H., Adjaye, J. and Herwig, R. 2009. In silico identification of a core regulatory network of OCT4 in human embryonic stem cells using an integrated approach, *BMC Genomics* 10:314.

Chew, J. L., Loh, Y. H., Zhang, W., Chen, X., Tam, W. L., Yeap, L. S., Li, P., Ang, Y. S., Lim, B., Robson, P. and Ng, H. H. 2005. Reciprocal transcriptional regulation of Pou5f1 and Sox2 via the Oct4/Sox2 complex in embryonic stem cells, *Mol Cell Biol* 25: 6031–6046.

Chickarmane, V. and Peterson, C. 2008. A computational model for understanding stem cell, trophectoderm and endoderm lineage determination, *PLoS One* 3:e3478.

Chickarmane, V., Troein, C., Nuber, U. A., Sauro, H. M. and Peterson, C. 2006. Transcriptional dynamics of the embryonic stem cell switch, *PLoS Comput Biol* 2:e123.

Davidson, E. H., Rast, J. P., Oliveri, P., Ransick, A., Calestani, C., Yuh, C. H., Minokawa, T., Amore, G., Hinman, V., Arenas-Mena, C., Otim, O., Brown, C. T., Livi, C. B., Lee, P. Y., Revilla, R., Rust, A. G., Pan, Z., Schilstra, M. J., Clarke, P. J., Arnone, M. I., Rowen, L., Cameron, R. A., McClay, D. R., Hood, L. and Bolouri, H., 2002. A genomic regulatory network for development, *Science* 295:1669–1678.

Delbrück, M. 1949. Unités biologiques douées de continuité génétique, *Editions du Centre National de la Recherche Scientifique* 33–35.

Demongeot, J., Kaufman, M. and Thomas, R. 2000. Positive feedback circuits and memory, *C R Acad Sci III* 323:69–79.

Gillespie, D. 1977. Exact stochastic simulation of coupled chemical reactions, *J Phys Chem* 81:2340–2361.

Gillespie, D. T. 2000. The chemical Langevin equation, *J Chem Phys* 113:297–306 http://link.aip.org/link/?JCP/113/297/291.

Glass, L. 1975. Classification of biological networks by their qualitative dynamics, *J Theor Biol*:85–107.

Glass, L. and Kauffman, S. A. 1973. The logical analysis of continuous, non-linear biochemical control networks, *J Theor Biol* 39:103–129.

Guo, X. and Wang, X. F. 2009. Signaling cross-talk between TGF-beta/BMP and other pathways, *Cell Res* 19:71–88.

Hanna, J., Saha, K., Pando, B., J. van Zon, Lengner, C. J., Creyghton, M. P., van Oudenaarden, A. and Jaenisch, R. 2009. Direct cell reprogramming is a stochastic process amenable to acceleration, *Nature* 462:595–601.

Huang, S., Eichler, G., Bar-Yam, Y. and Ingber, D. E. 2005. Cell fates as high-dimensional attractor states of a complex gene regulatory network, *Phys Rev Lett* 94:128701.

Ivanova, N., Dobrin, R., Lu, R., Kotenko, I., Levorse, J., DeCoste, C., Schafer, X., Lun, Y. and Lemischka, I. R. 2006. Dissecting self-renewal in stem cells with RNA interference, *Nature* 442:533–538.

Kalir, S. and Alon, U. 2004. Using a quantitative blueprint to reprogram the dynamics of the flagella gene network, *Cell* 117:713–720.

Kalmar, T., Lim, C., Hayward, P., Munoz-Descalzo, S., Nichols, J., Garcia-Ojalvo, J. and Martinez Arias, A. 2009. Regulated fluctuations in nanog expression mediate cell fate decisions in embryonic stem cells. *PLoS Biol* 7:e1000149.

Kim, J., Chu, J., Shen, X., Wang, J. and Orkin, S. H. 2008. An extended transcriptional network for pluripotency of embryonic stem cells, *Cell* 132:1049–1061.

Kirkpatrick, S., Gelatt, C. D., Jr. and Vecchi, M. P. 1983. Optimization by simulated annealing, *Science* 220:671–680.

Klipp, E. 2005. *Systems Biology in Practice: Concepts, Implementation and Application.* Wiley-VCH Verlag GmbH & Co. KGaA, Weinheim.

Klipp, E., Liebermeister, W., Helbig, A., Kowald, A. and Schaber, J. 2007. Systems biology standards — the community speaks, *Nat Biotechnol.* 25:in press.

Kopp, J. L., Ormsbee, B. D., Desler, M. and Rizzino, A. 2008. Small increases in the level of Sox2 trigger the differentiation of mouse embryonic stem cells, *Stem Cells* 26:903–911.

Lillacci, G. and Khammash, M. 2010. Parameter estimation and model selection in computational biology, *PLoS Comput Biol* 6:e1000696.

Loh, Y. H., Wu, Q., Chew, J. L., Vega, V. B., Zhang, W., Chen, X., Bourque, G., George, J., Leong, B., Liu, J., Wong, K. Y., Sung, K. W., Lee, C. W., Zhao, X. D., Chiu, K. P., Lipovich, L., Kuznetsov, V. A., Robson, P., Stanton, L. W., Wei, C. L., Ruan, Y., Lim, B. and Ng, H. H. 2006. The Oct4 and Nanog transcription network regulates pluripotency in mouse embryonic stem cells, *Nat Genet* 38:431–440.

Loh, Y. H., Zhang, W., Chen, X., George, J. and Ng, H. H. 2007. Jmjd1a and Jmjd2c histone H3 Lys 9 demethylases regulate self-renewal in embryonic stem cells, *Genes Dev* 21:2545–2557.

MacArthur, B. D., Ma'ayan, A. and Lemischka, I. R. 2009. Systems biology of stem cell fate and cellular reprogramming, *Nat Rev Mol Cell Biol* 10:672–681 http://dx.doi.org/610.1038/nrm2766.

MacArthur, B. D., Please, C. P. and Oreffo, R. O. 2008. Stochasticity and the molecular mechanisms of induced pluripotency, *PLoS One* 3:e3086.

Mendes, P. and Kell, D. 1998. Non-linear optimization of biochemical pathways: Applications to metabolic engineering and parameter estimation, *Bioinformatics* 14:869–883.

Milo, R., Shen-Orr, S., Itzkovitz, S., Kashtan, N., Chklovskii, D. and Alon, U. 2002. Network motifs: Simple building blocks of complex networks, *Science* 298:824–827.

Moles, C. G., Mendes, P. and Banga, J. R. 2003. Parameter estimation in biochemical pathways: A comparison of global optimization methods, *Genome Res* 13:2467–2474.

Newman, M. E. J. 2003. The structure and function of complex networks, *SIAM Review* 167–256.

Niwa, H. 2007. How is pluripotency determined and maintained? *Development* 134:635–646.

Pan, G., Li, J., Zhou, Y., Zheng, H. and Pei, D. 2006. A negative feedback loop of transcription factors that controls stem cell pluripotency and self-renewal, *FASEB J* 20:1730–1732.

Rodda, D. J., Chew, J. L., Lim, L. H., Loh, Y. H., Wang, B., Ng, H. H. and Robson, P. 2005. Transcriptional regulation of nanog by OCT4 and SOX2. *J Biol Chem* 280:24731–24737.

Rosenfeld, N., Elowitz, M. B. and Alon, U. 2002. Negative autoregulation speeds the response times of transcription networks, *J Mol Biol* 323:785–793.

Sanchez, L. and Thieffry, D. 2003. Segmenting the fly embryo: A logical analysis of the pair-rule cross-regulatory module, *J Theor Biol* 224:517–537.

Singh, S. K., Kagalwala, M. N., Parker-Thornburg, J., Adams, H. and Majumder, S. 2008. REST maintains self-renewal and pluripotency of embryonic stem cells, *Nature* 453:223–227.

Sisson, S. A., Fan, Y. and Tanaka, M. M. 2007. Sequential Monte Carlo without likelihoods, *Proc Natl Acad Sci USA* 104:1760–1765.

Srinivas, M. and Patnaik, L. M. 1996. On modeling genetic algorithms for functions of unitation, *IEEE Trans Syst Man Cybern B Cybern* 26:809–821.

Takahashi, K. and Yamanaka, S. 2006. Induction of pluripotent stem cells from mouse embryonic and adult fibroblast cultures by defined factors, *Cell* 126:663–676.

Thieffry, D. and Thomas, R. 1998. Qualitative analysis of gene networks, *Pac Symp Biocomput* 77–88.

Valdimarsdottir, G. and Mummery, C. 2005. Functions of the TGFbeta superfamily in human embryonic stem cells, *APMIS* 113:773–789.

Waddington, C. 1957. *The Strategy of the Genes*. George Allen & Unwin, London.

Wang, J., Rao, S., Chu, J., Shen, X., Levasseur, D. N., Theunissen, T. W. and Orkin, S. H. 2006. A protein interaction network for pluripotency of embryonic stem cells, *Nature* 444:364–368.

Xu, N., Papagiannakopoulos, T., Pan, G., Thomson, J. A. and Kosik, K. S. 2009. MicroRNA-145 regulates OCT4, SOX2 and KLF4 and represses pluripotency in human embryonic stem cells, *Cell* 137:647–658.

Yamanaka, S. 2009. Elite and stochastic models for induced pluripotent stem cell generation, *Nature* 460:49–52.

Zhou, Q., Chipperfield, H., Melton, D. A. and Wong, W. H. 2007. A gene regulatory network in mouse embryonic stem cells, *Proc Natl Acad Sci USA* 104:16438–16443.

Heterogeneity and Flexibility of Stem Cell Fate Decisions

2

A Conceptual Perspective

Ingmar Glauche and Ingo Roeder

2.1. Introduction

In 1868, the German biologist Ernst Haeckel used the term "Stammzelle" (German for *stem cell*) to denote the unicellular ancestor from which all multicellular organisms developed during evolution. In analogy to this evolutionary perspective, he proposed to call the fertilized egg — as the founder cell of an organism — a *stem cell* as well (Ramalho-Santos and Willenbring, 2007). Around 1900, the term *stem cell* was first used to characterize a progenitor cell that gives rise to different types of blood cells. However, the existence of an undifferentiated hematopoietic stem cell (HSC) that is capable of self-renewal (i.e. maintaining or even regenerating its undifferentiated state) and of differentiation (i.e. producing differentiated cells) could not be demonstrated experimentally for several decades. In the 1960s, Till and McCulloch proved for the first time that there are undifferentiated bone marrow cells with the potential to develop into spleen colonies of irradiated mice. It could be demonstrated that these colonies itself contain differentiated cells but also cells retaining the potential to generate secondary spleen colonies (Till and McCulloch, 1961; Becker *et al.*, 1963; Siminovitch *et al.*, 1963).

Meanwhile, stem cells are not only known for the hematopoietic system but could also be identified in other tissues like the skin (Alonso and Fuchs, 2003), the gut (Potten and Morris, 1988), the liver (Theise, 2003), and even in neural tissues such as the brain (Temple, 2001). All these different types of tissue-resident stem cells, which are also called adult or somatic stem cells, are typically found within a particular tissue or organ. There, the stem cells guarantee the sustained function of the tissue by the maintenance of their own population and by the generation of the specialized cell types which are needed for replacement and repair (Potten and Loeffler, 1990; Lemischka, 2004). Generally, their differentiation potential is narrowed to the cell types of the particular tissue.

Over the last decades, further types of stem cells have been identified. Most prominently, embryonic stem (ES) cells are derived from the cells of the inner cell mass of a blastocyst and are maintained in *in vitro* cultures (Evans and Kaufman, 1981; Martin, 1981; Thomson and Marshall, 1998). These cells are characterized by their ability to differentiate into all cell types of the organism. However, they represent a fixation of a transient cell stage (corresponding to the cells of the inner cell mass of the blastocyst) and therefore do not have a stable *in vivo* counterpart. Although embryonic stem cells have a superior differentiation potential compared to tissue stem cells, their application for medical therapy imposes a number of technical problems and ethical questions (de Wert and Mummery, 2003).

Very recently, a new type of stem cells — denoted as *induced pluripotent* stem (iPS) cells — has been described. iPS cells show ES cell-like characteristics and can be obtained by reprogramming tissue stem cells or even fully differentiated cells, e.g. by a targeted over-expression of only a few specific transcription factors, such as Sox2, Oct4, c-Myc, and Klf4 (Takahashi and Yamanaka, 2006; Okita *et al.*, 2007).

Unlike ES and iPS cells, which are defined *in vitro*, the prospective isloation of tissue stem cells is still an experimental challenge. Over the last decades, many sophisticated methods have been developed to identify stem cells based on their expression of characteristic cell surface molecules (Goodell, 2002; Bonnet, 2002; Wognum *et al.*, 2003; Kiel *et al.*, 2005; Challen *et al.*, 2009), their typical mRNA expression patterns (Ivanova *et al.*, 2002; Ramalho-Santos *et al.*, 2002) or their precise location within a certain tissue (Blanpain and Fuchs, 2006; van der Flier and Clevers, 2009). However, these techniques only enrich for stem cells and it is not (yet) possible to prospectively decide whether a particular cell will act as a stem cell in a functional assay or not. Therefore, the "gold standard" for defining tissue stem cells is a functional characterization of evaluating the cell's ability for long-term tissue regeneration *in vivo*.

Based on this functional characterization, tissue stem cells are formally defined as undifferentiated cells being able to self-renew their own population[1] and to guarantee the long-term production of differentiated cells. Although this definition appears simple and straightforward, it imposes some difficulties when applied to a single cell. It has been correctly pointed out by Loeffler and Roeder (2002) that the above definition refers to general capabilities of the cell type in question but requires

[1] Within the definition the terms *self-renewal* and *self-maintenance* are often used interchangeably, although they might refer to slightly different phenomena (Loeffler and Potten, 1997). Self-maintenance generally refers to the property of a stem cell population to preserve their functionality to act as stem cells, which is the general case in a homeostatic situation. In contrast, self-renewal does also include the repopulation of a tissue and the reestablishment of the corresponding stem cell population after serious stress.

a high level of flexibility in their actual use. It is obvious that a single cell is unable to fulfil all these demands simultaneously. It might well be the case that under certain environmental conditions, just one of the capabilities is used (e.g. the differentiation into a large number of functional cells) whereas other capabilities remain unused (e.g. the self-renewal ability). Furthermore, the use of these potentials might change over time, thus including reversible developments. In this sense, it appears far more plausible that stem cell characteristics are maintained on the cell population level, i.e. by the concerted action of multiple cells guaranteeing a robust population behavior, while cellular fate is hardly predictable on the individual cell level. Following this argument, it is evident that any functional definition of stem cells generally refers to a population of multiple, heterogeneous cells. We, therefore, argue against the view on stem cells as individual and isolated entities with fixed characteristics and suggest to perceive *stemness* as population property instead.

Nevertheless, the question remains: how does the robust population behavior of stem cells originates from the level of potentially heterogeneous and flexible single-cell constituants? Within this chapter we illustrate different aspects of flexibility in single-cell decisions and demonstrate how heterogeneous, although self-organizing stem cell populations show stable and robust behavior on the tissue level.

2.2. Stem Cell Fate Decisions

Starting out from the above definiton of stem cell populations, there are two principle levels on which stem cells have to "make a decision" about their future development. Namely, these cells have to decide on whether they retain their stem cell characteristics or whether they contribute to the pool of differentiating cells. In the latter case, the multi-potent cells have to decide to which functional cell type they are committing. Although this categorization is simplistic, there are several hints that the distinction of stem cell fate decisions between *self-renewal vs. differentiation* and *lineage specification* is appropriate. A discussion of the different conceptual approaches to these cell fate decisions is provided in the following section.

2.2.1. *Models of stem cell self-renewal*

2.2.1.1. *Entity-based models*

Entity-based models of stem cell organization emerged from the classical perception of stem cells as well-defined and clearly distinguishable objects. These entities are characterized by a set of fixed developmental options describing the transition between different states. Cells with similar states are summarized into so-called "compartments" that are commonly envisioned to contain populations of cells with equal characteristics like stem cells, progenitors, or more mature cell types. The transition rules can potentially be influenced by extrinsic signals like feedback regulation from mature cell stages mediated by growth factors or microenvironmental cues. However, the stem cell status is viewed as a property that can be kept or lost, but never regained. In this sense, the flow through different subsequent compartments is generally perceived as being unidirectional (Weissman, 2000; Prudhomme *et al.*, 2004).

Entity-based models can either be described on the single-cell level by defining transition propabilities for each cell depending on their affiliation to a certain compartment or, (assuming homogeneous cell populations of sufficient size) by population-based approaches in which individual stochastic decisions are replaced by averaged, deterministic rate constants.

A typical example of this model class is the mathematical model that accompanied the first experimental demonstration of the self-renewing ability of hematopoietic progenitor cells by Till and McCulloch in the 1960s (Till *et al.*, 1964). This model assumes that within the stem cell compartment, each cell can either divide symmetrically (giving rise to two stem cells, corresponding to a self-renewal of the population) or differentiate (Figure 2.1). These options are realized with fixed probabilities denoted as p and q, respectively. Even without the assumption of feedback regulation or microenvironment interaction, the authors could explain the frequency distribution of spleen colonies (CFU-S colonies) in secondary recipients.

Picking up on this idea of stochastic transitions between rather distinct cell populations, in which a probability is assigned to each developmental option, the

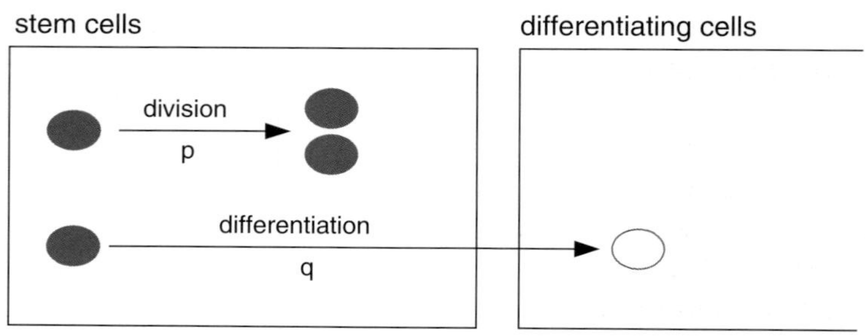

Figure 2.1. Compartment model of hematopoietic self-maintenance. A simple compartment model was proposed by Till *et al.* (1964). Cells in the stem cell compartment (dark grey) are either transferred into the compartment of differentiating cells (grey circles) with rate q or they undergo division with rate p producing two daughter stem cells. In the stable situation, the rates for differentiation and division are equal $p = q$.

mathematical concept was extended for the hematopoietic system (Vogel *et al.*, 1969) and for other tissues (Loeffler and Grossmann, 1991). In a simple, albeit elegant approach, Abkowitz *et al.* used such a model to access the frequency and replication kinetics of hematopoietic stem cells *in vivo* as well as the evolution of myeloproliferative disorders (Abkowitz *et al.* 1996, 2000; Catlin *et al.* 2005). Besides the developmental options of symmetric division and differentiation, the authors also accounted for the process of apoptosis.

Feedback regulations can be introduced for both levels of description by a dynamic adaption of either transition probabilities balancing self-renewal and differentiation in the stochastic (single-cell–based) approaches or by the corresponding rate constants in the deterministic (population-based) approaches. In addition to the evaluation of the steady-state situation, such feedback regulations allow the analysis of the system response to various perturbations, as they occur after damage or injury. Sophisticated examples for these kinds of models have been developed by Mackey and coworkers. Under the assumptions of delayed feedbacks from later cell stages on the hematopoietic stem cell compartment, the authors explained typical oscillation patterns occurring in a number of specific malignancies of the hematopoietic system (Mackey and Glass, 1977; Mackey, 1978; Hearn *et al.*, 1998; Colijn and Mackey, 2005a,b). Even more extensive are the multi-compartment models of later stages of hematopoiesis by Loeffler *et al.*, as there are models on granulopoiesis (Reincke *et al.*, 1984; Wichmann *et al.*, 1984; Scholz *et al.*, 2005) and erythropoiesis (Loeffler *et al.*, 1989).

2.2.1.2. *Symmetric and asymmetric divisions*

Although asymmetric cell divisions (i.e. one stem cell giving rise to one stem and one differentiated cell, Figure 2.2b) have not been directly observed in hematopoiesis, it was already recognized by Till *et al.* (1964) that the possibility of such an event could be integrated into their early concept of separating division and differentiation by assuming that a symmetric division event is followed by the differentiation of just one daughter cell. This view of separating division and differentiation and thus avoiding the notion of *functionally* asymmetric divisions (Figure 2.2c) was also supported by single-cell experiments using bcl-2 transfected FDCP-mix cells (Fairbairn *et al.*, 1993) as well as by further conceptual works (Potten and Loeffler, 1990; Loeffler and Roeder, 2002; Roeder and Lorenz, 2006). In contrast, and driven by the observation of functionally asymmetric divisions in other stem cell systems [e.g. in invertebrates (Betschinger and Knoblich, 2004) or in non-hematopoietic tissues like the nervous system (Goetz and Huttner, 2005) and the epidermis (Lechler and Fuchs, 2005)], the concept of asymmetric cell divisions is still strongly promoted in hematopoiesis

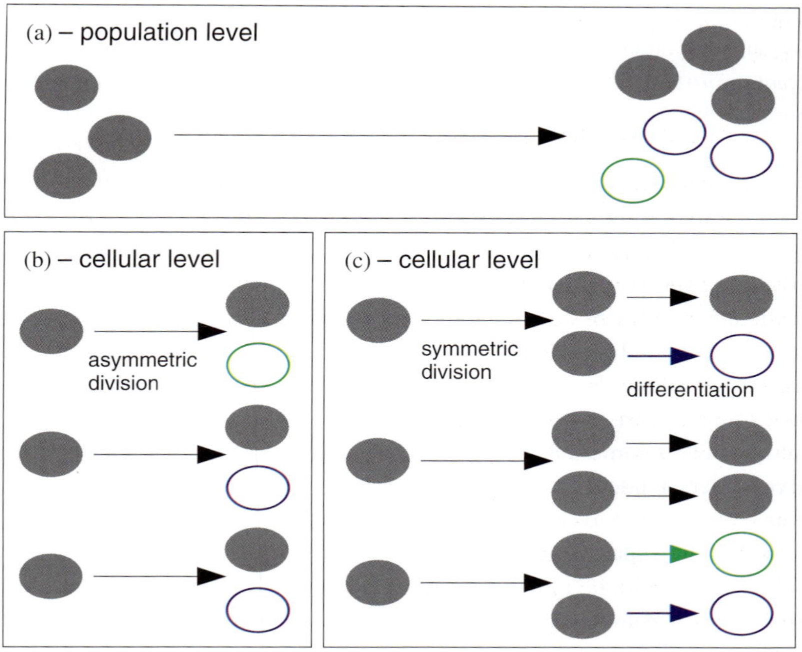

Figure 2.2. Symmetric and asymmetric divisions. (a) On the population level, the number of stem cells (dark grey) is maintained, while at the same time a fraction of differentiating cells (colored circles) is generated. (b) On the cellular level, these results can be obtained by assuming functionally asymmetric divisions of the stem cells in which one daughter cell remains a stem cell whereas the other daughter cell differentiates. (c) Alternatively, the findings on the population level can be obtained if the processes of division and differentiation are decoupled. Thus, each stem cell division leads to the generation of two closely similar daughter cells (symmetric division) of which some might undergo further differentiation (colored arrows).

and is present in many debates about organizational principles (Morrison *et al.*, 1997; Takano *et al.*, 2004; Ho and Wagner, 2007). On a simplified level, it is often assumed that only by means of asymmetric divisions the balance between self-renewal and differentiation can be established. However, to guarantee functional tissues, it is sufficient to maintain this balance on the population level (Figure 2.2a) and not necessarily on the single-cell level. This argument is in line with the current lack of evidence for a functionally asymmetric division event in hematopoiesis. Moreover, the assumption of strictly asymmetric divisions of HSCs on the single-cell level is not sufficient to explain the observed robust, dynamic response of the hematopoietic system to injury or damage as a loss in stem cell number cannot be compensated on the basis of asymmetric cell divisions only.

2.2.1.3. *Models of adaptive and self-organizing stem cell systems*

It became increasingly evident that an entity-based view on stem cells is not appropriate to account for all facets of "stemness" including the degree of heterogeneity within stem cell populations, the reversibility of cellular developments, and the adaptiveness to changing demands. Motivated by, but also stimulating these discussions, a number of stem cell models have been established, representing stem cells as individual objects with inheritable and dynamically regulated properties which only exhibit stem cell functionality in the population context. Individual cellular properties are continuously changing under the influence of environmental signals and are not restricted by the boundaries of an abstract compartmentalization. For example, within such a concept, the long-term repopulation ability of a cell does not vanish suddenly (classically described by the cell's transits from the stem cell to a progenitor compartment), but the loss of function is gradual. This also implies the relativistic standpoint that the "quality" of a stem cell always has to be appraised relative to other (competing) cells, i.e. in a scenario with many competitor cells, a particular cell might not contribute to long-term repopulation, whereas in a different situation with few competitors, it could do so.

The interactions between stem cells and their microenvironment (modeled either by local interaction rules or by global information and feedback loops) allow for an adaptive control of the balance between self-renewal and differentiation, generally referred to as self-organization. The future development of each cell depends on its current state *as well as* the state of the surrounding environment, and can only be predicted in a probabilistic sense. Single-cell–based models are particularly suited to describe and analyze the heterogeneity in clonal development as it can be observed in chimerism studies and clonal tracking experiments where the progeny of particular (stem) cells can be followed e.g. by using retroviral markers (Jordan *et al.*, 1995; Drize *et al.*, 1997; Gerrits *et al.*, 2010). Moreover, the single-cell–based structure allows the study of tissue-level consequences resulting from pertubations on the individual cell level or from changes of the local cell–cell or cell–microenvironment interaction rules (Galle *et al.*, 2005).

Building on the early conceptual works by Potten and Loeffler (1990; 1997) which placed stem cell properties in the population context, the idea of "stemness" as a systemic property was formalized by Loeffler and Roeder (2002). Inspired by the special role of the hematopoietic niche for the maintenance of HSCs, their model concept assumes that stem cell development is directly influenced by each cell's local

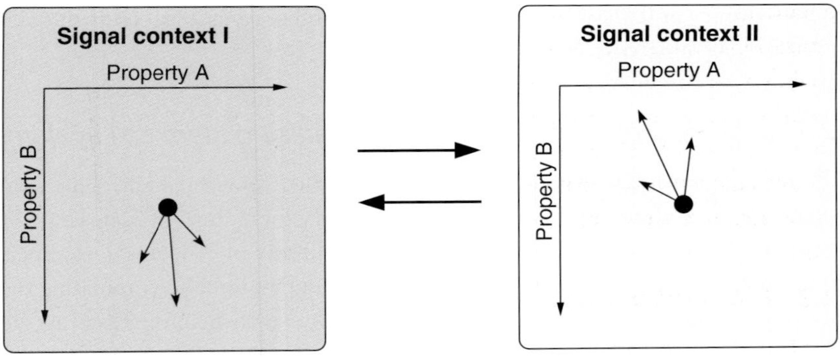

Figure 2.3. Influence of the signaling context on cellular development. Two abstract cellular properties (named A and B) are under the governance of two distinct signaling contexts (named I and II). The signaling contexts, between which the cells can actively change, modulate the properties A and B in a contrary fashion.

microenvironment, or more generally speaking, by the signaling context to which a cell is presently exposed to. This simple assumption is outlined in Figure 2.3 in which two abstract cell properties (A and B) are modulated in a contrary fashion within two distinct signaling contexts. Subsequently, a mathematical formalization of this concept has been developed by Roeder and Loeffler (2002) to construct a single-cell–based model of hematopoietic stem cell organization in which the cellular development is directly regulated by the cell's microenvironment. The assumptions of reversible binding and detachment from a niche-like environment place the cells under the governance of two antagonistic regimes which is sufficient to explain self-renewal and differentiation of HSCs as a self-organizing process.

Another elegant approach to model stem cells as self-organizing, adaptive systems has been presented by Kirkland (2004). This work demonstrates that stem cell organization can be sufficiently described in a continuum space, explicitly excluding any kind of compartmentalization in discrete subpopulations. In this concept, cells are defined by continuous variables describing functional and phenotypic properties (such as repopulation potential or multipotency). Based on this idea, cells can hardly be classified as being either "stem cells" or "non-stem cells". The variables change according to a set of probability density functions which themselves depend on the current state of the cell: cells with higher "stemness" cycle slower and have a smaller probability for differentiation as compared to cells with reduced "stemness".

A formal foundation for the concept of perceiving cells as reactive agents, which underlies all the above mentioned models, was recently provided by Theise and d'Inverno (2004). The authors argue that critical phenomena of the system only emerge if a sufficient number of reactive agents are present. Moreover, self-organization of a complex adaptive system requires interactions between the agents as well as a certain degree of non-determinism. The authors demonstrate that the perception of stem cells as reactive agents is a possible tool to relate the phenomena on the single-cell level to the resulting population dynamics of stem cells. In this sense, the characteristics that define "stemness" on the population level can possibly be translated to a set of rules on the single-cell level.

2.2.2. *Intracellular models of lineage specification*

Apart from the fate decision between self-renewal vs. differentiation, differentiating cells have to decide to which lineage they are committing. It is long recognized that this process of *lineage specification* in adult stem cell systems has a close counterpart in embryology where a zygote gives rise to a complete organism composed of a multitude of different tissues. In 1956, Conrad Waddington pictured this decision sequence by a ball rolling down an incline with different branching ridges and valleys (Waddington, 1956; 1957) (Figure 2.4). Whereas the valleys correspond to different developmental options, the ridges represent the distinctions between tissue types. This neat visualization of the process of lineage specification in development received a comeback in the last years (Ingber, 2003; Goldberg *et al.*, 2007) and is now commonly used especially in discussions about reprogramming and induced pluripotency (Ng *et al.*, 2008; Goldmann, 2008; Yamanaka, 2009).

Although the building principles of a lineage specification process are already captured by this early visualization of a developmental landscape, the picture deserves a critical interpretation and discussion. Starting from the general assumption that the phenotype of a cell is the direct result of the activity of the cell's genes and gene products, it follows that lineage specification (i.e. the change in the cell's phenotype) is ultimately linked to a change in the underlying molecular expression pattern. However, a number of questions remain: Which molecular components interact with each other? How are these interactions characterized? And how can a robust and self-stabilizing expression pattern of the above mentioned components be achieved?

An increasing number of theoretical works on the dynamics of such molecular regulatory systems focus on a conceptual understanding of lineage specification. It is already evident that the "epigenetic landscape" proposed by Waddington (1956) does not correspond to physical structures or potential energy landscapes as they are known from physics. Rather, they represent unstable (ridges) and stable (valleys) states of the cell's underlying complex interaction network (Ingber, 2003; Ao, 2009). The resulting landscape visualizes the probability for finding a cell in a certain

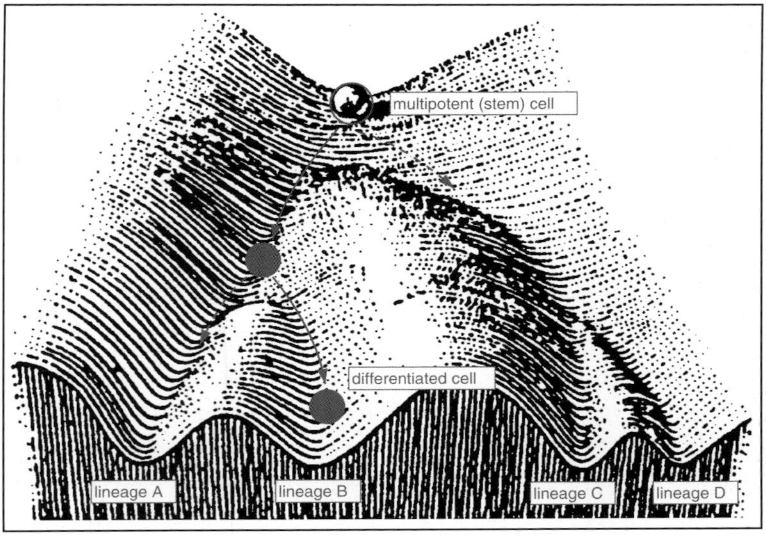

Figure 2.4. Lineage specification in the concept of state space. The "epigenetic landscape" as suggested by Waddington (1956; 1957) provides the background to understand lineage specification as a decision process in a high-dimensional landscape that arises as a consequence of the complex intracellular interaction network. A possible development is shown by the red arrows, and alternative options are indicated by the dashed arrows. Figure modified from Waddington (1957).

state (with higher probabilities in the valleys), therefore providing very helpful and intuitive features for an interpretation of the underlying network dynamics (Huang *et al.*, 2007; Cao and Liang, 2008).

2.2.2.1. *The concept of state space*

The network dynamics (i.e. temporal changes in gene expression and the convergence towards steady states) are determined by the particular arrangement and strength of the network interactions. Assume for a simplified example that gene A strongly represses gene B and that genes C and D are mutual activators. Given that these interactions are fixed based on the promoter structure of the genes, only certain expression patterns are possible while others are not. For example, A and B cannot both be activated simultaneously if one inhibits the other; conversely, C and D must both turn on if they are mutual activators. Interactions like these limit the number of self-adapting expression states and determine how they evolve over time. Given a certain stimulus, the internal state of a cell might transiently shift through various intermediate stages before it finally stabilizes in one of the self-adapting expression states (Preisler and Kauffman, 1999).

In a more formal description, every cellular component (i.e. gene transcript, protein, soluble factor etc.) within a certain cell can be represented as a dimension in a high-dimensional *state space*. In such a state space, each cell can be identified by a vector $S = \{S_1, S_2, \ldots, S_n\}$ in which each component S_i represents the concentration of a particular molecular component X_i. Given the general consensus that the phenotypic appearance of a cell is the result of the molecular interactions between all the genes, their transcripts, proteins, signaling pathways and other intracellular components, this also implies that the dimensions of the state space are not independent of each other. Such a complex, high-dimensional system of interactions is most conveniently represented by a complex network in which the nodes correspond to the different molecular components X_i and the edges characterize their mutual interactions x_{ij}. It is now the supportive and repressive nature of these mutual interactions x_{ij} leading to the emergence of patterns in which certain components X_i are expressed while others are silenced. The ability to maintain such an internal equilibrium is termed *homeostasis* whereas the stabilized, self-adapting patterns are referred to as *attractors* of the network. On the phenotypic level, such a stable attractor can be associated with a certain cell type (e.g. stem cells, mature cell types) being characterized by a typical expression of molecular components which enables the cell to fulfill specific functional tasks (Kauffman, 1993; Huang *et al.*, 2005).

In the above interpretation, the picture by Waddington (Figure 2.4) is a representation of the intracellular attractor landscape and the developmental options between them. However, there is a fundamental question which is not captured by the sketch: Is this attractor landscape fixed or is it changing in the course of differentiation? In terms of the state space formulation, the cell's intrinsic state is characterized by the vector of concentrations S_i for each of the intracellular components. However, it is the the matrix of all potential interactions x_{ij} that actually shapes the attractor landscape and that determines how the cell state S_i develops within the state space. As any specific representation of a developmental landscape is a simplification of a highly complex process, there is no unique answer on whether this landscape is constant in time or not. In fact, the answer depends on the assumptions and interpretations made for the underlying molecular network. Two contrasting perspectives are outlined below:

- The *local view* considers only a limited number of important network components for which the dynamic behavior is subject to a quantitative analysis. The abundance of other components, which are considered constant over time or change on long time scales only, are not explicitly described as nodes of the network. They are represented as (constant) parameters. As an example, one might think of transcription rates which can be described as functions of DNA methylation patterns. If such patterns change on long time scales, the depending transcription rates change slowly over time. As a consequence of this interpretation, quantitative (and qualitative) changes of the network interactions ultimately alter the network dynamics and potentially modify the attractor landscape.

As for the example of bistable switches, which is illustrated in more detail below, it is the modification of a parameter (such as a transcription rate), which drives the system from a state with just one attractor towards a state with multiple attractors (Gardner *et al.*, 2000; Roeder and Glauche, 2006). Such a parameter that is able to induce a qualitative alteration of the attractor landscape is called a *bifurcation* parameter.

Furthermore, it could be shown by studying Boolean dynamics on larger networks that modifications of the interaction rules can (although not necessarily have to) lead to dramatic effects on the attractor landscape (Wuensche, 2004). On the basis of these findings, the processes of differentiation and lineage specification, which are evidently coupled to specific epigenetic changes and are closely regulated from external signals, can be interpreted as alterations of the developmental landscape.

- In contrast to the local view, the *global view* represents each and every relevant molecular component (genes, transcripts, proteins, receptors etc.) within the underlying network structure. In this framework all (local view) parameters influencing the system behavior are considered additional nodes of the network graph. In principle, this approach can be applied to all external influences (like external cytokines, nutrition, temperature etc.) leading to a significantly increased dimension of the state space vector S.

  It is the idea of the global view that interactions between any of the network components are solely characterized by physical and chemical properties and do not involve any temporal dependence. As all external influences and changes [even those which act on slow time scales (e.g. epigenetic remodeling)] are explicitely represented in the network structure, the interaction rates represent absolute and conserved properties. This, in turn, leads to a fixed developmental landscape in which all possible attractors exist simultaneously. However, if the number of interacting components is large, the resulting landscape can be extremely diverse and the attractors are possibly scattered over the whole state space. This view closely resembles the rigorous definition of a cellular state space first formulated by Kauffman (1969; 1993).

Coming back to the developmental landscape as proposed by Waddington (cf. Figure 2.4), this picture can be interpreted in two different ways. In the *local view*, the attractor landscape changes as the interaction parameters are functions of the time of differentiation and the developmental landscape has to be interpreted as a time-ordered sequence of snapshots rather than a fixed landscape.

In contrast, a *global view* would imply that the landscape remains fixed and only the trajectory of the cell in the simplified, two-dimensional state space is a function of time. In this sense, there is a coexistence of multiple attractors corresponding to the multipotent and the subsequently restricted developmental states, and differentiation is illustrated as a transition from one attractor towards the next attractor potentially involving the decision between more than one option. It should be clearly stated that any low-dimensional representation (in terms of an attractor landscape) is insufficient to capture the complex, high-dimensional state space in the global view.

Taken together, it needs to be emphasized that the two different views on developmental landscapes are not mutually exclusive. In fact, the local view appears as a low-dimensional projection of the global view, where some dimensions of the global view (i.e. concentrations of certain molecular components) are collapsed into interaction parameters of the network such as transcription and inhibition rates. The preference for one or the other view should depend on the particular problem in question.

2.2.2.2. *Fluctuations and robustness*

Even within a stable attractor of the state space, the concentration of the intracellular component S_i might fluctuate around the theoretical steady state. Such fluctuations occur naturally and are the results of small copy numbers, degradation and diffusion processes within the cells (intrinsic noise), or spatial arrangements and influences of the local microenvironment (extrinsic noise) (Elowitz *et al.*, 2002; Kaern *et al.*, 2005; Raser and O'Shea, 2005). Noise effects can be further enhanced in nonlinear systems (Metzner *et al.*, 2009).

To study the system dynamics, one needs to distinguish whether the "shape" of the attractor and the noise amplitude are suited to induce a change of the attractor state (see also Section 2.3). Generally, small fluctuations (in most cases) do not lead to a change between available attractors and thus do not significantly change the overall expression pattern and the cell's phenotype (Huang, 1999). This feature is commonly referred to as *robustness*. In contrast, fluctuations of sufficient amplitude (e.g. induced by extrinsic effects) might contribute to phenotypic variability. In fact, such noise-induced fluctuations are a possible stimulus to induce changes of the cell state and to shift the cell into a different attractor (i.e. into another phenotype) (Hasty *et al.*, 2000; Tian and Burrage, 2006; Suel *et al.*, 2007; Losick and Desplan, 2008).

In the concept of developmental landscapes, as illustrated in Figure 2.4, the fluctuations do (in principle) enable the cells to move from one valley (i.e. attractor) into another. The probability for such state changes depends predominantly on the noise amplitude and on the "location" of the alternative attractor. As outlined above, in the global view, the picture of a low-dimensional developmental landscape is oversimplifying as the alternative attractors might be distant and separated through a rugged developmental landscape and are therefore practically not reachable. In the local view, this translates into the question of whether a certain alternative attractor coexists for a given parameter configuration or whether it needs to be created by a changing parameter configuration, e.g. encoding epigenetic remodeling.

2.2.2.3. *Reduction of complexity*

The complexity of the state space representation of a cell, including the characterization of all the mutual interactions between the cellular components, makes a comprehensive description of the cell's intrinsic dynamics in the *global view* numerically challenging, or even impossible. However, it is not just the feasibility, but also the availability of the particular data which limits a global approach. Although the availability of high-throughput measurements like mRNA microarrays allows the simultaneous determination of the expression of thousands of genes in a cell population, these methods are rarely adequate to characterize the complex interactions between them. Consequently, a dynamic picture of the cells behavior in a global state space cannot be obtained.

In order to reduce this level of complexity, the *local view* is suggested as an appropriate alternative to describe the phenomena of intracellular regulation and lineage specification. In this context, one may distinguish between two principle approaches: a *phenomenological* approach and a *functional* approach. Figure 2.5 outlines these principle perspectives. Whereas the cell in the upper panel is depicted with different sets of competing and potentially interacting genes and gene products as well as their mutual interactions (illustrating the *global view*), the snapshots in the lower panel refer to low-dimensional projections (*local view*). The *phenomenological approach* summarizes the complex regulations within sets of coregulated genes into different major components, thus allowing the study of the dynamic interactions between them. Although this approach misses the molecular

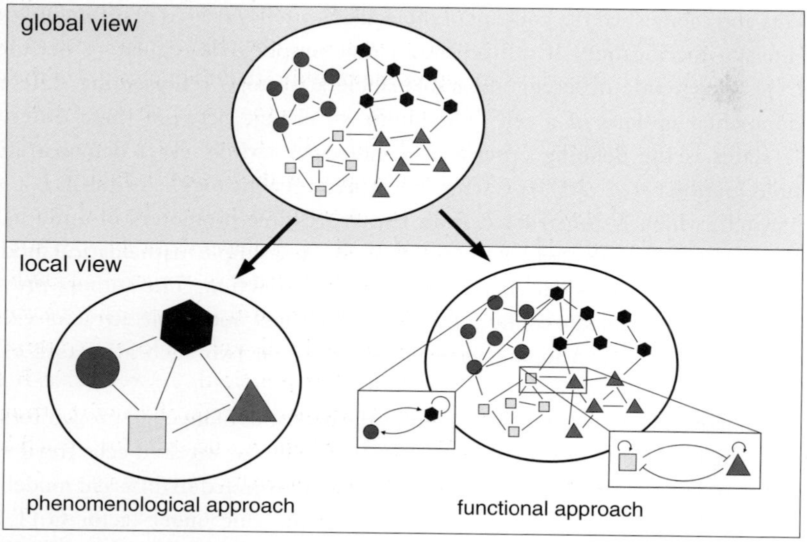

Figure 2.5. Levels of description. The cell in the upper panel reflects the entire state space of a cell in the *global view* (grey shaded area) with a multitude of interacting molecular components (e.g. genes, transcripts, proteins, receptors etc.). The representations below illustrate different variants of the *local view*. The left arm outlines the *phenomenological* approach in which complex regulations within coregulated gene sets (marked with identical colors) are summarized into different major components. In contrast, the *functional* approaches (shown for the right cell) focus on interactions between certain key regulators for which details of the molecular interactions are known.

details of the interactions within those regulatory complexes, it is sufficient to address the general principles of developmental phenomena like lineage specification and the temporal ordering of events. In contrast, the *functional approach* focuses on a detailed description of the interactions between a few key regulators, for which details of the molecular regulation are available. Although this process can elucidate the role of molecular switches for certain intracellular decisions, it captures just a brief sequence out of the complex and temporally extended differentiation dynamics and fails to embed these actions into a greater context.

2.3. Bistable Switches: Antagonistic Interactions between PU.1 and GATA-1

It was not until the 1990s that the advances in the characterization of molecular interactions led to the identification of certain key regulatory elements, which in turn opened the gate for mathematical modeling approaches of specific biological systems and established the concept of molecular switches related to differentiation and lineage specification. It can be shown that simple gene regulatory networks (GRN) can generate different, separable stable attractors, representing different developmental options of a cell. The ability to change between these different stable states is the defining criteria of a molecular switch. As a demonstrative example, Gardner *et al.* (2000) presented a synthetic genetic toggle-switch in *E. Coli*. The switch, which is constructed from two repressible promoters in a mutually inhibitory network and for which the authors provide a simple mathematical model, could be flipped by transient chemical or thermal induction. The authors proved the existence of parameter configurations in which just one stable attractor exists (monostability) or, for changing parameters, in which two such attractors exist (bistability).

With particular focus on hematopoiesis, the decisions of myeloid progenitors to differentiate either into erythroid/megakaryocytic cells or myeloid cells [reviewed in Cantor and Orkin (2002); Tenen (2003)] has been subjected to different modeling approaches. It is the mutual regulation between the zinc finger factor GATA-1 and the Ets-family transcription factor PU.1 that supports the idea of a molecular switch. Whereas increased levels of GATA-1 are required for the differentiation and maturation of erythroid/megakaryocytic cells, upregulation of PU.1 supports the development of myeloid and lymphoid cells. It has been demonstrated experimentally that there are physical interactions between GATA-1 and PU.1 that, by the formation of heterodimers, induce a mutual inhibition and, therefore, favors one lineage choice at the expense of the other (erythroid/megakaryocyte vs. myeloid) (Du *et al.*, 2002; Nerlov *et al.*, 2000; Rekhtman *et al.*, 2003; Voso *et al.*, 1994; Zhang *et al.*, 1999, 2000). Additionally, for both GATA-1 and PU.1, it has been shown that they are able to stimulate their own transcription through an auto-catalytic process (Chen *et al.*, 1995; Okuno *et al.*, 2005; Tsai *et al.*, 1991).

Based on these reported interactions, we have previously proposed a mathematical formalization of this molecular switch (Roeder and Glauche, 2006). Figure 2.6 illustrates the different types of interactions between PU.1 and GATA-1 that have been used for the particular model. Translating this network graph into a set of ordinary differential equations revealed a variety of possible dynamical phenomena. Whereas in the case of low transcriptional activity the system is characterized by a stable state in which both transcription factors are equally expressed, increasing the transcriptional activity of the auto-regulation (i.e. parameter s) of GATA-1 and PU.1 leads to the emergence of different stables states that are characterized by the dominance of either one of the two factors, thus revealing the typical properties of a molecular switch.

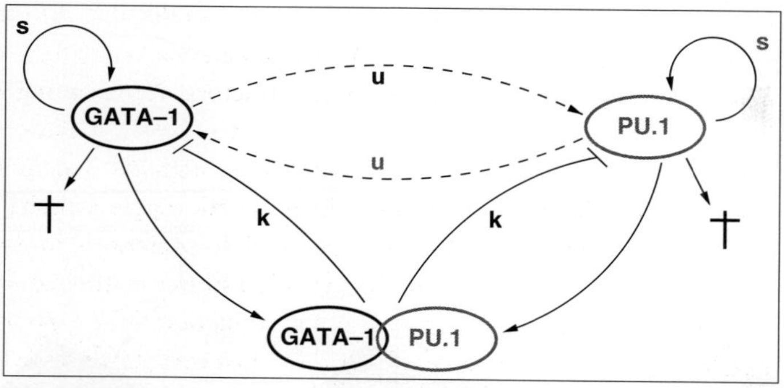

Figure 2.6. Interaction network for transcription factors GATA-1 and PU.1. The sketch shows a possible configuration of the interaction network between GATA-1 and PU.1 as discussed by Roeder and Glauche (2006) for a completely symmetrical system with the following parameters: s — transcription rate of the auto-regulation, k — inhibition of transcription by the heterodimer, u — mutual activation (this process is most likely indirect). Crosses indicate decay. As shown by Roeder and Glauche (2006), this interaction network can be translated in a set of dimensionless ordinary differential equations.

This qualitative change in system behavior is commonly depicted in a so-called bifurcation diagram. The bifurcation diagram shows the position of any fixed points of the system as a function of a regulating (i.e. bifurcation) parameter. Figure 2.7 provides the particular bifurcation diagram for the discussed example of the PU.1/GATA-1 switch. For different values of the bifurcation parameters (here, transcription rate $s = 2.5, 4, 6$), a sequence of corresponding attractor landscapes is shown in (Figures 2.7b–2.7d), illustrating the qualitative change in the number of attractors. Whereas (for low transcription rates) only one non-trivial[2] attractor exists for s (indicating the coexpression of both PU.1 and GATA-1), the system reaches a bistable regime for higher values of s (considering non-trivial steady states only) where the system can "choose" between either of two attractors (in which either PU.1 or GATA-1 is upregulated and the opposing TF is downregulated). In the latter situation (Figures 2.7c–2.7d), switching between the attractors has to be facilitated either by (intrinsic) noise or by external perturbations. In this interpretation, the bistable switch can be mapped onto the developmental landscape outlined in Figure 2.4, in which the different attractors are separated by the (conquerable) ridges in between.

[2]In the sense of non-zero concentrations.

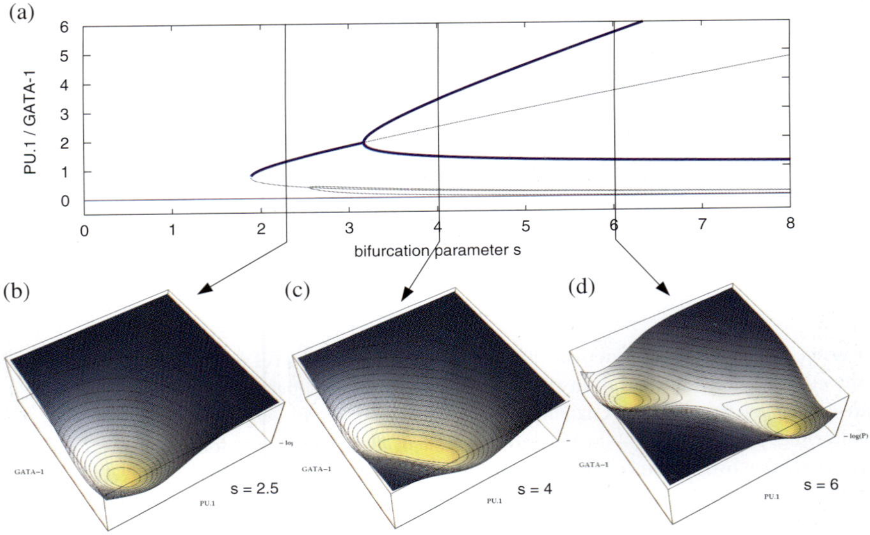

Figure 2.7. Bifurcation diagram for a bistable switch. (a) Stable fixed points for the PU.1-GATA-1 model introduced by Roeder and Glauche (2006) are indicated in blue as functions of the bifurcation parameter *s* (grey lines indicate unstable fixed points, thin blue line is the trivial fixed point at the origin). (b–d) A sequence of attractor landscapes for different values of the bifurcation parameter *s*. The *z*-axis corresponds to the negative logarithm of the probability for finding a cell at a certain position in the state space of PU.1 and GATA-1 concentrations. The attractor regions are indicated by the color yellow.

The existence of the two described regimens is well motivated by a number of findings that agree upon a general phenomenology of the lineage specification process: Whereas the undifferentiated state is characterized by a low level coexpression of lineage specific and potentially antagonistic genes in early progenitor cells, this coexpression, commonly referred to as *priming*, disappears in the course of differentiation when certain lineage restricted genes are upregulated while others are downregulated (Cross and Enver, 1997; Hu *et al.*, 1997; Enver and Greaves, 1998; Akashi *et al.*, 2003). It appears that the regulated maintenance of the priming state is a candidate mechanism for the molecular representation of multipotency. Such a "promiscuous" coexpression of genes specific for alternative lineage fates preserves the potential for different developmental options and is no longer maintained once the cells undergo terminal differentiation.

Within the picture of a bistable switch, differentiation can be described by a change of the bifurcation parameter, such as a general increase in transcriptional activity. At a certain point, a system bifurcation occurs and the system can no

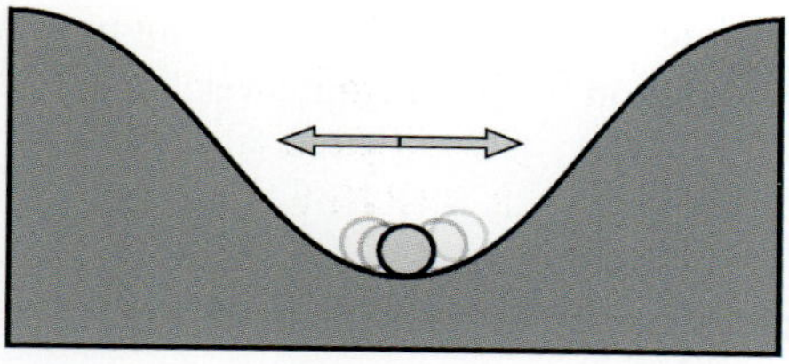

(a)

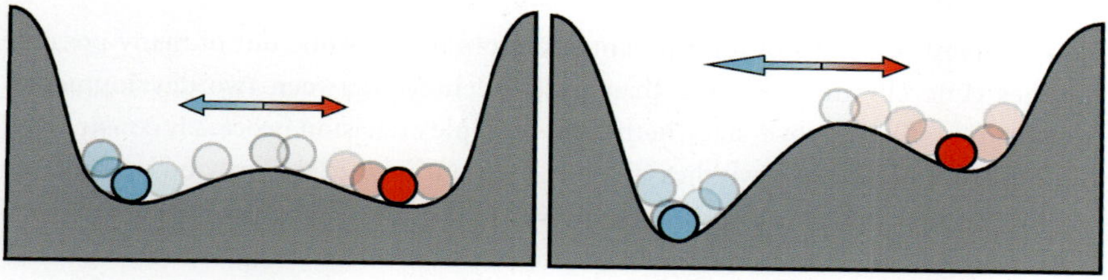

(b) (c)

Figure 2.8. Functional noise in bistable switches. (a) The situation of a monostable attractor. Albeit a certain level of noise exist, the system will ultimately return towards the unitary attractor. In (b) and (c), two bistable attractor landscapes are depicted. Whereas in the symmetric case (b), transitions between the two attractors are equally likely, this symmetry is no longer maintained in (c) resulting in a higher transition probability from the right to the left attractor (as compared to the opposite direction).

longer remain in the coexpression state. Instead, either of the two factors will be upregulated at the expense of the other. The final outcome of this decision process depends on (1) the level of (functional) noise within the system, and (2) the asymmetry of the switch. The two dimensions are both outlined in Figure 2.8. Starting out from the coexpression state (Figure 2.8a) the attractors for either the PU.1 high or GATA-1 high state are separated for increasing values of the bifurcation parameter. Initially, the system intrinsic noise is sufficient to allow changes between the attractors; however, this option vanishes as the attractors become more isolated Figure 2.8b. The noise dependence is even more obvious for the asymmetrical scenario Figure 2.8c. In this case, the probabilities for changes between the attractors are different, leading to the accumulation of cells in either of the two attractors. This behavior illustrates that the functionality of the switch requires a certain level of noise (or perturbation). However, the functional regulation of noise amplitudes and their role in developmental decisions is not completely understood and is subject to extensive experimental and theoretical research (Swain *et al.*, 2002; Raser and O'Shea, 2005; Chang *et al.*, 2008). As we demonstrate in the next section, the underlying stochasticity on the single-cell level does not prevent the establishment of robust and stable behavior on the population level.

2.4. Higher-Dimensional Decisions and Population Heterogeneity

2.4.1. *Higher-dimensional decisions*

Commitment decisions of multipotent ancestors towards one out of many possible lineages generally involve more than just the choice between two developmental options. It is still controversial whether this complex decision process is constructed as a sequence of bistable switches (like the ones outlined in the previous section) or whether special higher-dimensional switches are required (Cinquin and Demongeot, 2005; Foster *et al.*, 2009). In either case, the two-dimensional state space (e.g. PU.1 vs. GATA-1) has to be replaced by a higher-dimensional state space representing the complexity of several developmental options. In such a state space, the uncommitted ("priming") state is characterized by the low level coexpression of multiple, potentially antagonistic transcription factors, while the committment decision is associated with the dominance of one factor (or several similarly acting factors). It has been shown (on a conceptual level) by Roeder and Loeffler (2002) that the interplay of two, rather antagonistic environments is sufficient to explain this bivalence of different developmental options.

These studies have been motivated by observations in the hematopoietic system, suggesting that maintenance and self-renewal of HSCs is sensitively dependent on signals from so called stem cell niches. It is the protective action of this particular local microenvironment that preserves a fraction of cells in an uncommitted, mostly quiescent state with exceptionally high repopulation potential. In contrast, cells that are not under the tight regulation of such niches progressively lose their repopulation potential and undergo terminal differentiation.

Assuming that the process of lineage specification can be described by the competition between different developmental programs, the regulation of the process can be explained equally well in the context of two antagonistic control regimes. Illustrating this principle, we characterize each cell by their *lineage propensities* for the development into one out of N different potential lineages. Furthermore, we assume two different regulatory regiments, a *regressive* and a *dissipative* one. Whereas the *regressive* control regime maintains a balance between the individual lineage propensities, the *dissipative* control regime leads to the dominance of one propensity at the expense of the others.

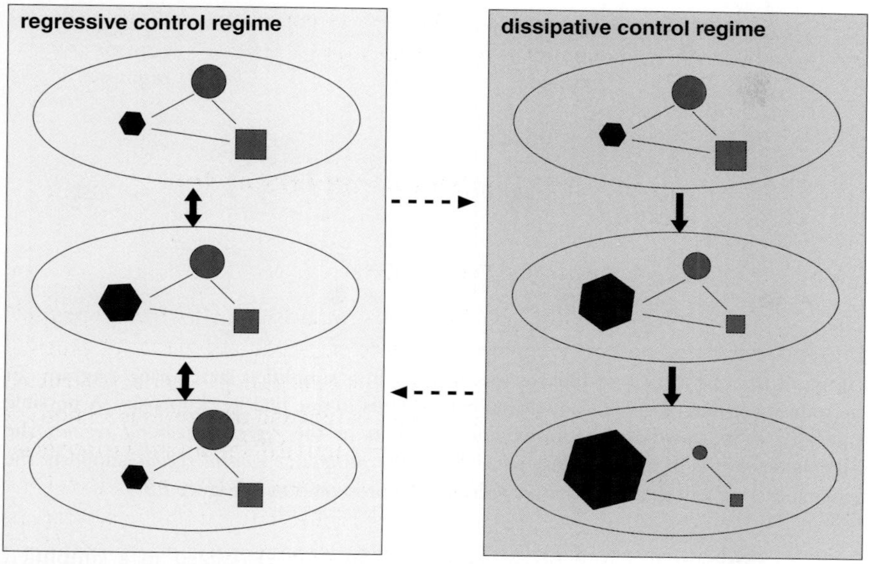

Figure 2.9. Lineage specification under two control regimes. Simplified representations of the state space S are depicted for a particular cell. Coregulated clusters of molecular components within the cell are summarized in a single, generic measure termed *lineage propensity* (indicated by the symbols). In the *dissipative control regime*, there is a clear tendency for the upregulation of one (in this case the "blue") lineage propensity at the expense of the others (directed movement in the state space). In contrast, the *regressive control regime* supports a leveling of the expression states of the three propensities with moderate fluctuations and reversible changes (undirected movement in the state space). The cells obey the principle ability to change between the two control regimes (dashed arrows).

The scheme in Figure 2.9 illustrates the action of the different control regimes. The individual lineage propensities (depicted by the colored symbols) comprise the whole spectrum of signals and regulators that are specifically expressed for one common lineage fate. This means that the action of different control regimes is not described on the level of individual genes but summarized as the combined expression of coregulated, lineage specific components (i.e. genes, transcription factors, surface factors, receptors and signaling pathways) that are representative for a certain cell fate (cf. the *phenomenological approach* described above). The level of the particular lineage propensity represents the potential of a cell to develop into the corresponding lineage. Changes between the control regimes are required in order to allow for an adaptive and dynamically stabilized system. However, such changes do also imply that reversible developments in the process of lineage specification are generally possible.

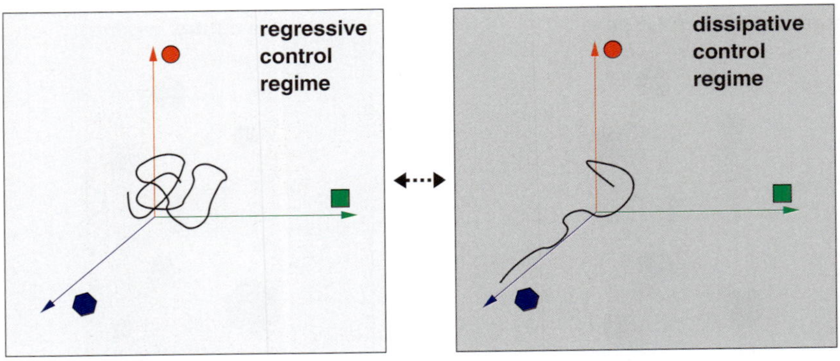

Figure 2.10. Dynamics of lineage specification in a simplified state space diagram. In the state space diagram, each dimension corresponds to one lineage propensity. A possible trajectory is indicated by the black line. Whereas in the *regressive control regime*, the propensities are kept at an almost equal level, the *dissipative control regime* supports the dominance of one of them. This is an alternative representation to Figure 2.9.

The differential action of the control regimes is visualized in a simplified state space diagram (Figure 2.10). For an example of three competing lineages, which are assigned to the axes of a three-dimensional diagram, each point in the state space refers to a triplet of lineage propensities. The temporal development is consequently depicted by a trajectory as a sequence of such points. In this state space, the *regressive control regimes* leads to trajectories close to the origin, characteristic for low and balanced lineage propensities, whereas the *dissipative control regime* drives the trajectory along one of the axis, indicating a dominating lineage propensity while the others are suppressed. The general concept of different competing cell fates can be easily extended to higher numbers of lineages. However, one has to keep in mind that this model is just a phenotypic representation of the underlying interaction network for which many details are not yet known.

2.4.2. *Population behavior*

Upon integration of the outlined model of higher-dimensional switches in a large number of individual cells, it is possible to study the dynamic effects of the lineage specification process on the population level. For a particular example, we start out with a population of 1000 cells for which three different lineage propensities are initially balanced. This corresponds to a population of uncommitted progenitor cells that are equally likely to contribute to three different lineages among their progeny. As the cells proliferate and undergo terminal differentiation, the initial balanced situation is lost and the descendents of the original cells convert into one of the

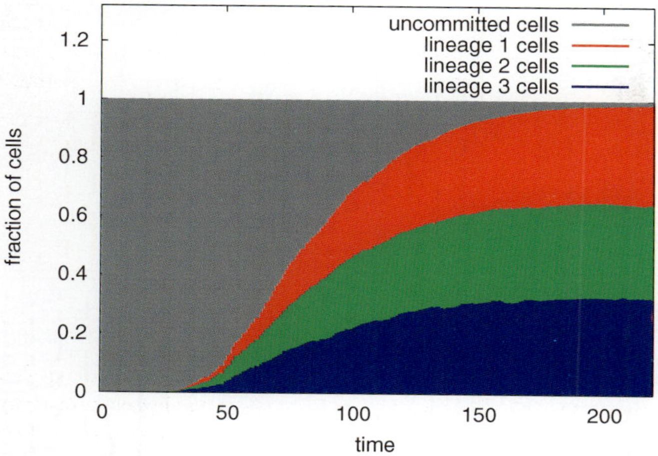

Figure 2.11. Lineage specification on the population level. Starting out from a population of uncommitted cells with balanced initial lineage propensities (shown in grey), the process of lineage specification among the individual progenitor cells leads to upregulating one lineage propensity at the expense of the others. Over time, this process results in the emergence of committed cells (indicated by the colors red, green, blue).

three lineages [by randomly upregulating one lineage propensity at the expense of the others; for details see Glauche *et al.* (2007)]. A typical time course of this overall commitment is illustrated in Figure 2.11. Starting out from a population of uncommitted cells (shown in grey), the process of lineage specification leads to the emergence of increasing fractions of cells committed to either of the three lineages (indicated by the colors red, green, blue).

Upon repetition of the *in silico* experiment, the emergence of three almost equally sized populations of "red", "green" and "blue" cells is robust and reproducible. However, looking on the commitment decisions of individual cells, it is obvious that the robust population behavior appears as a result of a very heterogeneous and flexible cellular development. We have recently introduced the notion of a "cellular genealogy" to summarize the development of an individual cell and its complete progeny into a tree-like structure (Glauche *et al.*, 2009). These cellular genealogies are well suited to incorporate further cellular attributes (like morphology, state of lineage commitment, expression of cell surface markers) in the divisional context and to illustrate and quantify their mutual relations.

For the example of a differentiating, *in silico* cell culture illustrated in Figure 2.11, we provide a selection of cellular genealogies originating from 6 of the 1000 initial cells in Figure 2.12. The color coding reflects the cell-intrinsic state

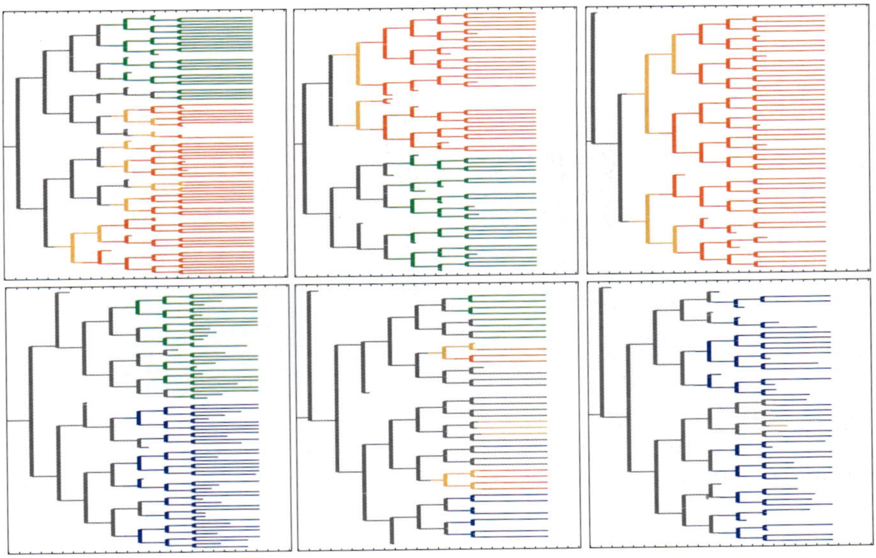

Figure 2.12. Cellular genealogies. Selection of six typical cellular genealogies for the differentiation process illustrated in Figure 2.11. Starting from the initial cells on the left side of each subfigure, the corresponding progeny is arranged in the branches. Color coding corresponds to the actual cell-intrinsic state of lineage commitment and indicates whether the cell is still uncommitted (grey) or whether one of the lineage propensities is already dominating (committed state, indicated by the red, green, blue coloring).

of lineage commitment (i.e. the lineage propensities) and indicates whether the cell is still uncommitted or whether one of the lineage propensities is already dominating. It should be emphasized that none of these individual cellular genealogies is identical to each other (nor to the remaining 994 cellular genealogies not shown here), indicating that the individual lineage specification is heterogenous, albeit leading to a stable and robust population behavior.

We furthermore point out that the progeny of some of the initial cells only contributes to one lineage, while some contribute to two or even three lineages. A histogram, indicating the fraction of cells contributing to each possible combination of lineage fates, is provided in Figure 2.13a. It is obvious that for the particular scenario, all developmental options are realized with very similar fractions. However, we point out that depending on the particular parameter choice, alternative scenarios are also possible, e.g. with dominating contributions to just one lineage fate or to a set of cellular genealogies, of which most realizations contribute to all three possible lineages.

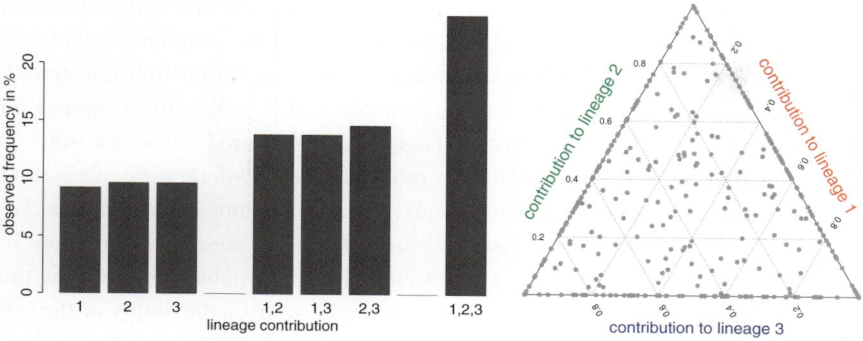

(a) (b)

Figure 2.13. Heterogeneity of single cell fates. (a) Histogram indicating the fraction of initial cells contributing to different combinations of lineages. (b) Ternary plot showing the mixture of lineages for each cellular genealogy. Cellular genealogies with contribution to just one lineage are mapped onto the corners of the triangle. Cellular genealogies with contribution to two out of three lineages are shown on the axes.

In addition the ternary plot in Figure 2.13b shows the fractions of "red", "green" and "blue" cells to which each of the individual cellular genealogies contributes. As the cellular genealogies with contribution to just one lineage are mapped onto the corners of the triangle and the cellular genealogies with contribution to two out of three lineages are shown on the axes, the plot is most illustrative to vizualize the heterogeneity among the cellular genealogies contributing to all three possible lineages. Here, again, a wide spectrum of possible outcomes is realized in the *in silico* simulations.

2.5. Summary

The characterization of individual cells in terms of their state space is an appropriate representation of their intracellular repertoire. Within this conceptual framework, it is possible to address different features such as cell fate decisions and cellular heterogeneity in a conceptual and mathematical consistent way. As a complete representation of a cell's molecular repertoire (including concentrations of different types of transcripts, proteins, surface factors, epigenetic configurations) is experimentally not feasible at the moment, we described different approaches for reducing this complexity to biologically meaningful and numerically practicable dimensions.

Taking the most simple example of two interacting transcription factors, it has been demonstrated that their mutual regulation is sufficient to generate several, qualitatively different system states. For the particular interactions between transcription factors PU.1 and GATA-1, we showed that the variation of an overall transcriptional activity shifts the system from a monostable regime (one non-trivial, stable fixed point) to a bistable regime (two non-trivial, stable fixed points), in which either of the two transcription factors is upregulated while the other is downregulated. Such qualitative changes in the attractor configuration of a system are commonly perceived as potential representations of molecular switches. We also point out that the actual decision for one of the two coexisting attractors in the bistable regime is potentially a stochastic, albeit adjustable process. It is the advantage of such a fluctuation-driven system to allow for the flexibility of the cell fate decision in response to environmental signals.

Although the individual cell fate decision on the single-cell level is no longer predictable, it is well possible to generate stable population behavior under the same conditions. Using the example of a simple, three-dimensional state space implemented for multiple cells we demonstrated a robust and reproducible differentiation pattern on the population level. However, the underlying stochastic model indicates that the individual cellular development is potentially very heterogeneous. We, therefore, argue that predictions about the future development of individual cells can only be made in a statistical sense and that stem cell populations naturally have a certain irreducible degree of heterogeneity. This heterogeneity might actually be a functional feature of any stem cell population as it guarantees for the coverage of the necessary functional spectrum. This notion is in very good agreement with the functional definition of stem cells which attributes "stemness" not to individual objects but to stem cell populations in general.

Acknowledgement

The authors thank Nico Scherf and Katrin Horn for assistance in preparing Figures 2.7, 2.8 and 2.13.

This research was supported by the European Commission project EuroSySTem (200270), and by the German Ministry for Education and Research (BMBF), MedSys-Verbundprojekt HaematoSys (0315452A).

Bibliography

Abkowitz, J. L., Catlin, S. N. and Guttorp, P. 1996. Evidence that hematopoiesis may be a stochastic process *in vivo*. *Nat Med* 2:190–197.

Abkowitz, J. L., Golinelli, D., Harrison, D. E. and Guttorp, P. 2000. *In vivo* kinetics of murine hemopoietic stem cells, *Blood* 96:3399–3405.

Akashi, K., He, X., Chen, J., Iwasaki, H., Niu, C., Steenhard, B., Zhang, J., Haug, J. and Li, L. 2003. Transcriptional accessibility for genes of multiple tissues and hematopoietic lineages is hierarchically controlled during early hematopoiesis, *Blood* 101:383–389.

Alonso, L. and Fuchs, E. 2003. Stem cells of the skin epithelium, *Proc Natl Acad Sci U S A* 100:11830–11835.

Ao, P. 2009. Global view of bionetwork dynamics: Adaptive landscape, *J Genet Genomics* 36:63–73.

Becker, A. J., McCulloch, E. A. and Till, J. E. 1963. Cytological demonstration of the clonal nature of spleen colonies derived from transplanted mouse marrow cells, *Nature* 197:452–454.

Betschinger, J. and Knoblich, J. A. 2004. Dare to be different: Asymmetric cell division in drosophila, c. elegans and vertebrates, *Curr Biol* 14:R674–R685.

Blanpain, C. and Fuchs, E. 2006. Epidermal stem cells of the skin, *Annu Rev Cell Dev Biol* 22:339–373.

Bonnet, D. 2002. Haematopoietic stem cells, *J Pathol* 197:430–440.

Cantor, A. B. and Orkin, S. H. 2002. Transcriptional regulation of erythropoiesis: An affair involving multiple partners, *Oncogene* 21:3368–3376.

Cao, Y. and Liang, J. 2008. Optimal enumeration of state space of finitely buffered stochastic molecular networks and exact computation of steady state landscape probability, *BMC Syst Biol* 2:30.

Catlin, S. N., Guttorp, P. and Abkowitz, J. L. 2005. The kinetics of clonal dominance in myeloproliferative disorders, *Blood* 106:2688–2692.

Challen, G. A., Boles, N., Lin, K. K.-Y. and Goodell, M. A. 2009. Mouse hematopoietic stem cell identification and analysis, *Cytometry A* 75:14–24.

Chang, H. H., Hemberg, M., Barahona, M., Ingber, D. E. and Huang, S. 2008. Transcriptome-wide noise controls lineage choice in mammalian progenitor cells, *Nature* 453:544–547.

Chen, H., Ray-gallet, D., Zhang, P., Hetherington, C. J., Gonzalez, D. A., Zhang, D. E., Moreau-gachelin, F. and Tenen, D. G. 1995. PU.1 (Spi-1) autoregulates its expression in myeloid cells, *Oncogene* 11:1549–1560.

Cinquin, O. and Demongeot, J. 2005. High-dimensional switches and the modelling of cellular differentiation, *J Theor Biol* 233:391–411.

Colijn, C. and Mackey, M. C. 2005a. A mathematical model of hematopoiesis–i. periodic chronic myelogenous leukemia, *J Theor Biol* 237:117–132.

Colijn, C. and Mackey, M. C. 2005b. A mathematical model of hematopoiesis: II. cyclical neutropenia, *J Theor Biol* 237:133–146.

Cross, M. A. and Enver, T. 1997. The lineage commitment of haemopoietic progenitor cells, *Curr Opin Genet Dev* 7:609–613.

de Wert, G. and Mummery, C. 2003. Human embryonic stem cells: Research, ethics and policy, *Hum Reprod* 18:672–682.

Drize, N., Chertkov, J., Sadovnikova, E., Tiessen, S. and Zander, A. 1997. Long-term maintenance of hematopoiesis in irradiated mice by retrovirally transduced peripheral blood stem cells, *Blood* 89:1811–1817.

Du, J., Stankiewicz, M. J., Liu, Y., Xi, Q., Schmitz, J. E., Lekstrom-Himes, J. A. and Ackerman, S. J. 2002. Novel combinatorial interactions of GATA-1, PU.1, and

C/EBPepsilon isoforms regulate transcription of the gene encoding eosinophil granule major basic protein, *J Biol Chem* 277:43481–43494.

Elowitz, M. B., Levine, A. J., Siggia, E. D. and Swain, P. S. 2002. Stochastic gene expression in a single cell, *Science* 297:1183–1186.

Enver, T. and Greaves, M. 1998. Loops, lineage, and leukemia, *Cell* 4:9–12.

Evans, M. J. and Kaufman, M. H. 1981. Establishment in culture of pluripotential cells from mouse embryos, *Nature* 292:154–156.

Fairbairn, L. J., Cowling, G. J., Reipert, B. M. and Dexter, T. M. 1993. Suppression of apoptosis allows differentiation and development of a multipotent hemopoietic cell line in the absence of added growth factors, *Cell* 74:823–832.

Foster, D. V., Foster, J. G., Huang, S. and Kauffman, S. A. 2009. A model of sequential branching in hierarchical cell fate determination, *J Theor Biol* 260: 589–597.

Galle, J., Loeffler, M. and Drasdo, D. 2005. Modeling the effect of deregulated proliferation and apoptosis on the growth dynamics of epithelial cell populations *in vitro*, *Biophys J* 88:62–75.

Gardner, T. S., Cantor, C. R. and Collins, J. J. 2000. Construction of a genetic toggle switch in *Escherichia coli*, *Nature* 403:339–342.

Gerrits, A., Dykstra, B., Kalmykowa, O. J., Klauke, K., Verovskaya, E., Broekhuis, M. J. C., de Haan, G. and Bystrykh, L. V. 2010. Cellular barcoding tool for clonal analysis in the hematopoietic system, *Blood* 115(13):2610–2618.

Glauche, I., Cross, M., Loeffler, M. and Roeder, I. 2007. Lineage specification of hematopoietic stem cells: Mathematical modeling and biological implications, *Stem Cells* 25:1791–1799.

Glauche, I., Lorenz, R., Hasenclever, D. and Roeder, I. 2009. A novel view on stem cell development: Analysing the shape of cellular genealogies, *Cell Prolif* 42:248–263.

Goetz, M. and Huttner, W. B. 2005. The cell biology of neurogenesis, *Nat Rev Mol Cell Biol* 6:777–788.

Goldberg, A. D., Allis, C. D. and Bernstein, E. 2007. Epigenetics: A landscape takes shape, *Cell* 128:635–638.

Goldmann, B. 2008. Smash the (cell) state! *Nature Reports Stem Cells*, doi:10.1038/stemcells.2008.115.

Goodell, M. A. 2002. Multipotential stem cells and 'side population' cells, *Cytotherapy* 4:507–508.

Hasty, J., Pradines, J., Dolnik, M. and Collins, J. J. 2000. Noise-based switches and amplifiers for gene expression, *Proc Natl Acad Sci U S A* 97:2075–2080.

Hearn, T., Haurie, C. and Mackey, M. C. 1998. Cyclical neutropenia and the peripheral control of white blood cell production, *J Theor Biol* 192:167–181.

Ho, A. D. and Wagner, W. 2007. The beauty of asymmetry: Asymmetric divisions and self-renewal in the haematopoietic system, *Curr Opin Hematol* 14:330–336.

Hu, M., Krause, D., Greaves, M., Sharkis, S., Dexter, M., Heyworth, C. and Enver, T. 1997. Multilineage gene expression precedes commitment in the hemopoietic system, *Genes Dev* 11:774–785.

Huang, S. 1999. Gene expression profiling, genetic networks, and cellular states: An integrating concept for tumorigenesis and drug discovery, *J Mol Med* 77:469–480.

Huang, S., Eichler, G., Bar-Yam, Y. and Ingber, D. E. 2005. Cell fates as high-dimensional attractor states of a complex gene regulatory network, *Phys Rev Lett* 94:128701.

Huang, S., Guo, Y.-P., May, G. and Enver, T. 2007. Bifurcation dynamics in lineage-commitment in bipotent progenitor cells, *Dev Biol* 305:695–713.

Ingber, D. E. 2003. Tensegrity II. How structural networks influence cellular information processing networks, *J Cell Sci* 116:1397–1408.

Ivanova, N. B., Dimos, J. T., Schaniel, C., Hackney, J. A., Moore, K. A. and Lemischka, I. R. 2002. A stem cell molecular signature, *Science* 298:601–604.

Jordan, C. T., Astle, C. M., Zawadzki, J., Mackarehtschian, K., Lemischka, I. R. and Harrison, D. E. 1995. Long-term repopulating abilities of enriched fetal liver stem cells measured by competitive repopulation, *Exp Hematol* 23:1011–1015.

Kaern, M., Elston, T., Blake, W. and Collins, J. 2005. Stochasticity in gene expression: From theories to phenotypes, *Nat Rev Genet* 6:451–464.

Kauffman, S. 1969. Homeostasis and differentiation in random genetic control networks, *Nature* 224:177–178.

Kauffman, S. A. 1993. *The Origins of Order: Self-Organization and Selection of Evolution*, Oxford University Press.

Kiel, M. J., Yilmaz, O. H., Iwashita, T., Yilmaz, O. H., Terhorst, C. and Morrison, S. J. 2005. Slam family receptors distinguish hematopoietic stem and progenitor cells and reveal endothelial niches for stem cells, *Cell* 121:1109–1121.

Kirkland, M. A. 2004. A phase space model of hemopoiesis and the concept of stem cell renewal, *Exp Hematol* 32:511–519.

Lechler, T. and Fuchs, E. 2005. Asymmetric cell divisions promote stratification and differentiation of mammalian skin, *Nature* 437:275–280.

Lemischka, I. 2004. (Post) genomic stem cell, in R. Lanza, J. Gearhart, B. Hogan, D. Melton, R. Pedersen, J. Thomson and M. West (eds.), *Handbook of Stem Cells*, Academic Press, Burlington, pp. 21–46.

Loeffler, M. and Grossmann, B. 1991. A stochastic branching model with formation of subunits applied to the growth of intestinal crypts, *J Theor Biol* 150:175–191.

Loeffler, M., Pantel, K., Wulff, H. and Wichmann, H. E. 1989. A mathematical model of erythropoiesis in mice and rats. Part 1: Structure of the model, *Cell Tissue Kinet* 22:13–30.

Loeffler, M. and Potten, C. S. 1997. *Stem Cells and Cellular Pedigrees — A Conceptual Introduction*. Academic Press, Cambridge, pp. 1–27.

Loeffler, M. and Roeder, I. 2002. Tissue stem cells: Definition, plasticity, heterogeneity, self-organization and models — a conceptual approach, *Cells Tissues Organs* 171:8–26.

Losick, R. and Desplan, C. 2008. Stochasticity and cell fate, *Science* 320:65–68.

Mackey, M. C. 1978. Unified hypothesis for the origin of aplastic anemia and periodic hematopoiesis, *Blood* 51:941–956.

Mackey, M. C. and Glass, L. 1977. Oscillation and chaos in physiological control systems, *Science* 197:287–289.

Martin, G. R. 1981. Isolation of a pluripotent cell line from early mouse embryos cultured in medium conditioned by teratocarcinoma stem cells, *Proc Natl Acad Sci U S A* 78:7634–7638.

Metzner, C., Sajitz-Hermstein, M., Schmidberger, M. and Fabry, B. 2009. Noise and critical phenomena in biochemical signaling cycles at small molecule numbers, *Phys Rev E* 80:021915.

Morrison, S. J., Shah, N. M. and Anderson, D. J. 1997. Regulatory mechanisms in stem cell biology, *Cell* 88:287–298.

Nerlov, C., Querfurth, E., Kulessa, H. and Graf, T. 2000. GATA-1 interacts with the myeloid PU.1 transcription factor and represses PU.1-dependent transcription, *Blood* 95:2543–2551.

Ng, R. K., Dean, W., Dawson, C., Lucifero, D., Madeja, Z., Reik, W. and Hemberger, M. 2008. Epigenetic restriction of embryonic cell lineage fate by methylation of elf5, *Nat Cell Biol* 10:1280–1290.

Okita, K., Ichisaka, T. and Yamanaka, S. 2007. Generation of germline-competent induced pluripotent stem cells, *Nature* 448:313–317.

Okuno, Y., Huang, G., Rosenbauer, F., Evans, E. K., Radomska, H. S., Iwasaki, H., Akashi, K., Moreau-Gachelin, F., Li, Y., Zhang, P., Gottgens, B. and Tenen, D. G. 2005. Potential autoregulation of transcription factor PU.1 by an upstream regulatory element, *Mol Cell Biol* 25:2832–2845.

Potten, C. S. and Loeffler, M. 1990. Stem cells: Attributes, cycles, spirals, pitfalls and uncertainties. lessons for and from the crypt, *Development* 110:1001–1020.

Potten, C. S. and Morris, R. J. 1988. Epithelial stem cells *in vivo*, *J Cell Sci Suppl* 10:45–62.

Preisler, H. and Kauffman, S. 1999. A proposal regarding the mechanism which underlies lineage choice during hematopoietic differentiation, *Leuk Res* 23:685–694.

Prudhomme, W. A., Duggar, K. H. and Lauffenburger, D. A. 2004. Cell population dynamics model for deconvolution of murine embryonic stem cell self-renewal and differentiation responses to cytokines and extracellular matrix, *Biotechnol Bioeng* 88:264–272.

Ramalho-Santos, M., Yoon, S., Matsuzaki, Y., Mulligan, R. C. and Melton, D. A. 2002. "stemness": Transcriptional profiling of embryonic and adult stem cells, *Science* 298:597–600.

Raser, J. M. and O'Shea, E. K. 2005. Noise in gene expression: Origins, consequences, and control, *Science* 309:2010–2013.

Reincke, U., Loeffler, M., Wichmann, H. E. and Harrison, B. 1984. The kinetics of granulopoiesis in long-term mouse bone marrow culture. Part I, *Int J Cell Cloning* 2:394–407.

Rekhtman, N., Choe, K. S., Matushansky, I., Murray, S., Stopka, T. and Skoultchi, A. I. 2003. PU.1 and pRB interact and cooperate to repress GATA-1 and block erythroid differentiation, *Mol Cell Biol* 23:7460–7474.

Roeder, I. and Glauche, I. 2006. Towards an understanding of lineage specification in hematopoietic stem cells: A mathematical model for the interaction of transcription factors GATA-1 and PU.1, *J Theor Biol* 241:852–865.

Roeder, I. and Loeffler, M. 2002. A novel dynamic model of hematopoietic stem cell organization based on the concept of within-tissue plasticity, *Exp Hematol* 30:853–861.

Roeder, I. and Lorenz, R. 2006. Asymmetry of stem cell fate and the potential impact of the niche observations, simulations, and interpretations, *Stem Cell Rev* 2:171–180.

Scholz, M., Engel, C. and Loeffler, M. 2005. Modelling human granulopoiesis under poly-chemotherapy with g-csf support, *J Math Biol* 50:397–439.

Siminovitch, L., McCulloch, E. A. and Till, J. E. 1963. The distribution of colony-forming cells among spleen colonies, *J Cell Physiol* 62:327–336.

Suel, G. M., Kulkarni, R. P., Dworkin, J., Garcia-Ojalvo, J. and Elowitz, M. B. 2007. Tunability and noise dependence in differentiation dynamics, *Science* 315:1716–1719.

Swain, P. S., Elowitz, M. B. and Siggia, E. D. 2002. Intrinsic and extrinsic contributions to stochasticity in gene expression, *Proc Natl Acad Sci U S A* 99:12795–12800.

Takahashi, K. and Yamanaka, S. 2006. Induction of pluripotent stem cells from mouse embryonic and adult fibroblast cultures by defined factors, *Cell* 126:663–676.

Takano, H., Ema, H., Sudo, K. and Nakauchi, H. 2004. Asymmetric division and lineage commitment at the level of hematopoietic stem cells: Inference from differentiation in daughter cell and granddaughter cell pairs, *J Exp Med* 199:295–302.

Temple, S. 2001. The development of neural stem cells, *Nature* 414:112–117.

Tenen, D. G. 2003. Disruption of differentiation in human cancer: AML shows the way, *Nat Rev Cancer* 3:89–101.

Theise, N. and d'Inverno, M. 2004. Understanding cell lineages as complex adaptive systems, *Blood Cells Mol Dis* 32:17–20.

Theise, N. D. 2003. Liver stem cells: Prospects for treatment of inherited and acquired liver diseases, *Expert Opin Biol Ther* 3:403–408.

Thomson, J. A. and Marshall, V. S. 1998. Primate embryonic stem cells, *Curr Top Dev Biol* 38:133–165.

Tian, T. and Burrage, K. 2006. Stochastic models for regulatory networks of the genetic toggle switch, *Proc Natl Acad Sci U S A* 103:8372–8377.

Till, J. E. and McCulloch, E. A. 1961. A direct measurement of radiation sensitivity of normal mouse bone marrow cells, *Radiat Res* 14:213–222.

Till, J. E., McCulloch, E. A. and Siminovitch, L. 1964. A stochastic model of stem cell proliferation, based on the growth of spleen colony-forming cells, *Proc Natl Acad Sci U S A* 51:29–36.

Tsai, S. F., Strauss, E. and Orkin, S. H. 1991. Functional analysis and *in vivo* footprinting implicate the erythroid transcription factor GATA-1 as a positive regulator of its own promoter, *Genes Dev* 5:919–931.

van der Flier, L. G. and Clevers, H. 2009. Stem cells, self-renewal, and differentiation in the intestinal epithelium, *Annu Rev Physiol* 71:241–260.

Vogel, H., Niewisch, H. and Matioli, G. 1969. Stochastic development of stem cells, *J Theor Biol* 22:249–270.

Voso, M. T., Burn, T. C., Wulf, G., Lim, B., Leone, G. and Tenen, D. G. 1994. Inhibition of hematopoiesis by competitive binding of transcription factor PU.1, *Proc Natl Acad Sci U S A* 91:7932–7936.

Waddington, C. H. 1956. *Principles of Embryology*. George Allen & Unwin, Ltd.

Waddington, C. H. 1957. *The Strategy of the Genes. A Discussion of Some Aspects of Theoretical Biology*. George Allen & Unwin, Ltd.

Weissman, I. L. 2000. Stem cells: Units of development, units of regeneration, and units in evolution, *Cell* 100:157–168.

Wichmann, H. E., Loeffler, M. and Reincke, U. 1984. The kinetics of granulopoiesis in long-term mouse bone marrow culture. Part II, *Int J Cell Cloning* 2:408–424.

Wognum, A. W., Eaves, A. C. and Thomas, T. E. 2003. Identification and isolation of hematopoietic stem cells, *Arch Med Res* 34:461–475.

Wuensche, A. 2004. *Modularity in Development and Evolution*, Chap. Basins of attraction in network dynamics: A conceptual framework for biomolecular networks. Chicago University Press, pp. 288–311.

Yamanaka, S. 2009. Elite and stochastic models for induced pluripotent stem cell generation, *Nature* 460:49–52.

Zhang, P., Behre, G., Pan, J., Iwama, A., Wara-Aswapati, N., Radomska, H. S., Auron, P. E., Tenen, D. G. and Sun, Z. 1999. Negative cross-talk between hematopoietic regulators: GATA proteins repress PU.1. *Proc Natl Acad Sci U S A* 96:8705–8710.

Zhang, P., Zhang, X., Iwama, A., Yu, C., Smith, K. A., Mueller, B. U., Narravula, S., Torbett, B. E., Orkin, S. H. and Tenen, D. G. 2000. PU.1 inhibits GATA-1 function and erythroid differentiation by blocking GATA-1 DNA binding, *Blood* 96:2641–2648.

Comparative Transcriptomics for Global Functional Profiling of Embryonic Stem Cells

3

Yu Sun and Ming Zhan

3.1. Introduction

Embryonic stem cells (ESCs) harvested from different species share similar pluripotent properties; they can replicate indefinitely and differentiate into all types of cells. However, major differences among ESCs isolated from different species have also been reported (Ginis *et al.*, 2004; Rao, 2004; Zhan *et al.*, 2005). For example, the LIF pathway is critical in maintaining the pluripotency of mouse ESCs, but it is not essential in humans to maintain ESC pluripotency (Ying *et al.*, 2003; Daheron *et al.*, 2004; Humphrey *et al.*, 2004). Identifying functional modules such as biological processes or pathways that show evolutionally conserved behavior will help us identify the core molecular mechanisms underlying the pluripotent property of ESCs.

In this chapter, we describe the findings and methodologies of large scale and global functional profiling of ESCs by comparative transcriptomics (Sun *et al.*, 2007). This analysis systematically surveyed all possible functional modules categorized by the Gene Ontology (GO) and Kyoto Encyclopedia of Genes and Genomes (KEGG), and all transcription factors and growth factors expressed in ESCs. All the expressed genes in a given functional module (instead of differentially expressed genes alone) were analyzed to explore, not only major but subtle yet important, transcriptional changes that occur upon ESC differentiation. As a result, evolutionarily conserved and divergent functional modules were identified from human and mouse ESCs. The conserved modules, showing positive cross-species correlation on the expression profile, represent core biological networks or molecular mechanisms fundamental for ESC development. The divergent modules, showing negative or no correlation on the expression profile, are suggestive of species-specific regulation in ESCs. Hundreds of functional modules, as defined by GO or KEGG, were associated with conserved expression patterns but bear no overt relationship to ESC development, suggestive of new mechanisms critical to ESC pluripotency. The comparison of ESCs across species and between human ESCs and embryonal carcinoma stem cells (ECCs) suggest that while pluripotency is an essential function in multi-cellular organisms conserved through evolution, mechanisms primed for differentiation are less conserved and contribute substantially to the differences among stem cells derived from different tissues or species.

3.2. Comparative Transcriptomics Analysis

In this study, we obtained the gene expression profile data of human ESCs (hESCs) from four hESC cell lines (BG01, BG01V, BG02, and BG03) and their early differentiated cell counterparts, embryoid bodies (EBs), using Illumina BeadArray. The array contained 23,584 probes, representing 20,692 unique genes. We obtained the gene expression profile data of mouse ESCs (mESCs) from the GEO database (GSE2375). The mESC dataset contained expression profiles of the R1 cell line and differentiated EB counterpart. From the hESC and mESC datasets, 5459 human–mouse orthologous genes that showed significant signal levels were selected for the analyses. The hESC samples had correlation coefficient values greater than 0.96, while the hEB samples had pair-wise correlation values greater than 0.92. The mESC and mEB samples had correlations greater than 0.99 and 0.93, respectively.

Many differentially expressed genes have been identified in ESCs in comparison with EBs and are found to be important for self-renewal or differentiation (Sperger *et al.*, 2003; Abeyta *et al.*, 2004; Brandenberger *et al.*, 2004; Miura *et al.*, 2004; Rao *et al.*, 2004; Richards *et al.*, 2004; Skottman *et al.*, 2005; Li *et al.*, 2006). However, small or subtle transcriptional changes can sometimes dramatically alter the flux of pathways and their overall active or inactive states (Hughes *et al.*, 2000; Mootha *et al.*, 2003; McCarroll *et al.*, 2004). A less than twofold increase in Oct3/4, for example, can lead to the differentiation of mESCs to primitive endoderm and mesoderm, while a repression of Oct3/4 results in the loss of pluripotency and differentiation to trophectoderm (Niwa *et al.*, 2000). It thus is important to capture such relatively small yet significant transcriptional changes, which are often overlooked by analyses focusing primarily on fold changes. To address the problem, our analysis included all expressed genes, regardless of their relative levels of changes, and analyzed the global trend of expression pattern across different functional sets in human and mouse ESCs. First, the expression change of a gene i is summarized as the ratio of average expression levels before and after differentiation (represented by x_i). For a functional module with n genes, the expression pattern of the module is represented as a vector of relative changes in expression across n genes (represented by $\mathbf{x} = [x_1, x_2, \ldots x_n]$). Next, the evolutionary conservation of the functional module on the expression pattern is determined by the Pearson correlation (r) between the vectors $\mathbf{x}_h$ (the expression profile vector in hESCs) and $\mathbf{x}_m$ (the expression profile vector in mESCs). The significance level of the observed Pearson correlation (r) from the n pairs of orthologous genes in the group is computed by treating $\frac{\sqrt{(n-2)}\times r}{\sqrt{(1-r^2)}}$ as coming from a t distribution with $n-2$ degree(s) of freedom. Therefore, a large group of genes with a relatively low correlation in the expression pattern could still be significant, and the method can thus help to reveal subtle changes and a global trend of transcriptomic conservation.

Our analysis first revealed that the global correlation of the expression pattern between human and mouse ESCs was 0.301, based on the 5459 expressed orthologous genes. The correlation was highly significant statistically ($P < 10^{-114}$). This indicates that despite profound differences between hESCs and mESCs, a certain number of orthologous genes behave similarly upon ESC differentiation, which represent core mechanisms for ESC fate control. A systematic survey across all functional modules categorized by the GO and KEGG databases, and examination of all expressed transcription factors and growth factors revealed three different types of functional modules: (1) the expression patterns were positively correlated between human and mouse ESCs ($r > 0$, $P < 0.05$); (2) the expression patterns were not correlated ($P \geq 0.05$); and (3) the expression patterns were negatively correlated ($r < 0$, $P < 0.05$). The functional modules of the first type were considered to be evolutionarily conserved, while those in the second and third types were evolutionarily divergent. Table 3.1 summarizes the results of this analysis. Out of the 903 GO-defined biological processes to which all human–mouse orthologous genes in our analysis were associated, 43.1% (i.e. 389 biological processes) were evolutionarily conserved. The other 56.9% (514 biological processes) were evolutionarily divergent, showing either a negative (0.07%) or no (56.2%) cross-species correlation in the expression pattern. The GO-defined molecular function and cellular component exhibited a similar distribution of conserved and divergent groups (Table 3.1). Among all KEGG-defined pathways to which the human–mouse orthologous genes of this analysis were associated, 14 showed a positively correlated expression pattern, 3 showed a negative correlation, while the remaining 88 showed a non-correlated expression pattern (Table 3.1). Figure 3.1 depicts expression patterns of some conserved and divergent functional modules that are important to ESC development. As shown, each column in the heatmap represents an orthologous gene expressed in hESCs (top row) and mESCs (bottom row), and the expression change of a gene is represented by different colors

Table 3.1. Number of functional modules showing positive, negative, or no correlation on the transcriptional changes upon ESC differentiation between human and mouse ($P \leq 0.05$).

Cross-species transcriptional correlation	Gene group	Biological process (GO)	Molecular function (GO)	Cellular component (GO)	Pathway (KEGG)
Positively correlated	Total	389	154	102	14
	Upregulated in hESCs	105	31	20	2
	Downregulated in hESCs	54	21	16	5
	Upregulated in mESCs	157	63	43	4
	Downregulated in mESCs	110	43	23	2
Not correlated	Total	508	258	121	88
	Upregulated in hESCs	135	67	28	4
	Downregulated in hESCs	113	85	39	10
	Upregulated in mESCs	265	129	84	8
	Downregulated in mESCs	333	138	47	5
Negatively correlated	Total	6	1	2	3

(green: downregulated in ESCs, red: upregulated in ESCs, black: no change between ESCs and EBs). The conserved modules (Figure 3.1a) were associated with high correlation values (r), in which many orthologous genes showed expression activities in the same direction (e.g. either upregulated in both species or downregulated in both). The divergent modules (Figure 3.1b) were associated with low or negative correlation values, in which most orthologous genes showed expression in opposite directions (e.g. upregulated in one species and downregulated in another).

3.3. Conserved Functional Modules

The conserved biological processes (GO terms) can be classified into several major functional categories. As shown in Figure 3.2, 13.8% of the conserved biological process were involved in development, such as cellular morphogenesis (97 orthologous genes, $r = 0.325$, $P < 0.001$), embryonic development (87 orthologs, $r = 0.318$, $P < 0.005$), pattern specification (50 orthologs, $r = 0.57$, $P < 10^{-5}$), and determination of left/right symmetry (6 orthologs, $r = 0.79$, $P < 0.05$). The conserved functional modules also included establishment and/or maintenance of chromatin architecture (1% of all conserved processes, 67 orthologs, $r = 0.505$, $P < 10^{-5}$; including chromosome remodeling), responses to stimulus (7.2%, including response to damaged DNA and unfolded proteins), cell proliferation and

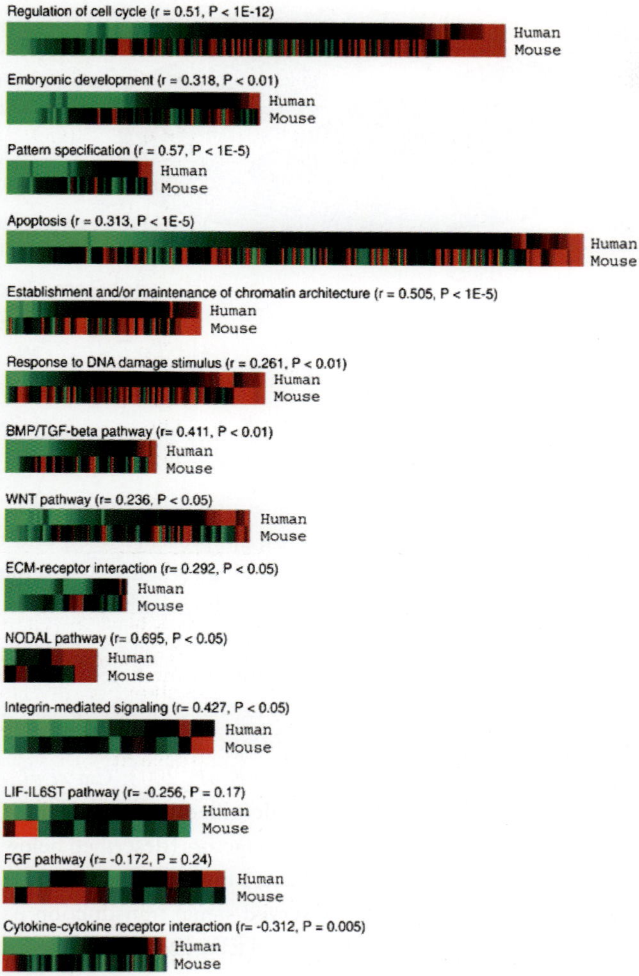

Figure 3.1. Expression patterns of selected (a) converged and (b) divergent biological processes and pathways. On the headmap of each module, the column represents an orthologous gene, expressed in hESC (top row) and in mESC (bottom row). The expression fold change of the gene is represented by different colors (green: downregulated in ESC, red: upregulated in ESC, black: no change). For (a) conserved modules (with high *r* values), many orthologous genes showed expression in the same direction (e.g. either upregulation in both species or downregulated in both) with similar expression fold changes between human and mouse. For (b) divergent modules (low or negative *r* values), most orthologous genes showed expression in opposite directions (e.g. upregulated in one species and downregulated in another) with dissimilar expression fold changes between human and mouse.

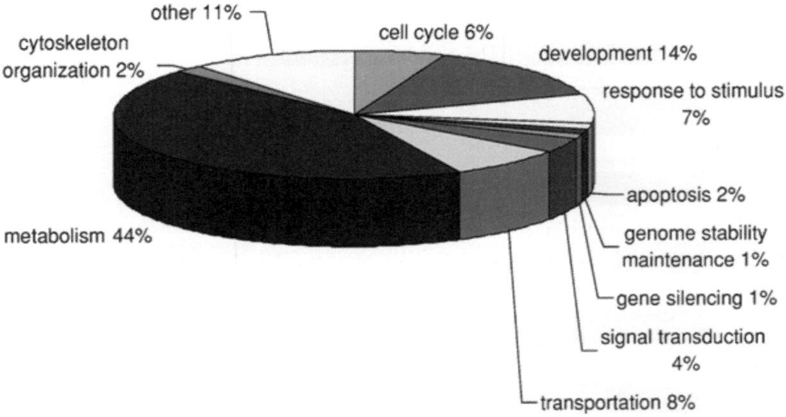

Figure 3.2. Summary of conserved biological processes showing correlated expression patterns between human and mouse ESCs.

cell cycle (6.2%), signal transduction (4.1%), apoptosis (1.8%), and metabolism (44.1%, including DNA replication and packaging) (Figure 3.2). Fourteen pathways defined by KEGG were conserved, including pathways such as TGF-β (52 orthologs, $r = 0.411$, $P < 0.005$), WNT (84 orthologs, $r = 0.236$, $P < 0.05$), Nodal (8 orthologs, $r = 0.695$, $P < 0.05$), and integrin-mediated signaling pathways (18 orthologs, $r = 0.427$, $P < 0.05$). These and many other conserved biological processes and pathways are presumably important for ESC development (Ivanova *et al.*, 2002; Ramalho-Santos *et al.*, 2002; Sato *et al.*, 2003; Cai *et al.*, 2004). The conserved biological processes and pathways identified, together with transcriptionally conserved transcription factors and growth factors (described below), represent core biological networks and mechanisms that are fundamental for ESC self-renewal and differentiation. Table 3.2a shows conserved signal transduction pathways and their expression in human and mouse ESCs and EBs. Details of some pathways and biological processes are described in Section 3.3.

3.3.1. *Cell cycle*

Cell cycle control mechanisms in ESCs differ from those in somatic cells (Fluckiger *et al.*, 2005). ESCs are primed for rapid cell proliferation. Both rodent and monkey ESCs achieve this by shortening the G1 phase and bypassing DNA damage G1 checkpoints (Fluckiger *et al.*, 2005). The 304 human–mouse orthologous genes involved in cell cycle progression showed a significantly conserved expression pattern ($r = 0.486$, $P < 10^{-18}$). In particular, 18 genes involved in the G1/S

Table 3.2. Transcriptionally (a) conserved and (b) diverged signaling transduction pathways.

(a)

Pathway	Correlation (r)	Ortholog pairs	P-value	Transcript changes	
				hESC	mESC
TGF-β signaling	0.41	52	0.001	Down	Up & down
Wnt signaling	0.24	84	0.015	Up & down	Down
Notch signaling	0.45	23	0.016	Down	None
Nodal signaling	0.7	8	0.028	Up	Up
ECM-receptor interaction	0.29	43	0.029	Down	Down
Integrin-mediated signaling	0.43	18	0.039	Down	Down

(b)

Pathway	Correlation (r)	Ortholog pairs	P-value	Transcript changes	
				hESC	mESC
Cytokine–cytokine receptor interaction	−0.31	69	0.005	Down	None
Phosphatidylinositol signaling system	0.19	39	0.123	Down	None
JAK–Stat signaling	0.14	53	0.157	Down	None
Toll-like receptor signaling	0.14	38	0.198	Down	None
Hedgehog signaling	0.17	25	0.204	Down	Down
FGF signaling	−0.17	19	0.24	None	Up
LIF signaling	−0.26	16	0.169	None	Down*
Insulin signaling	0.06	73	0.31	Down	Up

(*Continued*)

Table 3.2. (*Continued*)

Pathway	Correlation (r)	Ortholog pairs	P-value	Transcript changes	
				hESC	mESC
MAPK signaling	0.03	128	0.355	Down	None
Calcium signaling	−0.01	75	0.448	Down	None
Neuroactive ligand-receptor interaction	−0.01	58	0.472	Down	Down

*LIF pathway was downregulated in BG01, BG01V, BG02, but non-change in GB03.

transition of the mitotic cell cycle and 17 genes involved in the cell cycle arrest were highly conserved ($r = 0.623$ with $P < 0.005$, $r = 0.573$ with $P < 0.01$, respectively). Moreover, 171 additional genes involved in the regulation of cell cycle were also significantly conserved ($r = 0.51$, $P < 10^{-12}$) (Figure 3.1). The results suggest that these biological processes are fundamentally important to ESC self-renewal of different species.

3.3.2. *Genome stability*

Because of rapid DNA replication and the absence of cell cycle checkpoints, ESCs require extra mechanisms more sensitive than somatic cells to ensure genomic stability and to prevent the transmission of mutations to offspring cells (Burdon *et al.*, 2002). This may be achieved by active repression of spontaneous mutations, as well as the removal of damaged cells by apoptosis. Our analysis found that genes involved in responses to DNA-damaging stimuli and DNA repair had relatively low, yet significant cross-species correlation in their expression patterns ($r = 0.261$ and 0.269, respectively; $P < 0.01$ in both) (Figure 3.1). The evolutionary conservation of genes suggests their fundamental importance to ESC self-renewal. More specifically, the genes MLH1, MSH2, MSH3, MSH5, PMS2, and POLD3, involved in DNA mismatch repair and replication fidelity, were highly conserved ($r = 0.912$, $P < 0.01$) and thus comprised the core functional elements in DNA repair processes of ESCs. Apoptosis genes also showed a low yet highly significant cross-species correlation on the expression pattern (orthologs $= 198$, $r = 0.313$, $P < 10^{-5}$) (Figure 3.1). Moreover, genes involved in the response to unfolded proteins and telomere maintenance showed high levels of conservation ($r = 0.413$ and 0.802, respectively,

$P < 0.01$ in both). These results suggest that these biological processes critically contribute to ensure rapid yet accurate DNA replication and chromosome integrity during ESC self-renewal.

3.3.3. *Nodal pathway*

The Nodal pathway is implicated in early patterning of the embryo, left–right axis specification, and ESC development (Schier and Shen 2000; Besser 2004; Vallier *et al.*, 2004; Pfendler *et al.*, 2005). The expression profile of this pathway was highly conserved ($r = 0.695$, $P < 0.05$) (Figure 3.1). Among the nine human–mouse orthologous genes examined for this pathway, while SMAD2 and SMAD3 did not show significant transcriptional changes in both hESCs and mESCs, LEFTB and TDGF1 exhibited at least a twofold upregulation in both hESCs and mESCs. LEFTB is an antagonist and TDGF1 acts as a coreceptor for Nodal in Nodal signaling. The conserved expression pattern, especially for the coactivators of Nodal, TDGF1 and LEFTB, indicates that the Nodal pathway is fundamental for ESC pluripotency in different species.

3.3.4. *WNT pathway*

The pathway is actively involved in determining the fate of stem cells and their self-renewal (Aubert *et al.*, 2002; Tang *et al.*, 2002; Ding *et al.*, 2003; Reya *et al.*, 2003). The pathway exhibited a low yet significant level of cross-species correlation on the expression pattern ($r = 0.236$, $P < 0.05$) (Figure 3.1). Nineteen of the 84 orthologous genes examined for this pathway showed at least 1.5-fold expression changes upon ESC differentiation in both species. Among the 19 genes, 15 showed up- and downregulation in the same direction. The 15 conserved genes included WNT family members (WNT1, WNT2B, and WNT5B), frizzled receptors (FZD2, FZD6, and FZD8), the antagonist DKK1, and downstream members of the pathway (FRAT2, CTBP2, TP53, MYC, DVL3, RAC2, CCND2, and CCND3). The WNT family members and frizzled receptors, as well as DKK1, FRAT2, CTBP2, DVL3, RAC2, CCND2, and CCND3, were downregulated in both hESCs and mESCs, while TP53 and MYC were upregulated in ESCs. The observed evolutionary conservation of the expression pattern suggests that WNT signaling plays a fundamental role in maintaining ES cell pluripotency. However, it is important to note that four genes of the pathway showed transcriptional changes over 1.5-fold in opposite directions between hESCs and mESCs: CSNK2A2 (downregulated in hESCs but upregulated in mESCs), FZD9, SFRP1, and PRKCB1 (upregulated in hESCs but downregulated in mESCs). These genes represented divergent peaks on the overall conserved landscape of this pathway. This divergence may explain the differing results of the effects of WNT on hESC self-renewal (Dravid *et al.*, 2005).

3.3.5. *BMP/TGF-β pathway*

The pathway is involved in regulating cell proliferation and differentiation, mesoderm formation and patterning, and ESC development (Johansson and Wiles, 1995; Winnier *et al.*, 1995; Whitman, 1998; Goumans and Mummery, 2000; Massague *et al.*, 2000). The pathway showed a significant level of positive correlation on the expression pattern between hESCs and mESCs ($r = 0.411$, $P < 0.005$) (Figure 3.1). Eight of the 52 orthologous genes examined for this pathway showed at least 1.5-fold expression changes upon ESC differentiation, in which BMP4, BMP5, MADH6, ID2, and PITX2 were downregulated in both human and mouse ESCs, while THBS2, MYC, and GDF3 were upregulated in both species. This conserved expression pattern of ligands, intermediate signal transducers, and target genes suggests that BMP/TGF-β signaling is fundamentally important for ESC self-renewal and differentiation. However, there were five genes showing divergent expression patterns: INHBA, MADH7, THBS1, and MAPK1 were upregulated in mESCs but downregulated in hESCs; INHBE was downregulated in mESCs while they were upregulated in hESCs. The pathway therefore presents a mosaic structure with conserved functional cores and divergent domains, similar to the WNT pathway. This is consistent with reports showing that BMP supports mESC self-renewal, but activin appears to do so in hESCs (Ying *et al.*, 2003; Chambers and Smith, 2004).

3.3.6. *NOTCH pathway*

The pathway helps to determine cell fate through cell–cell interactions (Artavanis-Tsakonas *et al.*, 1999) and is reportedly involved in hESC self-renewal (Rho *et al.*, 2006). Overall, the pathway showed a conserved expression pattern across species ($r = 0.451$, $P < 0.05$) (Figure 3.1). However, 12 of the 23 orthologous genes examined for this pathway showed less than 1.5-fold expression changes upon ESC differentiation in both human and mouse ESCs, which included the receptors NOTCH1 and NOTCH4, and the ligands DVL2 and JAG2. Only three genes exhibited an increase of more than 1.5-fold of expression changes in both human and mouse and positively correlated expression pattern: DVL3 (downregulated in ESCs), RBPSUH, and CTBP2 (both upregulated in ESCs). The subtle yet conserved expression pattern seems important for this pathway's role in ESC self-renewal.

3.3.7. *Intergrin-mediated pathway*

Integrins mediate cell adhesion to the extracellular matrix (ECM), control cell cycle progression, regulate WNT1 and BMP4 expression, and activate the mesodermal and neuroectodermal lineages for ESC differentiation (Czyz and Wobus, 2001; Walker and Assoian, 2005). The 18 human–mouse orthologous genes involved in the integrin-mediated signaling pathway showed conserved transcriptional changes during ESC differentiation ($r = 0.427$, $P < 0.05$) (Figure 3.1). These transcriptional conserved genes included integrin members (e.g. ITGAs, ITGBs) and intracellular signal transducers (e.g. CD47, GAB2, CIB1). Among the seven genes encoding integrins, six were either slightly downregulated in ESC or had little transcriptional changes, while ITGB5 was at least 1.5-fold upregulated in hESC but not in mESC. GAB2 and CD47 were at least 1.5-fold downregulated in both human and mouse ESCs. These results suggest that integrin-mediated signaling is mostly repressed in ESCs or activated in EBs and fundamental for ESC development in both human and mouse.

Many functional modules were conserved on the expression pattern but not highly differentially expressed, and thus not disclosed from previous studies based on differentially expressed genes alone. A total of 237 biological processes, 102 molecular functions, and 66 cellular components fell into this category of modules. The modules included embryonic morphogenesis (orthologous pairs = 37, $r = 0.42$, $P = 0.005$), gene silencing (orthologs = 8, $r = 0.67$, $P = 0.035$), cell cycle arrest (orthologs = 17, $r = 0.57$, $P = 0.008$), G1/S transition of mitotic cell cycle (orthologs = 18, $r = 0.62$, $P = 0.003$), and aging (orthologs = 8, $r = 0.70$, $P = 0.026$), which are known to be important for ESCs. Many other modules in this category however bear no overt relationship to ESC development. Further analysis of these modules may lead to new insights into mechanisms regulating ESC development.

3.4. Divergent Functional Modules

Divergent functional modules include those showing negative or no correlation on transcriptional changes during differentiation between human and mouse ESCs. Figure 3.1b shows expression patterns of some divergent functional modules important for ESC development. Table 3.2b shows divergent signal transduction pathways and their expression in ESCs and EBs of humans and mice.

The functional modules with negatively correlated expression patterns represented extreme cases of divergent modules. One such module is that involved in cytokine–cytokine receptor interactions. The 69 human–mouse orthologous genes examined for this module showed a significant level of negative correlation on the expression pattern across species ($r = -0.312$, $P < 0.005$) (Figure 3.1b). The genes CCL7, PDGFRA, IL1R1, CXCL1, and INHBA were downregulated in hESCs but upregulated in mESCs by at least 1.5-fold in both species, while the genes INHBE, CXCL12, and BMP2 were upregulated in hESCs but downregulated in mESCs by at least 1.5-fold in both species. Opposite expression patterns of cytokines in hESCs and mESCs were also observed by other studies (Ginis *et al.*, 2004). The negatively correlated expression pattern of cytokine–cytokine receptor interactions is suggestive of unique and distinct responses of hESCs and mESCs to growth factors.

The majority of functional modules we examined, including biological process, molecular functions, and cellular components defined by GO, and pathways defined by KEGG, showed no correlation on transcriptional changes during ESC differentiation across species (Table 3.1). From these non-conserved modules, we identified those processes that were significantly up- or downregulated in hESCs and mESCs. The results are shown in Table 3.1. Among the 508 non-conserved biological processes (56.2% of the total examined), 135 processes had the genes significantly upregulated and 113 processes significantly downregulated in hESCs. These significantly changed yet not conserved functional modules are suggestive of species-specific mechanisms controlling ESC self-renewal and differentiation.

For example, the expression pattern of the LIF pathway was not conserved, with insignificant cross-species correlation ($r = -0.256$, $P = 0.17$), but was significantly downregulated in hESCs although little changed in mESCs (Figure 3.1b). This result is consistent with the fact that the pathway is not essential in hESCs but is critical in mESCs for maintaining pluripotency (Ying *et al.*, 2003; Daheron *et al.*, 2004; Humphrey *et al.*, 2004). Among the 16 orthologous genes examined for this pathway, LIF was expressed at a much lower level in hESCs than in mESCs, while IL11, STAT6, STAT5B, SOCS1, and SOCS2 showed highly different transcriptional changes between hESCs and mESCs. Similarly, the expression pattern of the FGF pathway was not conserved, with insignificant cross-species correlation ($r = -0.172$, $P = 0.24$), but was significantly upregulated in mESCs, not in hESCs (Figure 3.1). The pathway is implicated in rodent ICM development and maintaining undifferentiated ESCs in culture (Lamb and Rizzino, 1998; Xu *et al.*, 2001; Goldin and Papaioannou, 2003). Among the 19 orthologous genes examined for this pathway, the MAPK inhibitors DUSP1, DUSP6, and DUSP9 were highly upregulated (over twofold) in mESCs but they were downregulated or unchanged in hESCs. The repressed expression of these genes in hESCs were also observed by a RT-PCR analysis (Rho *et al.*, 2006). These results suggest that the FGF pathway acts differently between human and mouse for ESC self-renewal. In addition, the non-conserved biological processes cytokine secretion and hormone biosynthesis were significantly upregulated in hESCs but not in mESCs. In contrast, the non-conserved biological processes base-excision repair, cytokinesis, and embryonic cleavage, as well as the insulin signaling pathway, were significantly upregulated in mESCs but not in hESCs (Table 3.2). It should be noted that since the non-conserved modules identified in this study were based only on human–mouse orthologous genes, not all species-specific biological mechanisms involved in ESC development were disclosed. Many other species-specific regulations could be present and activated by non-homologous genes in human or mouse ESCs.

The observed conservation and variation of functional modules at the gene transcriptional level are consistent with their evolutionary patterns at the gene structural level. Consistent with the conserved expression pattern of the NODAL pathway (Figure 3.1), NODAL and TDFG1 of the pathway are highly conserved in the gene and promoter structure between human and mouse (Zhan *et al.*, 2005), suggestive of conserved activation mechanisms of the pathway. Consistent with the divergent expression pattern of the FGF pathway (Figure 3.1), FGF4 of the pathway shows sequence divergence in the Sox2–Oct3/4 co-binding site of an enhancer (Zhan *et al.*, 2005). The Sox2–Oct3/4 co-binding site is critical to the self-renewal of mESCs; the sequence divergence of the co-binding site makes gene transcription less effective in humans than rodents (Lamb and Rizzino, 1998). Furthermore, consistent with the divergent expression pattern of the LIF pathway (Figure 3.1), LIFR and IL6ST (GP130) of the pathway are highly divergent in the gene and promoter structure between human and mouse (Zhan *et al.*, 2005), highlighting the fact that LIF signaling is required in mice but is not essential in humans for maintaining ESC pluripotency (Ying *et al.*, 2003; Daheron *et al.*, 2004; Humphrey *et al.*, 2004). It is reported that the collaboration between LIF signaling and the BMP/TGF-β pathway helps to sustain pluripotency in mESCs (Ying *et al.*, 2003). Since the LIF pathway is diverged and not fundamental, the collaboration is likely not fundamental in humans or other species either. Although the BMP/TGF-β pathway is transcriptionally conserved (Figure 3.1), BMP4 has different effects on mouse and human ESCs: it blocks ESC differentiation along the neuroectoderm default pathway in mESCs (Ying *et al.*, 2003) while it induces trophectroderm differentiation of ESCs in hESCs (Xu *et al.*, 2002). It is possible that the different impacts of BMP signaling on ESC development in humans and mice are due to the diverged LIF signaling that interacts with BMP signaling.

3.5. Conserved and Divergent Transcription Factors and Growth Factors

We further examined 304 transcription factors which are orthologous between human and mouse. Globally, the transcription factors presented a relatively low yet highly significant cross-species correlation on the transcriptional changes during ESC differentiation ($r = 0.378$, $P < 10^{-11}$), suggesting the presence of conserved core regulatory mechanisms in ESCs. Specifically, 199 of the transcription factors were conserved by showing up-or downregulation in the same direction between hESCs and mESCs, in which 48 genes had at least 1.5-fold transcriptional changes in both species. The remaining 105 transcription factors were divergent, showing up- or downregulation in opposite directions between hESCs and mESCs, in which nine genes exhibited at least 1.5-fold transcriptional changes upon differentiation in either or both species. Table 3.3a lists the 48 transcription factors with conserved expression patterns and at least 1.5-fold transcriptional changes. Among them, 18 transcription factors showed significant differential expression in both hESCs and mESCs ($P < 0.05$). These transcription factors, some of which are upregulated in both hESCs and mESCs while others are downregulated in both species, are functionally essential and are likely to be the key elements in the regulatory network controlling ESC self-renewal and differentiation. Among the 48 conserved transcription factors, 26 are implicated in development (Table 3.3a). Some of them are known to be important for ESC pluripotency, such as POU5F1, UTF1, MYC, and TP53, all of which were upregulated in ESC, and HAND1 and GATA6, both of which were downregulated in ESCs (Table 3.3a). The divergent transcription factors (Table 3.3b) may play different or species-specific roles in ESC development. Some of the factors are members of important signaling pathways: STAT6 (no significant change in hESCs but were upregulated in mESCs) and STAT5B (upregulated in hESCs but were downregulated in mESCs) are members of the JAK-STAT pathway involved in LIF signaling; CEBPB (no significant change in hESCs while upregulated in mESCs) is involved in IL6 signaling.

Similarly, we examined 61 orthologous growth factors. The growth factors showed an overall conserved expression pattern between hESCs and mESCs ($r = 0.375$, $P < 0.01$). Twenty-one of the growth factors exhibited negative correlation on transcriptional changes during ESC differentiation across species, with eight showing transcriptional changes of at least 1.5-fold (Table 3.4b). The remaining 40 growth factors were conserved. Among them, 10 factors showed at least a 1.5-fold change in transcription upon differentiation in both species (Table 3.4a). Seven of the 10 growth factors are known to be involved in development and morphogenesis. BMP4, BMP5 (both downregulated in ESCs), and GDF3 (upregulated in ESCs)

Table 3.3. Transcriptionally (a) conserved and (b) diverged transcription factors. The transcription factors showed at least 1.5-fold up- or downregulation in the same direction (conserved) or opposite directions (diverged) between human and mouse ES cells. *P*-values represent the significance of differential expression in hESCs or mESCs.

(a)

Transcription factors	*P*-value in hESC	*P*-value in mESC	Expression in hESC and mESC	Implicated functions
MXD4	0.0024	0.0175	Down	Cell proliferation
MYC	0.0555	0.0178	Up	
MYST2	0.0009	0.0062	Up	
TP53	0.0204	0.0008	Up	
ARNT2	0.1231	0.0302	Down	Development
EVX1	0.2256	0.0112	Down	
GATA3	0.0150	0.0426	Down	
GATA6	0.0017	0.0561	Down	
HAND1	0.0033	0.0329	Down	
HEY1	0.0032	0.0118	Down	
HLX1	0.0588	0.0008	Down	
HOXB5	0.0014	0.0012	Down	
HOXB6	0.0043	0.0049	Down	
HOXB9	0.0578	0.0050	Down	
MAFB	0.0802	0.1325	Down	
MEF2C	0.0209	0.0551	Down	
MSX1	0.0030	0.0384	Down	
MSX2	0.0050	0.0933	Down	
PITX1	0.0121	0.0234	Down	
PITX2	0.0054	0.1141	Down	
RAX	0.0252	0.0004	Down	
SIX3	0.2344	0.0487	Down	
SIX5	0.0985	0.0163	Down	
SOX10	0.1450	0.0121	Down	
SOX11	0.2070	0.2287	Down	
TCF21	0.1014	0.0078	Down	
ZIC1	0.1209	0.3692	Down	
ZHX1	0.0013	0.0361	Down	
TGIF	0.0143	0.0128	Up	
POU5F1	0.0070	0.0328	Up	
ATBF1	0.0112	0.0086	Down	Others
CREB3L1	0.0644	0.1096	Down	
DSIPI	0.0423	0.1123	Down	

(*Continued*)

Table 3.3. (*Continued*)

Transcription factors	*P*-value in hESC	*P*-value in mESC	Expression in hESC and mESC	Implicated functions
FOXA1	0.0273	0.0236	Down	
FOXF1	0.1172	0.0690	Down	
ISL1	0.0480	0.0877	Down	
NR2F1	0.1048	0.1984	Down	
SOX18	0.1016	0.0574	Down	
SOX7	0.0908	0.0705	Down	
SPI1	0.1408	0.0076	Down	
ZFHX4	0.0456	0.0730	Down	
ZNF42	0.3586	0.0206	Down	
MYB	0.0176	0.0600	Up	
MYCN	0.0032	0.0125	Up	
NFYB	0.0017	0.0132	Up	
POLR3K	0.0109	0.0519	Up	
POU2F1	0.0005	0.0870	Up	
UTF1	0.1303	0.0088	Up	

(b)

Transcription factors	*P*-value in hESC	*P*-value in mESC	Expression in hESC	Expression in mESC	Implicated functions
CBFA2T2	0.0619	0.0195	Down	Up	Cell proliferation
GTF2H1	0.0295	0.0076	Down	Up	
IRF1	0.0487	0.0778	Down	Up	
STAT5B	0.0030	0.0043	Up	Down	Signal transduction
STAT6	0.2758	0.0078	No change	Up	
CEBPB	0.2703	0.0162	No change	Up	Others
ZNF277	0.2153	0.0318	No change	Up	
TPBG	0.0576	0.0237	Down	Up	
FOXA2	0.1128	0.0786	Up	Down	

are, for example, critical members of the BMP signaling pathway. LEFTB and TDGF1 (both upregulated in ESCs) participate in NODAL signaling. On the other hand, many other conserved growth factors, as well as conserved transcription factors showed no obvious relationship with ESC development (Tables 3.3 and 3.4). These factors provide a basis for new hypotheses about mechanisms underlying ESC development.

Table 3.4. Transcriptionally (a) conserved and (b) diverged growth factors. The growth factors showed at least 1.5-fold up- or downregulation in the same direction (conserved) or opposite directions (diverged) between human and mouse ES cells. *P*-values represent the significance of differential expression in hESCs or mESCs.

(a)

Growth factors	hESC *P*-value	mESC *P*-value	Expression in ESC	Implicated functions
SCGF	0.0133	0.0468	Down	Cell proliferation
BMP4	0.0146	0.0600	Down	Development/ morphogenesis
BMP5	0.0067	0.0080	Down	
IGF2	0.0003	0.0563	Down	
MDK	0.0116	0.1218	Down	
GDF3	0.0039	0.0169	Up	
LEFTB	0.0089	0.0463	Up	
TDGF1	0.0021	0.0214	Up	
MST1	0.0543	0.0655	Down	Others
FBS1	0.0424	0.0233	Down	

(b)

Growth factors	hESC *P*-value	mESC *P*-value	Expression in hESC	Expression in mESC	Implicated functions
AREG	0.0805	0.2423	Down	No change	Cell–cell signaling
CXCL12	0.0417	0.0188	Up	Down	
BMP2	0.0783	0.0231	Up	Down	Development
FGF13	0.0212	0.0067	Up	Down	
DTR	0.0573	0.1424	Down	No change	
INHBA	0.3262	0.0837	No change	Up	
S100A6	0.0443	0.0243	Down	Up	
INHBE	0.2185	0.0598	No change	Down	Others

3.6. Functional Profiling of Embryonal Carcinoma Stem Cells

NTera2 is a human embryonal carcinoma stem cell (hECC) line with less differentiation potential than hESCs. Comparing the expression profile of NTera2 and hESCs helps identify the "stemness" factors controlling stem cell development.

We identified 2818 upregulated and 2924 downregulated genes (FDR < 0.05) from the determined expression profiles of NTera2. The upregulated genes in undifferentiated NTera2 cells included marker genes of undifferentiated hESCs such as POUT5F1, TDGF1, DNMT3B, CD24, LIN28, CD9, PUM1, PUM2, GJA1, and TERF1. The downregulated genes included HAND1, IGF2, and other markers known to be involved in differentiation. Among the 238 biological processes (GO) that were enriched in upregulated genes, 124 processes (i.e. 52.1% of the total) were shared with the 229 biological processes that were enriched in upregulated genes in hESCs. Among the 98 biological processes that were enriched in downregulated genes, 11 processes (i.e. 11.2% of the total) were shared with the 68 downregulated processes of hESCs. Therefore, there are more biological processes in common between the hESCs and NTera2 cells in the undifferentiated state than the differentiated state. A similar pattern was also observed with the GO-defined molecular functions and cellular components, and KEGG-defined pathways. These results suggest that hESCs and hECCs are more similar in mechanisms activated for self-renewal, but less similar in mechanisms activated for differentiation (or suppressed for maintaining pluripotency). Thus, hESCs and hECCs differ mainly in their differentiation potentials while sharing many characteristics of pluripotency and expressing similar "stemness" markers. This is consistent with the experimental observations that hESCs can differentiate to cells forming mesoderm, endoderm, and ectoderm, while NTera2 cells mainly differentiate into neuroectoderm cells (Pera *et al.*, 1989; Walsh and Andrews, 2003; Schwartz *et al.*, 2005). Our data indicated that biological processes related to embryonic development were significantly downregulated and thus repressed in hESCs, but not downregulated in NTera2 undifferentiated cells. Moreover, about 30% of the downregulated biological processes in hESCs were related to development, whereas only ~7% of the downregulated processes in undifferentiated NTera2 cells were related to development. The results highlight the fact that NTera2 cells possess less differentiation potential than ESCs.

Among the 124 functional modules that were enriched in the upregulated genes of both hESCs and hECCs, 63 showed a conserved expression pattern between hESCs and mESCs. These functional modules represent core molecular mechanisms that are activated for self-renewal in stem cells of different kinds. To achieve indefinite self-renewal and maintain the pluripotent state, stem cells are primed for rapid and accurate DNA replication by shortening the cell cycle, relaxing G1/S checkpoints, strengthening DNA repair, enhancing the response to unfolded proteins, and activating apoptosis. In addition, metabolic processes related to nucleotide synthesis, DNA replication, and DNA packaging are also activated (Burdon *et al.*, 2002). All these biological processes or related pathways were conserved on the expression pattern between hESCs and mESCs and common between hESCs and hECCs (Figures 3.1a and 3.2). In addition, telomere maintenance and aging-related processes were also conserved and upregulated in both hESCs and hECCs, keeping with the fact that stem cells have a longer half-life than somatic cells.

3.7. Summary and Future Prospects

In this chapter, we describe results of a comparative transcriptomics study for global functional profiling of ESCs. The cross-species correlation of transcriptional changes during ESC differentiation is shown to be a powerful predictor of molecular mechanisms that are fundamental for ESC development. Much of our current knowledge on ESC-related pathways and factors are supported or reaffirmed by comparative transcriptomics analysis that justify the prediction. About 27% of biological processes showed significant up- or downregulation and conserved transcriptional changes during ESC differentiation, and are thus deemed to be functionally essential. The analyses thus provide a focused list of important pathways or functional modules for in-depth investigation. Between different stem cell types (e.g. hESCs and hECCs), there are more similarities on the mechanisms activated for self-renewal than those activated for differentiation. Within the same stem cell type, more upregulated modules are conserved than downregulated modules. Specifically, in hESCs, 43.7% upregulated biological processes (105 out of 240) were conserved, while 32.3% downregulated biological processes (54 out of 167) were conserved. In mESCs, 37.2% upregulated biological processes were conserved, while 19.5% downregulated biological processes were conserved. Our analysis thus demonstrate that while pluripotency (as an essential function in multicellular organisms) is conserved through evolution, the mechanisms primed for differentiation are less conserved and contribute (in a larger extent) to the differences among stem cells derived from different tissues or species. Because biological mechanisms primed for differentiation are evolutionarily less conserved, ESC lines derived from different species (e.g. hESC, mESC) or isolated from different tissues (e.g. ESCs, ECCs etc.) likely adapt distinctive mechanisms activated for differentiation.

Bibliography

Abeyta, M. J., Clark, A. T., Rodriguez, R. T., Bodnar, M. S., Pera, R. A. and Firpo, M. T. 2004. Unique gene expression signatures of independently-derived human embryonic stem cell lines. *Hum Mol Genet* 13:601–608.

Artavanis-Tsakonas, S., Rand, M. D., and Lake, R. J. 1999. Notch signaling: Cell fate control and signal integration in development. *Science* 284:770–776.

Aubert, J., Dunstan, H., Chambers, I. and Smith, A. 2002. Functional gene screening in embryonic stem cells implicates Wnt antagonism in neural differentiation. *Nat Biotechnol* 20:1240.

Besser, D. 2004. Expression of nodal, lefty-a, and lefty-B in undifferentiated human embryonic stem cells requires activation of Smad2/3. *J Biol Chem* 279:45076–45084.

Brandenberger, R., Wei, H., Zhang, S., Lei, S., Murage, J., Fisk, G. J.,, Li, Y., Xu, C., Fang, R., Guegler, K., Rao, M. S., Mandalam, R., Lebkowski, J. and Stanton, L. W. 2004. Transcriptome characterization elucidates signaling networks that control human ES cell growth and differentiation. *Nat Biotechnol* 22:707–716.

Burdon, T., Smith, A. and Savatier, P. 2002. Signalling, cell cycle and pluripotency in embryonic stem cells. *Trends Cell Biol* 12:432–438.

Cai, J., Weiss, M. L. and Rao, M. S. 2004. In search of "stemness". *Exp Hematol* 32:585–598.

Chambers, I. and Smith, A. 2004. Self-renewal of teratocarcinoma and embryonic stem cells. *Oncogene* 23:7150–7160.

Czyz, J. and Wobus, A. 2001. Embryonic stem cell differentiation: The role of extracellular factors. *Differentiation* 68:167–174.

Daheron, L., Opitz, S. L., Zaehres, H., Lensch, W. M., Andrews, P. W., Itskovitz-Eldor, J. and Daley, G. Q. 2004. LIF/STAT3 signaling fails to maintain self-renewal of human embryonic stem cells. *Stem Cells* 22:770–778.

Ding, S., Wu, T. Y. H., Brinker, A., Peters, E. C., Hur, W., Gray, N. S. and Schultz, P. G. 2003. Synthetic small molecules that control stem cell fate. *PNAS* 100:7632–7637.

Dravid, G., Ye, Z., Hammond, H., Chen, G., Pyle, A., Donovan, P., Yu, X. and Cheng, L. 2005. Defining the role of Wnt/beta-catenin signaling in the survival, proliferation, and self-renewal of human embryonic stem cells. *Stem Cells* 23:1489–1501.

Fluckiger, A. C., Marcy, G., Marchand, M., Negre, D., Cosset, F. L., Mitalipov, S., Wolf, D., Savatier, P. and Dehay, C. 2005. Cell-cycle features of primate embryonic stem cells. *Stem Cells* 24:547–556.

Ginis, I., Luo, Y., Miura, T., Thies, S., Brandenberger, R., Gerecht-Nir, S., Amit, M., Hoke, A., Carpenter, M. K., Itskovitz-Eldor, J. and Rao, M. S. 2004. Differences between human and mouse embryonic stem cells. *Dev Biol* 269:360–380.

Goldin, S. N. and Papaioannou, V. E. 2003. Paracrine action of FGF4 during perimplantation development maintains trophectoderm and primitive endoderm. *Genesis* 1:40–47.

Goumans, M. J. and Mummery, C. 2000. Functional analysis of the TGFbeta receptor/Smad pathway through gene ablation in mice. *Int J Dev Biol* 44:253–265.

Hughes, T. R., Marton, M. J., Jones, A. R., Roberts, C. J., Stoughton, R., Armour, C. D., Bennett, H. A., Coffey, E., Dai, H., He, Y. D., Kidd, M. J., King, A. M., Meyer, M. R., Slade, D., Lum, P. Y., Stepaniants, S. B., Shoemaker, D. D., Gachotte, D., Chakraburtty, K., Simon, J., Bard, M. and Friend, S. H. 2000. Functional discovery via a compendium of expression profiles. *Cell* 102:109–126.

Humphrey, R. K., Beattie, G. M., Lopez, A. D., Bucay, N., King, C. C., Firpo, M. T., Rose-John, S. and Hayek, A. 2004. Maintenance of pluripotency in human embryonic stem cells is STAT3 independent. *Stem Cells* 22:522–530.

Ivanova, N. B., Dimos, J. T., Schaniel, C., Hackney, J. A., Moore, K. A. and Lemischka, I. R. 2002. A stem cell molecutar signature. *Science* 298:601.

Johansson, B. M. and Wiles, M. V. 1995. Evidence for involvement of activin A and bone morphogenetic protein 4 in mammalian mesoderm and hematopoietic development. *Mol Cell Biol* 15:141–151.

Lamb, K. A. and Rizzino, A. 1998. Effects of differentiation on the transcriptional regulation of the FGF-4 gene: Critical roles played by a distal enhancer. *Mol Reprod Dev* 51:218–224.

Li, H., Liu, Y., Shin, S., Sun, Y., Loring, J. F., Mattson, M. P., Rao, M. S. and Zhan, M. 2006. Transcriptome coexpression map of human embryonic stem cells. *BMC Genomics* 7:103.

Massague, J., Blain, S. W. and Lo, R. S. 2000. TGFbeta signaling in growth control, cancer, and heritable disorders. *Cell* 103:295–309.

McCarroll, S. A., Murphy, C. T., Zou, S., Pletcher, S. D., Chin, C. S., Jan, Y. N., Kenyon, C., Bargmann, C. I. and Li, H. 2004. Comparing genomic expression patterns across species identifies shared transcriptional profile in aging. *Nat Genet* 36:197–204.

Miura, T., Luo, Y., Khrebtukova, I., Brandenberger, R., Zhou, D., Thies, R. S., Vasicek, T., Young, H., Lebkowski, J., Carpenter, M. K. and Rao, M. S. 2004. Monitoring early differentiation events in human embryonic stem cells by massively parallel signature sequencing and expressed sequence tag scan. *Stem Cells Dev* 13:694–715.

Mootha, V. K., Lindgren, C. M., Eriksson, K. F., Subramanian, A., Sihag, S., Lehar, J., Puigserver, P., Carlsson, E., Ridderstrale, M., Laurila, E., Houstis, N., Daly, M. J., Patterson, N., Mesirov, J. P., Golub, T. R., Tamayo, P., Spiegelman, B., Lander, E. S., Hirschhorn, J. N., Altshuler, D. and Groop, L. C. 2003. PGC-1alpha-responsive genes involved in oxidative phosphorylation are coordinately downregulated in human diabetes. *Nat Genet* 34:267–273.

Niwa, H., Miyazaki, J. and Smith, A. G. 2000. Quantitative expression of Oct-3/4 defines differentiation, dedifferentiation or self-renewal of ES cells. *Nat Genet* 24:372–376.

Pera, M. F., Cooper, S., Mills, J. and Parrington, J. M. 1989. Isolation and characterization of a multipotent clone of human embryonal carcinoma cells. *Differentiation* 42:10–23.

Pfendler, K. C., Catuar, C. S., Meneses, J. J. and Pedersen, R. A. 2005. Overexpression of nodal promotes differentiation of mouse embryonic stem cells into mesoderm and endoderm at the expense of neuroectoderm formation. *Stem Cells Dev* 14:162–172.

Ramalho-Santos, M., Yoon, S., Matsuzaki, Y., Mulligan, R. C. and Melton, D. A. 2002. "Stemness": Transcriptional Profiling of Embryonic and Adult Stem Cells. *Science* 298:597–600.

Rao, M. 2004. Conserved and divergent paths that regulate self-renewal in mouse and human embryonic stem cells. *Dev Biol* 275:269–286.

Rao, R. R., Calhoun, J. D., Qin, X., Rekaya, R., Clark, J. K. and Stice, S. L. 2004. Comparative transcriptional profiling of two human embryonic stem cell lines. *Biotechnol Bioeng* 88:273–286.

Reya, T., Duncan, A. W., Ailles, L., Domen, J., Scherer, D. C., Willert, K., Hintz, L., Nusse, R. and Weissman, I. L. 2003. A role for Wnt signalling in self-renewal of haematopoietic stem cells. *Nature* 423:409.

Rho, J. Y., Yu, K., Han, J. S., Chae, J. I., Koo, D. B., Yoon, H. S., Moon, S. Y., Lee, K. K. and Han, Y. M. 2006. Transcriptional profiling of the developmentally important signalling pathways in human embryonic stem cells. *Hum Reprod* 21:405–412.

Richards, M., Tan, S. P., Tan, J. H., Chan, W. K. and Bongso, A. 2004. The transcriptome profile of human embryonic stem cells as defined by SAGE. *Stem Cells* 22:51–64.

Sato, N., Sanjuan, I. M., Heke, M., Uchida, M., Naef, F. and Brivanlou, A. H. 2003. Molecular signature of human embryonic stem cells and its comparison with the mouse. *Dev Biol* 260:404.

Schier, A. F. and Shen, M. M. 2000. Nodal signalling in vertebrate development. *Nature* 403:385.

Schwartz, C. M., Spivak, C. E., Baker, S. C., McDaniel, T. K., Loring, J. F., Nguyen, C., Chrest, F. J., Wersto, R., Arenas, E., Zeng, X., Freed, W. J. and Rao, M. S. 2005. NTera2: A model system to study dopaminergic differentiation of human embryonic stem cells. *Stem Cells Dev* 14:517–534.

Skottman, H., Mikkola, M., Lundin, K., Olsson, C., Stromberg, A. M., Tuuri, T., Otonkoski, T., Hovatta, O. and Lahesmaa, R. 2005. Gene expression signatures of seven individual human embryonic stem cell lines. *Stem Cells* 23:1343–1356.

Sperger, J. M., Chen, X., Draper, J. S., Antosiewicz, J. E., Chon, C. H., Jones, S. B., Brooks, J. D., Andrews, P. W., Brown, P. O. and Thomson, J. A. 2003. Gene expression patterns in human embryonic stem cells and human pluripotent germ cell tumors. *Proc Natl Acad Sci U S A* 100:13350–13355.

Sun, Y., Li, H., Liu, Y., Shin, S., Mattson, M. P., Rao, M. S. and Zhan, M. 2007. Cross-species transcriptional profiles establish a functional portrait of embryonic stem cells. *Genomics* 89:22–35.

Tang, K., Yang, J., Gao, X., Wang, C., Liu, L., Kitani, H., Atsumi, T. and Jing, N. 2002. Wnt-1 promotes neuronal differentiation and inhibits gliogenesis in P19 cells. *Biochem Biophys Res Commun* 293:167.

Vallier, L., Reynolds, D. and Pedersen, R. A. 2004. Nodal inhibits differentiation of human embryonic stem cells along the neuroectodermal default pathway. *Dev Biol* 275:403–421.

Walker, J. L. and Assoian, R. K. 2005. Integrin-dependent signal transduction regulating cyclin D1 expression and G1 phase cell cycle progression. *Cancer Metastasis Rev* 24:383–393.

Walsh, J. and Andrews, P. W. 2003. Expression of Wnt and Notch pathway genes in a pluripotent human embryonal carcinoma cell line and embryonic stem cell. *Apmis* 111:197–210; discussion 210–191.

Whitman, M. 1998. Smads and early developmental signaling by the TGFbeta superfamily. *Genes Dev* 12:2445–2462.

Winnier, G., Blessing, M., Labosky, P. A. and Hogan, B. L. 1995. Bone morphogenetic protein-4 is required for mesoderm formation and patterning in the mouse. *Genes Dev* 9:2105–2116.

Xu, C., Inokuma, M. S., Denham, J., Golds, K., Kundu, P., Gold, J. D. and Carpenter, M. K. 2001. Feeder-free growth of undifferentiated human embryonic stem cells. *Nat Biotechnol* 19:971–974.

Xu, R. H., Chen, X., Li, D. S., Li, R., Addicks, G. C., Glennon, C., Zwaka, T. P. and Thomson, J. A. 2002. BMP4 initiates human embryonic stem cell differentiation to trophoblast. *Nat Biotechnol* 20:1261–1264.

Ying, Q. L., Nichols, J., Chambers, I. and Smith, A. 2003. BMP induction of Id proteins suppresses differentiation and sustains embryonic stem cell self-renewal in collaboration with STAT3. *Cell* 115:281–292.

Zhan, M., Miura, T., Xu, X. and Rao, M. S. 2005. Conservation and variation of gene regulation in embryonic stem cells assessed by comparative genomics. *Cell Biochem Biophys* 43:379–405.

A Common Integrative Nuclear Signaling Module for Stem Cell Development

4

Michal K. Stachowiak, Ewa K. Stachowiak, John M. Aletta, and Emmanuel S. Tzanakakis

Ontogeny requires the coordinated regulation of multi-gene programs by a plethora of extracellular and intracellular signals. As a result of gene regulation, stem cells alternate between states of self-renewal and differentiation. Disruption of this regulation may cause oncogenic transformation in which stem cells are "arrested" in the proliferative state. The emerging principles of systems biology suggest that computational network modules are responsible for the integration of epigenetic signals that determine specific developmental stages. Modules that control entry into cell cycle and promote perpetual self-renewal of the stem cells have been identified. In this chapter, we summarize the current understanding of a novel Feed-Forward-And-Gate network module that affects the differentiation of stem cells and neural progenitor cells. In the center of this module is the new gene-controlling mechanism — integrative nuclear fibroblast growth factor receptor (FGFR)-1 signaling (INFS). Enhancement of the INFS may be used to re-establish coordinated gene regulation to activate cell differentiation for regenerative purposes or to provide interruption of cancer stem cell proliferation.

4.1. How can Ontogenesis be Reproducible in the Stochastic Environment?

Neurogenesis, like embryological ontogeny, is the product of the inherited genomic blueprint, as well as extracellular and intracellular environmental factors (referred to as epigenetics). The former is developed and perfected over millions of years of evolution and the latter is subject to stochastic events. Nevertheless, every successive generation exhibits a reproducible reconstruction of the organism with relatively infrequent errors. To understand how the remarkable fidelity of ontogeny is enforced, one needs to focus on the mechanisms and the principles that operate at the interface of genomic and epigenetic information.

As the embryological plan unfolds, stem-like cells engage in mitotic self-renewal or produce developmentally restricted and progressively more specialized cell progeny. Some examples include (1) the totipotent cells of the morula that generate entire embryos and associated extra-embryonic tissues, (2) the pluripotent embryonic stem cells of the blastocyst inner mass that are responsible for generating an entire embryo, and (3) the multipotent stem cells which generate endoderm-mesoderm- or ectoderm-derived tissues. This chapter focuses on ectoderm-derived neural stem cells which generate cellular components of the central nervous system, neurons, astrocytes, and oligodendrocytes.

The developmental decisions for each of these different cell types employ multi-gene programs that are guided by a plethora of environmental signals (i.e. external growth factors, neurotransmitters, hormones, immobile cell surface or extracellular matrix-associated molecules, intracellular signaling cascades and metabolic factors). Signals may direct the cells toward different, mutually exclusive developmental fates. How are those cellular signals computed and integrated such that the specific multi-gene developmental programs are effectively engaged? How can stochastic molecular events lead to the reproducible coordination of gene activities that enable neuronal development?

Systems biology addresses such questions by postulating computational modules embedded in the intricately complex biochemical pathways in terms of digital switches for activation or suppression of gene programs. In this manner, epigenetic information may be objectively quantified. For example, molecules (i.e. cyclins and partner kinases) known to carry out specific biochemical reactions have been identified to also have surprising global functions as molecular "switches" governing cell proliferation. Cyclin E/CDK2 reside at the nexus of complex mitotic, metabolic and signaling pathways (Neganova and Lako, 2008). The accumulation of cyclin E/CDK2 serves as a switch that triggers cell entry into the "S" phase. Similarly, the self-renewal of stem cells is maintained through the promotion of proliferation and the prevention of differentiation. This special state is enabled by a self-organized network of three transcription factors Oct3/4, Sox2 and Nanog which form positive feedback loops (Niwa, 2007) (see Chapter 1). In recent studies, the presence of Oct3/4 and Sox2, as well as two additional factors KlF4 and c-Myc, were found to be sufficient to establish pluripotency in fibroblast nuclei (Takahashi and Yamanaka, 2006). Hence, targeting the self-renewal gene regulatory module offers a promising technological advancement that is now widely replicated in many laboratories.

Studies from our group have revealed a novel universal regulatory module referred to as an integrative nuclear FGFR1 signaling (INFS), which in a manner analogous to the above mentioned modules, integrates diverse epigenetic signals (Stachowiak *et al.*, 2003a; 2007) (Figure 4.1). However, unlike the cell cycle and pluripotency modules, the INFS stimulates the cell's exit from the mitotic cycle and differentiation, thus complementing the stem cell developmental mechanisms (strategies). This new mechanism is the focus of our chapter.

In the center of the INFS module lie proteins that bear historical names based on originally perceived functions such as the fibroblast growth factor receptor (FGFR)-1 and its ligand fibroblast growth factor (FGF)-2.

4.2. The Curious Evolution of Fibroblast Growth Factors Family: Extracellular Cell–Cell and Intracellular Cytoplasmic Nuclear Messengers

The first discovered member of the fibroblast growth factor (FGF) family was a basic FGF, FGF-2, purified from bovine brain by Denis Gospodarowicz (Gospodarowicz *et al.*, 1982; 1987). The initial studies tested the actions of FGF as an external protein in fibroblast cell cultures, hence the factor's name. There are now over 20 members of the mammalian FGF family that share similar sequences and three-dimensional structures (Ornitz and Itoh, 2001). FGFs are not found in single-celled organisms but are common to the diverse multicellular animals (e.g. *C. elegans*, drosophila, zebrafish, frog, chicken, mouse and human), suggesting that the generation of tissues and organs with specialized cells requires FGFs.

In the mammalian lineage, the common primordial FGF gene, FGF13, is thought to be the source of various forms of FGFs (Itoh and Ornitz, 2011). Currently it is clear that many different cell types can elaborate various FGFs. Of particular interest during the evolution of the FGF family is the inclusion or exclusion of a cleavable hydrophobic signal peptide (SP) domain and an acquisition of a positively charged nuclear localization signal (NLS). FGFs with an SP and those without a cleavable hydrophobic signal sequence, but which may externalize from injured cells or by a phosphorylation-dependent mechanism (Ebert *et al.*, 2010), interact with plasma membrane FGF receptors (Ornitz and Itoh, 2001). Factors like FGF1 (acidic) and FGF2 (basic) that have lost the SP sequence and are produced in the cytosol retain a NLS, and thus appear to have evolved to act inside the cell nucleus. Indeed, the translation of FGF2 mRNA yields 18-kDa FGF2 and additional higher molecular weight (HMW) isoforms (21–24 kDa) (Florkiewicz *et al.*, 1991; Stachowiak *et al.*, 1994b; Joy *et al.*, 1997; Powell and Klagsbrun, 1991; Stachowiak *et al.*, 2003a). All HMW forms of FGF2 contain a classical NLS within

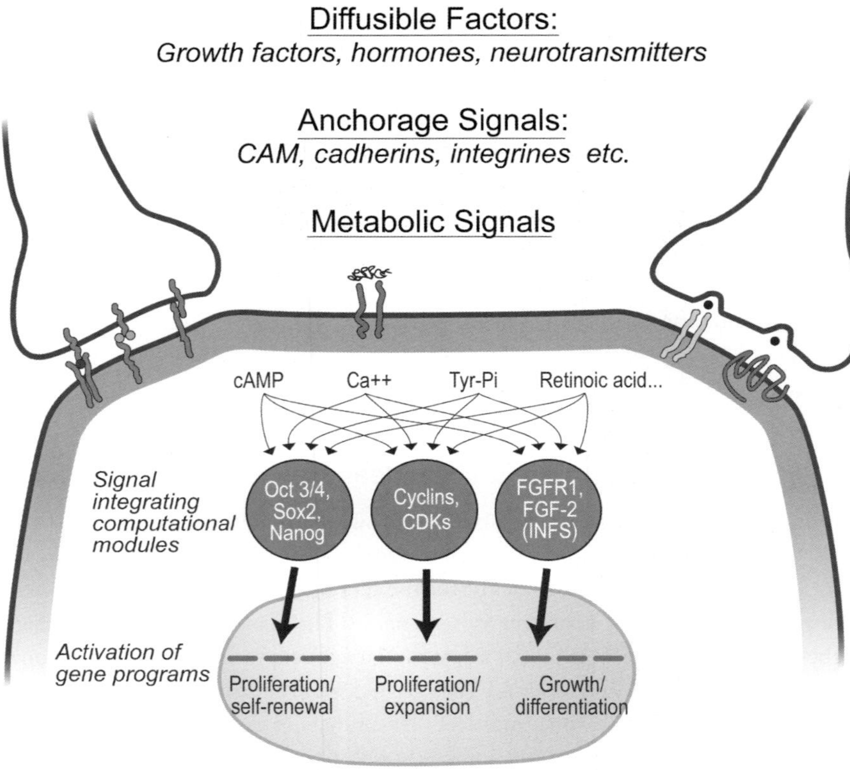

Figure 4.1. Integrative signaling modules in stem cell development. Self-organizing pluripotency network for cell renewal involves positive feedback interactions among transcriptional factors Oct3/4, Sox2 and Nanog. Cell cycle proteins (cyclins and cyclin-dependent kinases) integrate mitogenic signals. Cell differentiation is mediated by integrative nuclear FGFR1 signaling (INFS).

extended N-terminal sequences, thus enabling nuclear import (Florkiewicz *et al.*, 1991). The 18-kDa FGF2 also contains a non-classical C-terminal NLS (Sheng *et al.*, 2004). Whilst some tissues and cells display small amounts of 18-kDa FGF2 at the cell surface and in the extracellular matrix (Schechter, 1992; Vlodavsky *et al.*, 1991), both high- and low-molecular weight FGF2s in the brain and cultured cells are found almost exclusively in the nuclear fraction (Stachowiak *et al.*, 2003a) (Figure 4.2a).

Studies in our and other laboratories have revealed an association between nuclear FGF2 accumulation and (i) the growth of glial cells, (ii) the growth, differentiation, and functional activation of neurons, and (iii) the activation of

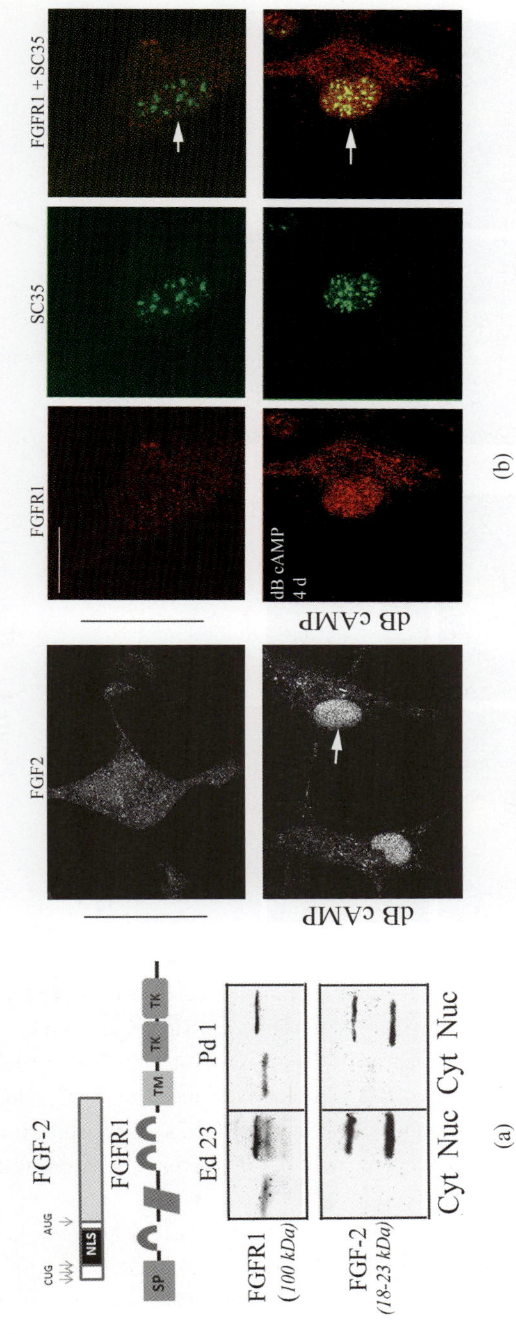

Figure 4.2. *(Continued)*

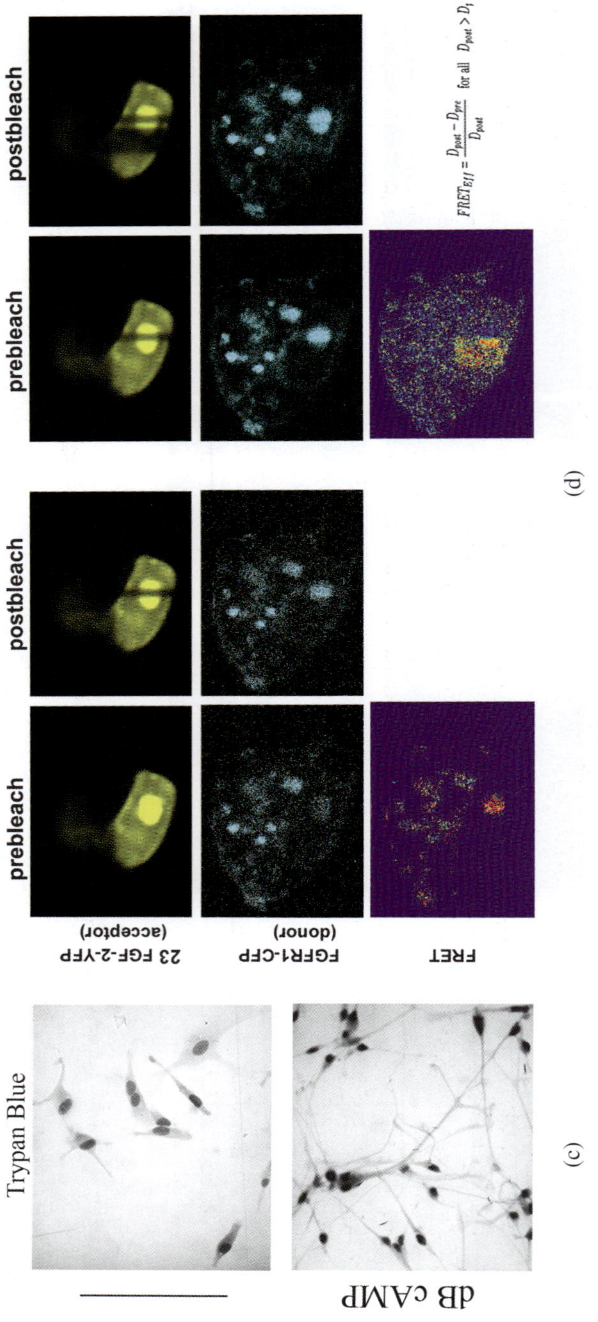

Figure 4.2. (*Continued*)

Figure 4.2. Structure, subcellular localization and interaction of nuclear FGF2 and FGFR1. (a) Top: Schematic structure of FGF2 and FGFR1 molecules. Low molecular weight (LMW, 18 kD) FGF2 is generated from typical AUG translational start codon. Higher molecular weight forms (HMW, 21–23 kDa) are generated from additional CUG codons and include nuclear localization signal (NLS). Western blot analyses of FGF2 and FGFR1 in Nuc and Cyt fractions of the whole rat brain at embryonic day 23 and postnatal day 1 (Stachowiak *et al.*, 2003a). (b–c) Non-differentiated proliferating HNPCs treated with 0.1-mM dBcAMP to induce their neuronal differentiation (Stachowiak *et al.*, 2003a). cAMP-treated cells display nuclear accumulation of FGF2 and FGFR1 ("B" at 24 hours) followed by neuronal differentiation ("C" at 48 hours) (Stachowiak *et al.*, 2003a). (b) Cells were fixed with 4% paraformaldehyde and immunostained with FGF2 Ab or co-immunostained with FGFR1 McAb6 (+ anti-mouse-Alexa 488) and rabbit antibody against splicesome assembly factor SC-35 (+ goat-anti rabbit-CY3). Merged images: FGFR1 and SC35 immunoreactive pixels co-localize (yellow) within SC35-rich nuclear foci (arrows pointing to nucleus) (Hu *et al.*, 2004; Peng *et al.*, 2002; Stachowiak *et al.*, 2003a; Stachowiak *et al.*, 1996a; Stachowiak *et al.*, 1996b). Antibody against SC-35 labels nuclear speckle-like domains which correspond to sites of RNA Pol II-mediated transcription and co-transcriptional, pre-mRNA processing (Blencowe *et al.*, 1994). In proliferating HNPC, little or no co-localization between SC35 speckles and FGFR1 grains was observed. In contrast, co-localization of SC35 and FGFR1 was observed in several larger aggregates of dB–cAMP-treated differentiating cells as shown in this figure. Analyses of consecutive, confocal sections demonstrated the FGFR1 sites within the interior of the SC35 speckles (not shown) (Stachowiak *et al.*, 2003a). (c) After 48 hours, incubation cells were fixed and permeabilized with 0.5% Triton X100 and their morphology were revealed at lower magnification by staining with Trypan Blue (Stachowiak *et al.*, 2003a). (d) Interaction of FGFR1 with FGF2 in the nucleus. TE671 cells were transfected with FGFR1–CFP (receptor fused to CFP) + 23-kDa FGF2–YFP (ligand fused to YFP). FRET was measured using acceptor bleaching. Bleaching of the 23-kDa FGF2–YFP energy acceptor increased the fluorescence emitted by the FGFR1–CFP energy donor in the nucleus and the nucleolus, as illustrated by two sequential acceptor bleachings of adjacent nuclear regions. In contrast, no detectable FRET (interaction) occurred between co-transfected nuclear 18-kDa FGF2 and FGFR1 (not shown). Thus, while HMW (23 kDa) FGF2 serves as a ligand for nuclear FGFR1, LMW (18 kDa) FGF2 does not interact with nuclear FGFR1. Size bar: 25 um. Color scale of FRET efficiency: 0.500 = 50%. The efficiency of energy transfer between nuclear 23-kDa FGF2 and FGFR1 was 35–40%. Figures 4.2a–4.2c were reprinted from *J. Neurochemistry* 84, pp. 1299–1300, 2003, with permission of International Society of Neurochemistry. Figure 4.2d was reprinted from *Molec. Biol. Cell* 20, pp. 2409, 2009.

*CFP: cyan fluorescent protein, Cyt: cytoplasmic, FRET: fluorescence resonance energy transfer, HNPC: human neural progenitor cells, Nuc: nuclear, SP: signal peptide, TK: split tyrosine kinase domain, TMD: transmembrane domain, YFP: yellow fluorescent protein.

neuroendocrine adrenal medullary cells. Nuclear accumulation of FGF2 is induced by various extracellular factors and major intracellular signaling pathways unrelated to the FGFs (Joy *et al.*, 1997; Peng *et al.*, 2001; Peng *et al.*, 2002; Stachowiak *et al.*, 1994b; Stachowiak *et al.*, 1997b) (see also Clarke *et al.*, 2001; Dono and Zeller, 1994; Grothe *et al.*, 1991; Klimaschewski *et al.*, 1999; Matsuda *et al.*, 1992; Matsuda *et al.*, 1994; Woodward *et al.*, 1992).

Studies have shown that FGF2 does not need to be secreted to stimulate the differentiation of avian Schwann cells (Sherman *et al.*, 1993). The intracellular-expressed nuclear HMW FGF2 effectively stimulates cellular growth (Arese *et al.*, 1999), neuronal differentiation and associated gene activities without surface FGF receptor activation (Peng *et al.*, 2001; Peng *et al.*, 2002; Stachowiak, 2009). Also, when added directly to the isolated nuclei, FGF2 stimulates rRNA synthesis (Bouche *et al.*, 1987) and affects gene transcription *in vitro* (Nakanishi *et al.*, 1992). Thus, FGF2, now appears to be a cytoplasmic–nuclear signaling factor rather than an extracellular-acting protein as initially proposed by the experiments of Gospodarowicz. We have proposed the concept of an "intracrine" growth factor signaling molecule which translocates from the cytosol to the nucleus (without externalization) to induce the cellular responses listed above (Stachowiak *et al.*, 1997a). The interest in FGF2 as a nuclear signaling factor has increased greatly with the observation that nuclear accumulation is highly regulated and correlates with the transition of cells between major developmental stages and with the discovery of nuclear FGF receptors (Bryant and Stow, 2005; Stachowiak *et al.*, 2003a; Stachowiak *et al.*, 2007).

4.2.1. *The discovery of the nuclear FGF receptor-1*

FGFs are known to interact with two main classes of receptors: (1) high affinity tyrosine kinase receptors which bind FGFs with picomolar affinity and are thought to mediate FGF cellular responses, and (2) low affinity receptors which bind FGFs with nanomolar affinity and are characterized by associated heparan sulfate proteoglycans (HSPGs) (Szebenyi and Fallon, 1999). Four genes encode the high-affinity FGFRs with numerous additional variants arising due to differential splicing. FGFR1–4 share a similar structure, consisting of an N-terminal signal peptide (SP), which directs FGFR synthesis to the endoplasmic reticulum (ER), two or three immunoglobulin-like domains, which contain the ligand-binding site, a single transmembrane domain (TMD), a split tyrosine kinase domain, and a C-terminal domain (see Figure 4.2a). Important differences among FGFRs reside in the TMD. Typically, the TMD of type-1 transmembrane proteins contains an

α-helix of approximately 30 amino acids with more than 11 consecutive hydrophobic amino acid residues. These non-polar amino acid residues are oriented outside the α-helical core and interact with the lipid bilayer, thereby conferring stability to the peptide in the membrane (Creighton, 1984). The TMD of FGFR4 has an exclusively α-helical conformation typical to many other type-1 tyrosine kinase receptors (Myers *et al.*, 2003). The TMDs of FGFR2 and FGFR3 have shorter α-helices as well as β-sheet regions. FGFR1 possesses an atypical TMD consisting of short stretches of hydrophobic amino acids interrupted by hydrophilic, polar amino acids, and a predicted β-sheet conformation (Myers *et al.*, 2003). Such a disruption of the hydrophobic region by polar amino acids has been shown (in other proteins) to make their membrane association less stable, allowing proteins to be released from the membrane (Eisenberg *et al.*, 1984).

High-affinity FGF2 binding sites formed by FGFR1 have been found in the nuclei of different cell types (Stachowiak *et al.*, 1996a; 1996b). The nuclear presence of FGFR1 was demonstrated with an array of different FGFR1 antibodies in different cell types, including non-transformed cells (Maher, 1996; Reilly and Maher, 2001), cancer cell lines (Bryant and Stow, 2005; Peng *et al.*, 2001; Stachowiak *et al.*, 2003a; Stachowiak *et al.*, 1996a; 1996b), cells found in the rat and mouse brains (Clarke *et al.*, 2001; Gonzalez *et al.*, 1995; Fang *et al.*, 2005; Leadbeater *et al.*, 2006; Stachowiak *et al.*, 2003a) (Figure 4.2a) and cochleo-vestibular ganglion cells (Bilak *et al.*, 2003). Importantly, FGFR1 nuclear accumulation has been observed *in vivo* in both developing neurons and glial cells (Clarke *et al.*, 2001; Fang *et al.*, 2005; Gonzalez *et al.*, 1995; Horbinski *et al.*, 2002; Klimaschewski *et al.*, 1999; Leadbeater *et al.*, 2006).

The molecular weight of nuclear FGFR1 corresponds to that of its full-length, cell surface counterpart and it can undergo autophosphorylation and phosphorylate other proteins in an FGF2-stimulated manner (Stachowiak *et al.*, 1996a; 1996b). One simple explanation for the nucleo-cytoplasmic localization of FGFR1 is through potential alternate splicing events that remove either the transmembrane domain (TMD) or the signal peptide (SP) domain. However, this explanation seems unlikely due to the lack of potential splice sites surrounding these areas in the FGFR1 gene, and is further discounted by the observation that transfection of cells with full-length FGFR1, including the receptor with various tags, leads to the accumulation of recombinant FGFR1 in the nucleus (Peng *et al.*, 2002; Puchacz *et al.*, 1993; Stachowiak *et al.*, 1997a).

The presence of FGFR1 along with FGF2 in the nuclei of a developing brain is illustrated on Figures 4.2a and 4.2b. The ligand that interacts with the nuclear FGFR1 is HMW (23 kDa) FGF2. Co-immunoprecipitation experiments show that a 23-kDa FGF2 ligand interacts with FGFR1 in the nucleus and cytoplasm whereas, an 18-kDa FGF2 interacts with FGFR1 only in the cytoplasm. The intranuclear interaction between FGFR1 and 23-kDa FGF2 in intact cells is verified by FRET analyses (Dunham-Ems *et al.*, 2009) (Figure 4.2d). In cells transfected with FGFR1 fused to cyan fluorescent protein (FGFR1-CFP) and 23-kDa FGF2 fused to yellow fluorescent protein (23 FGF-2-YFP), bleaching of the 23 FGF-2-YFP acceptor increases the fluorescence emitted by the FGFR1-CFP donor in the nucleus as well as the nucleolus (Figure 4.2d). In contrast, no detectable FRET occurs between co-transfected nuclear 18-kDa FGF2 and FGFR1 (not shown). Thus, while HMW (23 kDa) FGF2 serves as a ligand for nuclear FGFR1, LMW (18 kDa) FGF2 does not interact with nuclear FGFR1.

4.2.2. *What is the source of nuclear FGFR1?*

Biotinylation of the cell surface receptors of glioma or TE671 medulloblastoma cells does not lead to the appearance of biotinylated FGF receptor in cell nuclei (Carpenter, 2003; Peng *et al.*, 2002; Stachowiak *et al.*, 1997b), indicating that nuclear FGFR1 is not derived from the cell surface. Furthermore, in many cells, the binding of extracellular FGF to its surface receptors does not elicit nuclear FGFR1 accumulation (Carpenter, 2003; Horbinski *et al.*, 2002; Moffett *et al.*, 1996; Moffett *et al.*, 1998a; Moffett *et al.*, 1998b; Peng *et al.*, 2001; Peng *et al.*, 2002; Stachowiak *et al.*, 2003a; Stachowiak *et al.*, 1994a; Stachowiak *et al.*, 1996a; 1996b; Stachowiak *et al.*, 1994b; Stachowiak *et al.*, 1997b). Instead, nuclear FGFR1 accumulation is induced by a number of other factors (see Table 4.1) and is accompanied by an increase in receptor synthesis (Carpenter, 2003; Hu *et al.*, 2004; Myers *et al.*, 2003; Peng *et al.*, 2001; Peng *et al.*, 2002). Hence, we named it integrative nuclear R1 signaling (INFS) (Peng *et al.*, 2002; Stachowiak *et al.*, 2003a). In some cells, extracellular FGFs can stimulate the nuclear accumulation of FGFR1 (Maher, 1996). Whether this represents an internalization of the cell surface FGFR1 or newly synthesized ER-derived receptor remains to be further established.

FGFRs are type-1 transmembrane proteins, possessing N-terminal signal peptides that should predispose these receptors for their insertion into cellular membranes. However, cell biochemical fractionation demonstrated that a fraction of the cellular receptor is a soluble, nonmembrane-associated protein. How the receptor can enter the cytosol and undergo subsequent nuclear accumulation is an important, compelling question.

Pulse-chase experiments reveal that both cytosolic and nuclear FGFR1 represent newly synthesized full-length receptors — a notion supported by the fact that much of the nuclear FGFR1 is non-glycosylated. It appears that this receptor

Table 4.1. Stimuli causing nuclear accumulation of FGFR1 and FGFR2.

FGFR1	FGFR2
Angiotensin II (Stachowiak *et al.*, 1996a)	Prostaglandin (Marchetti *et al.*, 2006; Sabbieti *et al.*, 2005; Schmahl *et al.*, 2004)
Acetylcholine (Stachowiak *et al.*, 1996a)	
Cell depolarization (Peng *et al.*, 2002)	
Bone morphogenetic protein-7 (Horbinski *et al.*, 2002)	
EGF (Stachowiak *et al.*, 1997b)	
Retinoic acid (Kehoe, 2008)	
cAMP (Stachowiak *et al.*, 2003a)	
Protein kinase C (Peng *et al.*, 2002)	
Reduced cell contact (Stachowiak *et al.*, 1997a)	
Nerve growth factor (Stachowiak, 2009)	
Intracellular nuclear FGF2 (Peng *et al.*, 2002)	
Exogenous FGF2 (Maher, 1996)	

enters the cytoplasm or nucleus before Golgi processing and that the ER is one site at which FGFR1 is released into the cytosol (also see below).

Studies of FGFR1 fused to Enhanced Green Fluorescent Protein (FGFR1-EGFP) in live cells have greatly helped to elucidate FGFR1 processing and dynamics in live cells. A transfected full-length FGFR1-EGFP is able to enter cell nuclei similar to endogenous, non-tagged FGFR1 (Myers *et al.*, 2003; Peng *et al.*, 2002). Examination of FGFR1-EGFP mobility in live cells by fluorescence recovery after photobleaching (FRAP) analysis demonstrates a slow-moving population ($t_{1/2} = 69$ sec) associated with cellular membranes, and a fast-moving ($t_{1/2} = 0.2$ sec), cytosolic pool not associated with membranes (Dunham-Ems *et al.*, 2006). Studies using brefeldin A (BFA), an inhibitor of intracellular protein transport to the Golgi complex, show that the fast-moving receptor is transported out of the ER into the cytosol and then into the nucleus. This work further indicates that soluble and nuclear FGFR1 are not derived from the cell surface (Figure 4.3a).

In a fluorescence loss after photobleaching (FLIP) experiment, repeated photobleaching of FGFR1-EGFP in a small, defined nuclear region resulted in a loss of FGFR1-EGFP fluorescence within the entire nucleus, as well a delayed loss in the non-vesicular cytoplasm (Figure 4.3b). These experiments demonstrate that (i) "nuclear" and "cytosolic" FGFR1 are indeed present in distinct subcellular

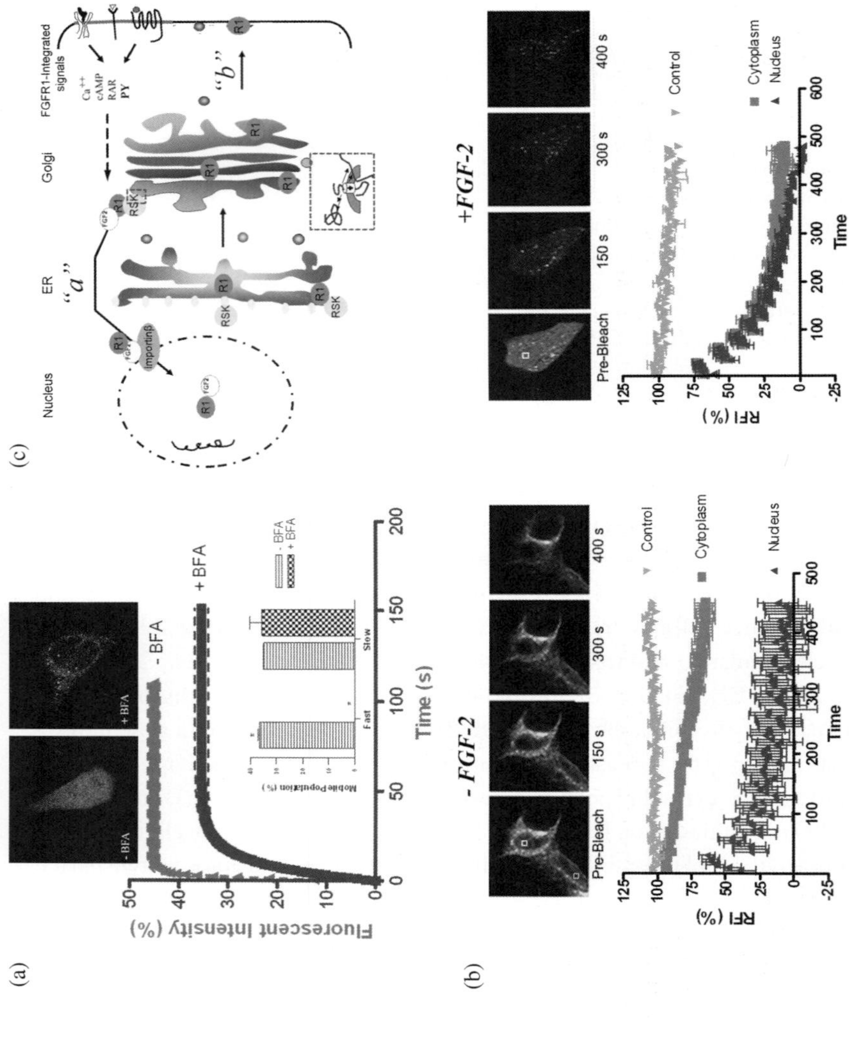

Figure 4.3. *(Continued)*

Figure 4.3. Cytoplasmic processing and nuclear targeting of FGFR1 (Fang *et al.*, 2005; Stachowiak *et al.*, 2003a). (a) ER to Golgi translocation is required for the generation of soluble fast mobility cytosolic FGFR1 and its nuclear accumulation. FGFR1–EGFP was co-transfected with pcDNA. Equal amounts (1 μg) of each plasmid were transfected. Cells were treated for 1 hour with Brefeldin A (BFA), 24 hours post transfection. Image shows a representative cell. FRAP measurements of FGFR1–EGFP–pcDNA3.1 compared to FGFR1–EGFP treated with 10-μg/ml BFA for 1 hour. Fluorescent intensity is representative of the mobile population. Data is the mean of (at least) 14 cells. The fast population is completely eliminated by BFA, leaving only one mobile population. (b) FGFR1 undergoes cytoplasmic–nuclear shuttling affected by cell stimulation — FLIP analysis (from Star Dunham-Ems — Ph.D. thesis, SUNY, Buffalo, 2006). The fluorescence images shown are prior to photobleaching (pre-bleach) and then at the indicated times, post-bleach. Five image scans were followed by repetitive bleaching (10 iterations), which was followed by image acquisition and repetitive bleaching; this sequence was repeated for 450 sec. The plane of focus was a cross-section through the cell. The graph indicates the FLIP curves where the fluorescence was quantified at the bleach point (nuclear region: indicated by square) at a distant cytoplasmic points (square) and at a non-bleached control cell. Graphs show FLIP analysis. The data are representatives from at least five different cells. (c) Model of nuclear trafficking of FGFR1. Cytoplasmic FGFR1 exists in three separate populations: (1) an immobile, newly synthesized ER population, (2) highly mobile, non-glycosylated, cytosolic population, and (3) a slowly diffusing, membrane receptor population (Dunham-Ems *et al.*, 2006). FGFR1 can enter either the (a) "nuclear pathway", where the non-glycosylated receptor is released from the ER into the cytosol and accumulates in cell nuclei, or the (b) "membrane pathway" in which FGFR1 is processed through the Golgi to the plasma membrane. Integration of diverse signals by the nuclear pathway: the activation of diverse classes of surface membrane receptors and their signaling cascades causes the association of the adaptor protein RSK1 with the cytoplasmic tail of newly synthesized FGFR1, and the subsequent release of FGFR1 into the cytoplasmic pool, before trafficking through the Golgi apparatus and to the cell surface. Once in the cytoplasm, FGFR1 indirectly complexes with importin β to facilitate nuclear import. In addition, activation of cell surface FGFR1 by FGF induces FGFR1 internalization, which is dependent on the function of the ARF6, Dynamin2 and Rab5 endocytic machinery, and is inhibited by surface E–cadherin adhesion complexes. Once internalized, FGFR1 may potentially be released from endosomes, or may be trafficked in a retrograde fashion to the ER/Golgi for cytoplasmic release via the RSK1-associated pathway (Stachowiak *et al.*, 2007). Figure 4.3a was reprinted from *Molec. Biol. Cell* 17, pp. 2230, 2006.

*FLIP: fluorescence loss induced by photobleaching, FRAP: fluorescence recovery after photobleaching.

compartments, (ii) nuclear FGFR1 remains in equilibrium with the cytosolic receptor, and (iii) cytoplasmic–nuclear transport limits the rate of nuclear FGFR1 accumulation. Importantly, in FGF-stimulated cells, depletion of cytosolic FGFR1-EGFP after nuclear bleaching was markedly accelerated, indicating that the cytosol to nucleus translocation can be increased.

The information on how FGFR1 protein may be extruded from the ER membrane comes from substitution experiments, in which the atypical (β-sheet containing polar amino acids) TMD of FGFR1 is replaced with the typical (α-helical, hydrophobic) TMD of FGFR4. The chimera FGFR1/R4 mutant is primarily associated with membranes and is absent from the cytosol and nucleus (Myers *et al.*, 2003). In contrast, FGFR1 mutants, in which hydrophobic TMD amino acids are replaced with increasing numbers of hydrophilic amino acids (Figure 4.3b), show reduced association with membranes and increased cytosolic/nuclear accumulation as TMD hydrophilicity is increased. Thus, the distinct TMD of FGFR1 appears to play a critical role in the weak association with cellular membranes and the intriguing FGFR1 accumulation in the cytosol and nucleus (Myers *et al.*, 2003).

Thus, a model emerges (Figure 4.3c) in which FGFR1 exists in three separate populations: (1) an immobile, newly synthesized ER population which is increased by brefeldin A treatment, (2) a newly synthesized, highly mobile, non-glycosylated, cytosolic population that is depleted by brefeldin A that resides outside the ER-Golgi lumen, and (3) a slowly diffusing, membrane receptor population. The newly synthesized FGFR1 can enter either the "membrane pathway" in which FGFR1 is processed through the Golgi to the plasma membrane, thus showing limited lateral mobility or the "nuclear pathway", in which a non-glycosylated receptor is released from the ER, producing highly mobile, cytosolic receptor molecules that accumulate in the cell nucleus. The release from the ER and entry into the nuclear pathway is favored by the interaction of FGFR1 with cytoplasmic proteins. One such interacting regulator is pp90 ribosomal S6 kinase (RSK)-1, which significantly influences the receptor's mobility and diffusion rate (Dunham-Ems *et al.*, 2006). The interaction occurs via association with the tyrosine kinase (TK) domain of FGFR1 (Fang *et al.*, 2005; Hu *et al.*, 2004). The kinase activity of RSK1 is responsible for redirecting FGFR1 synthesis from the Golgi glycosylation pathway into the cytosol, consistent with the known association of RSK1 with ER-attached polyribosomes (Angenstein *et al.*, 1998).

Striking support for nuclear FGFR1 import (Reilly and Maher, 2001) and the likely nuclear role of FGFR1 is revealed by the demonstration of a functional dependency on importin β, a component of multiple nuclear import pathways. Since importin β is a soluble protein that traffics between the cytoplasm and nucleus, the interaction with FGFR1 in this model must occur in the cytoplasm, perhaps after RSK1-associated cytoplasmic release (Figure 4.3c). The interaction between FGFR and importin β appears to be indirect as FGFR1 does not exhibit a detectable NLS. The fact that the FGF2 ligands possess NLS domains makes them potentially attractive linker/chaperone molecules.

4.3. FGFR1 as a Direct Global Gene Regulator

The nuclear targeting of FGFR1 suggests that FGFR1 mediates diverse extracellular and intracellular signal transduction directly to the genome. Immunoelectron microscopy of *in situ*-fractionated adherent cells finds 0.2-μm aggregates of FGFR1 within the interior of the nuclear matrix, suggesting that FGFR1 may exert its influence over a number of cellular processes by acting within the environment of the nuclear matrix (Stachowiak *et al.*, 1996b). Confocal microscopy of human neuronal progenitor cells stimulated with either cAMP, BMP7 or bovine adrenal medullary cells stimulated with cAMP or angiotensin II demonstrates that FGFR1 is localized to nuclear speckle domains containing snRNPs involved in RNA processing (Peng *et al.*, 2001) and the spliceosome assembly factor SC-35 (Stachowiak *et al.*, 2003a). These speckles are presumed to be equivalent to the nuclear matrix-attached granules observed in the EM (Fu and Maniatis, 1990; Wei *et al.*, 1999) and correspond to sites of RNA polymerase II-mediated transcription and co-transcriptional, pre-mRNA processing that are located on the periphery and in the interior of the speckles, respectively (Wei *et al.*, 1999). Consistent with this model, confocal microscopy showed the presence of BrUTP-labeled sites of transcription on the periphery of many FGFR1-rich speckles (Somanathan *et al.*, 2003). Furthermore, in cells stimulated with cAMP, FGFR1 concentrates within nuclear domains, showing high levels of RNA synthesis, revealed by an incorporation of 5'-fluorouridine (FU) (Figure 4.4a). In contrast, no significant co-localization of DNA replication sites with FGFR1 speckles was found arguing against a direct role in DNA replication (Somanathan *et al.*, 2003; Stachowiak *et al.*, 2003a).

The co-localization of FGFR1 with sites of RNA synthesis and processing indicated that the receptor could act as a direct and global transcriptional regulator, controlling the activities of genes located at different genomic locales. Indeed, the increases in the activities of FU-labeled transcription sites accompanying cell stimulation were diminished by transfection of a dominant negative tyrosine kinase-deleted FGFR1(TK−) mutant. FGFR1(TK−) blocks FGFR signaling by heterodimerization with endogenous receptors (Robinson *et al.*, 1995), and by

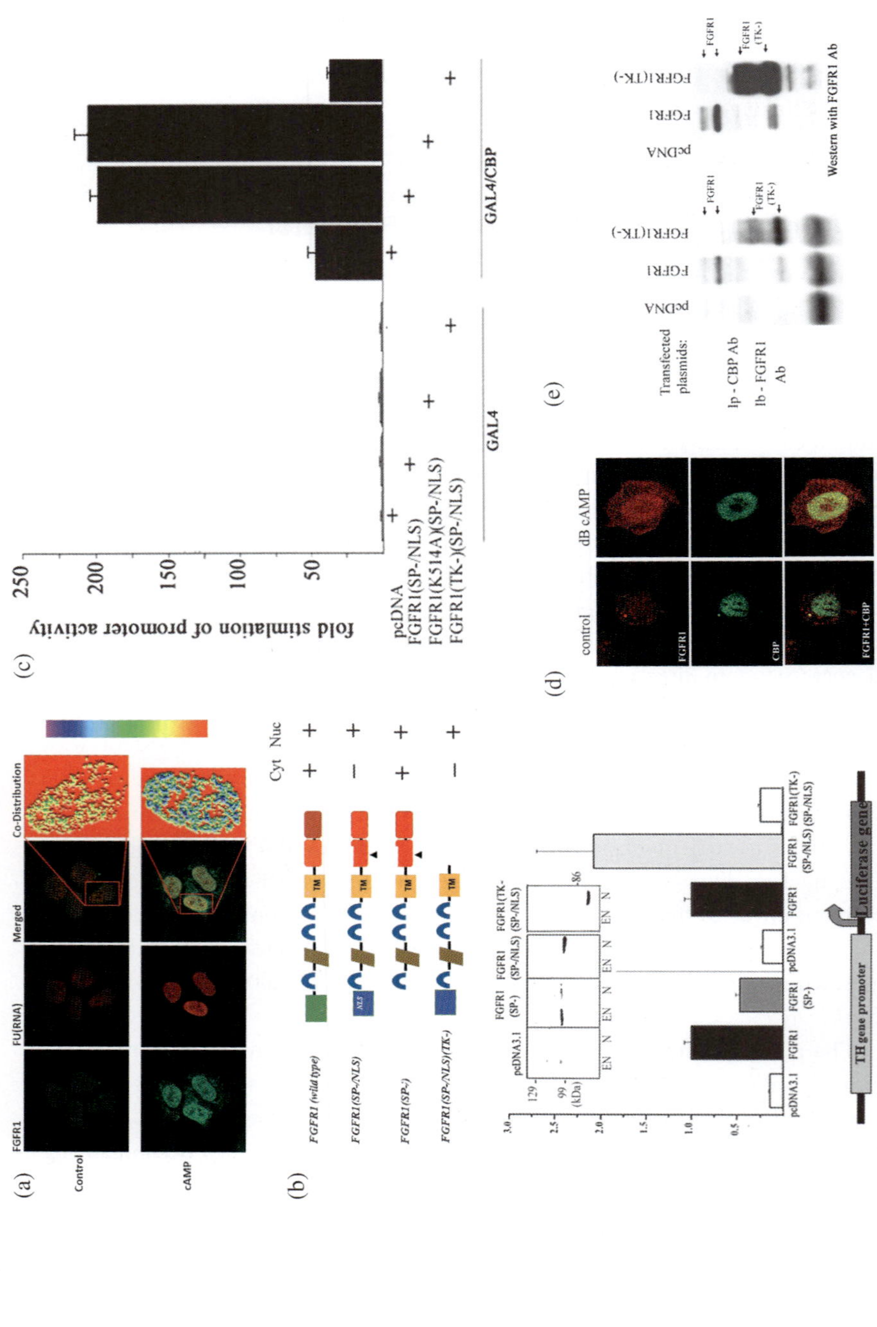

Figure 4.4. (*Continued*)

Figure 4.4. Nuclear FGFR1 as a transcriptional regulator in partnership with CBP (Peng *et al.*, 2001; Peng *et al.*, 2002). (a) Codistribution of nuclear FGFR1 and transcription sites. Neuroblastoma cells were incubated for 4 hours with or without 100-μM dBcAMP. The newly synthesized RNA cells were exposed to 0.5-mM 5'-FU for 15 minutes for labeling. FU was revealed with rat monoclonal anti-BrdU antibody + Cy3-conjugated secondary antibody. The endogenous FGFR1 was stained with McAb6 + Alexa 488-conjugated secondary antibody. Red (FU) and green confocal images were merged and the co-localized pixels were revealed. Codistribution image: colors indicate intensity of co-localized pixels (red: no co-localization; blue: highest intensity of co-localized pixels). (b) Nuclear FGFR1 stimulates transcription from TH gene promoter. TH promoter–luciferase reporter gene was co-transfected with control pcDNA3.1 vectors or vectors expressing wild type FGFR1, FGFR1(SP−) with depleted SP, FGFR1(SP−/NLS) with the SP replaced by NLS or FGFR12(SP−/NLS)(TK−) that lacks TK domain. Subcellular presence (+) or absence (−) of transfected receptor proteins is indicated and is shown on the western immunoblot. TH promoter was activated by cytoplasmic/nuclear FGFR1, non-membrane cytoplasmic/nuclear FGFR1(SP−), and by non-membrane exclusively nuclear FGFR1(SP−/NLS). Presence of TK domain is required for gene activation (Peng *et al.*, 2002). (c) Nuclear FGFR1 stimulates CBP-mediated transcription. The GAL4 promoter-luciferase reporter was co-transfected with GAL4/CBP or non-fused GAL4. GAL4/CBP-dependent transcription was stimulated by co-transfected, active R1(SP−/NLS) or kinase-inactive R1(K514A)(SP−/NLS) but not by kinase-deleted R(SP−/NLS)(TK−) (Fang *et al.*, 2005). (d) FGFR1 co-localizes with CBP in differentiating neural stem cells. Cultures of neural progenitor-like cells isolated from human umbilical cord blood were treated with 0.1-mM dB cAMP for 48 hours and co-immunostained with FGFR1 McAb6 (+ anti-mouse-Alexa 488) and with rabbit CBP (+ goat-anti rabbit-CY3). Confocal sections through the middle of the nuclei are shown. Merged images: FGFR1 and CBP immunoreactive pixels co-localize within CBP-rich nuclear foci (Fang *et al.*, 2005). (e) Both FGFR1 and TK-deleted FGFR1(TK−) bind to CBP. TE671 medulloblastoma cells were transfected with FGFR1 or FGFR1(TK−). Cell extracts were co-immunoprecipitated with CBP Ab and immunoblotted with FGFR1 McAb6 or were directly subjected to western analysis with FGFR1 McAb6. The interaction between CBP and transfected FGFR1 was further confirmed by CBP co-immunoprecipitation with FGFR1 antibody, in developing brain and by GST-CBP pull down of FGFR1 (Fang *et al.*, 2005). Figure 4.4b was reprinted from *J. Neurochemistry* 81, pp. 516, 2002, with permission of International Society for Neurochemistry. Figures 4.4c and 4.4d were reprinted from *J. Biological Chemistry* 280, pp. 28454–28456, 2005, with permission of American Society for Biochemistry and Molecular Biology Inc.

*FU: fluorouridine, NLS: nuclear localization signal, SP: signal peptide, TH: tyrosine hydroxylase, TK: tyrosine kinase.

forming inactive complex with CREB binding protein (CBP), a nuclear target of FGFR1 (see further below) (Fang *et al.*, 2005).

The FGF2 gene, located on human chromosome 4, was the first gene shown to be regulated by nuclear FGFR1 (Peng *et al.*, 2001). Tyrosine hydroxylase (TH), the rate-limiting enzyme in catecholamine biosynthesis, was also shown to be regulated by nuclear FGFR1 (Peng *et al.*, 2002). The TH gene is located on human chromosome 11 and is expressed specifically in catecholamine-producing cells. Co-transfection of FGFR1 (TK−) completely prevents the activation of transfected FGF2 and TH gene promoters, as well as their endogenous genes by angiotensin receptors, acetylcholine receptors, cell depolarization and by PKC and cAMP signaling pathways, thus demonstrating that FGFR1 is essential for the transcriptional activation of TH and FGF2 genes by various stimuli (Peng *et al.*, 2001; Peng *et al.*, 2002). In contrast, the extracellular FGFR antagonists do not prevent the same gene activations, suggesting that the promoter activation is mediated by intracellular, nuclear FGFR1 and the INFS mechanism (Peng *et al.*, 2001; Peng *et al.*, 2002). Indeed, the TH and FGF2 gene promoters, as well as other endogenous genes are activated by transfected wild type FGFR1, soluble (non-membrane) cytoplasmic/nuclear FGFR1(SP−) with a deleted signal peptide, and by soluble nuclear FGFR1(SP−/NLS), in which the signal peptide was replaced with a nuclear localization signal (Figure 4.4b shows activation of TH gene promoter). Thus, an increase in the nuclear FGFR1 content is sufficient to activate the TH and FGF2 genes. Importantly, blocking the cell surface FGFR and the extracellular ligand-induced FGFR internalization do not affect gene activation by transfected FGFR1 or by its nuclear ligand, HMW FGF2 (Peng *et al.*, 2001; Peng *et al.*, 2002). This supports the conclusion that the nuclear FGFR1 mediating gene transactivation via INFS is not a consequence of ligand-induced receptor internalization from the cell surface.

Although the activation of transcription by nuclear FGFR1 is prevented by the deletion of the receptor tyrosine kinase (TK) domain, the K514A mutation, which eliminates FGFR1 kinase activity (Bellot *et al.*, 1991) and receptor auto-phosphorylation, has little effect on transcription. Indeed, both FGFR1(K514A)(SP−/NLS) and FGFR1(SP−/NLS) transactivate the TH gene, as well as the co-transfected TH and FGF2 promoters, and CRE linked to a luciferase reporter (Fang *et al.*, 2005). Thus, activation of transcription by nuclear FGFR1 does not require receptor TK activity and, therefore, is distinct from the TK activity-dependent effects of plasma membrane FGFR1 (Mohammadi *et al.*, 1996).

Other genes thus far shown to be regulated by nuclear FGFR1 include neuron-specific enolase (NSE) (Stachowiak *et al.*, 2003b), neurofilament (NF)-l, c-Jun and cyclin D1 (Reilly and Maher, 2001). Different chromosomal locales of these nuclear

FGFR1-regulated genes further supports the role of INFS in global gene regulation (Stachowiak *et al.*, 2007).

4.3.1. *Transduction by nuclear FGFR1 involves the transcription co-activator, CREB Binding Protein (CBP)*

TH gene activation by nuclear FGFR1 is mediated by classical CRE. The promoters of other genes known to be regulated by nuclear FGFR1 contain CRE, AP1, and other cis-acting elements. Transfected nuclear FGFR1(SP−/NLS) transactivates CRE, as well as AP1-, retinoic acid receptor-, and NF-κB-binding elements (Fang *et al.*, 2005; Kehoe, 2008), all of which are known to be co-activated by CBP Goodman and Smolik, 2000). This raised the question of whether nuclear FGFR1 regulates transcription in cooperation with the CBP. The cooperation between these two nuclear proteins was documented by showing that co-transfected CBP and FGFR1(SP−/NLS) activate the TH and FGF2 gene promoters in a synergistic fashion, while depletion of endogenous CBP with antisense CBP RNA inhibits gene activation by nuclear FGFR1(SP−/NLS) (Fang *et al.*, 2005). The mechanisms of gene co-activation by nuclear FGFR1 and CBP were further elucidated using gal4-luciferase gene construct. In this artificial gene, the luciferase reporter is driven by a promoter with a DNA sequence that binds yeast transcription factor GAL4. Transfection of FGFR1 fused to DNA-binding fragment GAL4 (1-147 aa) failed to stimulate luciferase transcription, however, it did so in cooperation with co-transfected CBP (Fang *et al.*, 2005). Similarly, in a reverse experiment (shown in Figure 4.4c), transcription of gal4-luciferase driven by the chimera, GAL4-CBP, was augmented several-fold by co-transfected nuclear FGFR1(SP−/NLS) (Fang *et al.*, 2005). Deletion of the FGFR1 tyrosine kinase (TK) domain prevented the GAL4-CBP activation. Surprisingly, a point mutation (K514A) in FGFR1 TK which abolished TK activity did not prevent GAL4-CBP activation. Thus, the TK domain (not the TK activity) of the receptor is essential for the stimulation of transcription by a combination of FGFR1 and GAL4/CBP (Figure 4.4c). Further studies showed that the N-terminal domain of FGFR1 was also essential for transcriptional co-activation and cooperation with the CBP (Dunham-Ems *et al.*, 2009; Fang *et al.*, 2005).

Cross-linked chromatin immunoprecipitation experiments (Fang *et al.*, 2005) show that FGFR1 may associate with FGFR1-activated gene promoters (i.e. core FGF2 gene promoter) and induce the association of CBP with the promoter. Both kinase-active and -inactive receptors increase the binding of RNA Pol II and acetylation of histone H3 at the promoter. Thus, nuclear FGFR1 may stimulate the recruitment of CBP and RNA Pol II, as well as chromatin remodeling in a FGFR1 kinase activity-independent manner.

In the nucleus, CBP (von Mikecz *et al.*, 2000) and FGFR1 (Somanathan *et al.*, 2003; Stachowiak *et al.*, 2003a) are concentrated predominantly within splicing factor-rich nuclear speckle domains; double immunostaining revealed that the FGFR1- and CBP-labeled pixels were closely co-localized within the nuclear speckles in cAMP-differentiated human neural progenitor cells (see Figure 4.4d) and in developing mouse brain neurons (Fang *et al.*, 2005). Further support for a direct interaction between FGFR1 and CBP comes from experiments in which transfected FGFR1, FGFR1(TK−) or FGFR1(K514A) were co-immunoprecipitated with endogenous CBP (Figure 4.4e). These results indicate that the N-terminal domain of the receptor associates with CBP. Consistent with the nuclear trafficking of non-glycosylated FGFR1(Dunham-Ems *et al.*, 2006), CBP interacted predominantly with this form of nuclear FGFR1 (Fang *et al.*, 2005) (Figure 4.4e). An interaction between endogenous nuclear FGFR1 and CBP was seen in nuclear extracts of the developing rodent brain and was further confirmed by GST-CBP pull down of FGFR1 (Fang *et al.*, 2005).

4.3.2. *Transcription-gating function of CBP is regulated by nuclear FGFR1*

Gene activities in developing cells are affected by a variety of extracellular signals, transduced by diverse intracellular messengers including cAMP, Ca^{++}, NFkB, STATs or nuclear receptors. The transcriptional co-activators, CBP and its homolog p300, are signal integrators that function at the termini of all of these signal transduction pathways in which the co-activator interacts with diverse sequence-specific transcription factors (ssTFs), which includes CREB, AP1, STATs, NFkB, and retinoic acid receptors, and modulates the expression of genes involved in cell development and homeostasis (Chrivia *et al.*, 1993; Glass *et al.*, 1997; Heasley *et al.*, 1991; Janknecht and Hunter, 1996; Kwok *et al.*, 1998; Kwok *et al.*, 1994; Lundblad *et al.*, 1995). Mechanistically, CBP facilitates the formation and stabilization of the RNA Pol II holoenzyme (RPIIH) and is involved in histone acetylation, leading to chromatin remodeling, which is essential for initiation and continued transcription (Arany *et al.*, 1994; Boyes *et al.*, 1998; Chrivia *et al.*, 1993; Eckner *et al.*, 1996; Horvai *et al.*, 1997; Kraus *et al.*, 1999; Kwok *et al.*, 1998; Kwok *et al.*, 1994; Lundblad *et al.*, 1995).

In the nucleus, CBP is present in limiting quantities and, in addition, its activity is suppressed by other nuclear proteins. Ribosomal S-6 serine/threonine kinases RSK2 (Nakajima *et al.*, 1996) and RSK1 are known to bind to a region of CBP that is also recognized by the adenovirus oncoprotein E1A (Arany *et al.*, 1994). Like E1A, RSK also represses transcriptional activation of a number of cAMP or PKC responsive genes, including tyrosine hydroxylase (TH) (Nakajima *et al.*, 1996), FGF-2, and other specific cis-acting elements and repress cell differentiation. cAMP (Nakajima *et al.*, 1996), phorbol esters, and growth factors cause the dissociation of the CBP-RSK1 complex (Merienne *et al.*, 2001), necessary for CREB-mediated activation of genes such as TH (Nakajima *et al.*, 1996). Cell differentiation triggered by cAMP, PKC and other signaling pathways, promotes the dissociation of RSK-CBP complexes, allowing CBP and RSK to exert their functions as a RNA Pol II activator, HAT enzyme (CBP) and a ssTF/histone kinase (RSK) (Cheung *et al.*, 2000a; Nakajima *et al.*, 1996; Sassone-Corsi *et al.*, 1999). The enzymatic activities of RSK and CBP exert a coordinate action at the histone H3 tail. The phosphorylation facilitates acetylation; the combined phosphorylation–acetylation process may constitute an essential step in local chromatin remodeling (Cheung *et al.*, 2000b; Janknecht and Hunter, 1996; Kwok *et al.*, 1998; Wolffe and Pruss, 1996).

Not all CBP functions need be inhibited by RSK. In mitogen-stimulated cells, the RSK-CBP complex may promote a productive interaction between CBP and SRF or Elk-1, which bind to SRE. The interaction of CBP with RSK1 appears to be required for Ras-mediated stimulation of c-fos mRNA transcription (Nakajima *et al.*, 1996). CBP may also form complexes with RSK1 and simultaneously with ssTF such as ER81 (known to be activated by the CBP/RSK1 complex) (Wu and Janknecht, 2002). Thus, dynamic sequestering of CBP by RSK may allow channeling of the cell signals via mitogenic Ras for proliferation or via cAMP for differentiation (Figure 4.5d). Until recently, the signaling events regulating these critical CBP-RSK interactions to affect gene transcription and development have been obscure. INFS may provide such a mechanism (Fang *et al.*, 2005; Stachowiak *et al.*, 2007).

Triple immunostaining combined with confocal microscopy demonstrates that in proliferating human neural progenitor cells (which lack nuclear FGFR1), CBP-rich nuclear speckles contain RSK1 and nearly all of the CBP-immunoreactive pixels co-localize with RSK1-immunoreactive pixels (Fang *et al.*, 2005). The accumulation of nuclear FGFR1 induced by cAMP treatment correlated with the loss of CBP and RSK1 co-localized pixels together with the appearance of co-localized FGFR1 and CBP or FGFR1 and RSK1 pixels. Hence, in the nuclei with accumulated FGFR1, CBP and RSK1 no longer appear to interact, however instead, separate FGFR1-CBP and FGFR1-RSK1 mechanisms may be activated. This notion was further confirmed by immunoprecipitation experiments in which the CBP-RSK1 complex that forms in mitogen-stimulated proliferating cells is effectively disrupted by increasing levels of nuclear FGFR1 (i.e. FGFR1 transfection) (Figure 4.5a). Furthermore, the RSK1-CBP complex can be restored by increasing the cellular content of RSK1 (i.e. RSK1 transfection). The disruption of CBP-RSK1 interactions is accompanied by the generation of separate FGFR1 complexes with CBP and RSK1 (Fang *et al.*, 2005; Hu *et al.*, 2004). Thus, the FGFR1 that accumulates in the nucleus of stimulated cells is sufficient to disrupt the RSK1-CBP complex, resulting in the formation of individual FGFR1-CBP and FGFR1-RSK1 complexes.

The mutually exclusive interaction of FGFR1 and RSK1 with CBP was confirmed using GST-CBP-affinity selection assays. These assays showed that different CBP domains bind FGFR1 (CBP-CH1,KIX) and RSK1 (CBP-CH3) (Fang *et al.*, 2005) (Figure 4.5b). Thus, FGFR1 must disrupt CBP-RSK1 complex formation by a mechanism other than competition for a common CBP site. Further studies have shown that FGFR1 TK domain and CBP bind to a common region of RSK1 which may explain the disruption of CBP-RSK1 complexes by FGFR1 (Figure 4.5c) (Fang *et al.*, 2005; Hu *et al.*, 2004). The dissociation of the CBP-RSK1 complex by nuclear FGFR1 correlates with the activation of CBP-mediated transcription, whereas their reassociation is accompanied by the inhibition of CBP-mediated transcription (Fang *et al.*, 2005). Together, these studies show that CBP, FGFR1 and RSK1 are in a dynamic equilibrium, whereby an upregulation of nuclear FGFR1 disrupts the inactive CBP-RSK1 complex and activates transcription, whereas the upregulation of RSK1 reinstates the CBP-RSK1 complex and diminishes transcription (Figure 4.5d).

Nuclear FGFR1 can also partially activate CBP by a RSK1-independent mechanism since FGFR1 activated CBP (albeit to a lesser extent) under conditions (e.g. serum free cultures, deletion of RSK1 binding CBP region) where the interaction between CBP and RSK1 was prevented (Fang *et al.*, 2005). Furthermore, the FGFR1 TK binding to RSK1 also increases RSK1 activity and its stimulation of CREB-mediated transcription. However, unlike the activation of CBP, the FGFR1 activation of RSK1 requires FGFR1 tyrosine kinase activity.

Considered together, these findings frame a threefold mechanism for nuclear FGFR1 activation of transcription (Figure 4.5a).

(1) In proliferating cells, CBP is found in a complex with RSK1. As FGFR1 accumulates in the nucleus, the TK domain binds to the N-terminal kinase domain of RSK1 and disrupts the CBP-RSK1 complex.
(2) The N-terminal domain of another FGFR1 molecule interacts with the N-terminal domain of CBP. The FGFR1-CBP complex activates transcription,

utilizing the N-terminal portion of the CBP in a process that leads to the recruitment of RNA Pol II and histone acetylation (Fang *et al.*, 2005).

(3) RSK1-bound FGFR1 augments RSK1 activity, potentially resulting in the phosphorylation of ssTF and/or chromatin proteins.

4.3.3. *Dynamic, stochastic model of INFS gene regulation*

The canonical view of the principal mechanism of transcriptional regulation involves the recruitment of preformed stable protein holo-complexes to promoters. An alternative model relies on an emerging stochastic description of nuclear function, which proposes that gene activation is dependent upon a probabilistic interaction of highly dynamic molecules (Misteli, 2001). Biophotonic tools (FRAP, FRET) provided us with novel insights into the dynamic nature and molecular proximity of gene regulation by the INFS mechanism (Dunham-Ems *et al.*, 2009; Dunham *et al.*, 2004). FRAP (Figure 4.6) revealed three nuclear FGFR1 populations (Dunham-Ems *et al.*, 2009): (1) an immobile FGFR1 "I" population associated with the nuclear matrix, (2) a hyperkinetic population ("F" fast; $t_{1/2} \leq 1$ sec) representing freely diffusing nucleoplasmic FGFR1 engaged in random molecular collisions, and (3) a hypokinetic population ("S" slow; $t_{1/2} = 10$–30 sec) which appears to represent chromatin associated FGFR1 (Dunham-Ems *et al.*, 2009). The transcriptional activation by cAMP or other stimuli is accompanied by the generalized conversion of immobile and hyperkinetic "F" nuclear FGFR1 into a hypokinetic, chromatin-binding FGFR1 and further decrease in the rate of its movement ($t_{1/2} = 100$–500 sec; Extra-Slow "ES" FGFR1) (Dunham-Ems *et al.*, 2009). These changes are triggered by the interaction of FGFR1 with CBP [kinetic changes are mimicked by transfected FGFR1-binding CBP (but not by nonbinding mutant) and are prevented by the depletion of endogenous CBP RNA (Dunham-Ems *et al.*, 2009)]. FGFR1 interaction with RSK1, does not generate additional hypokinetic FGFR1, however, RSK1 slows down the hypokinetic FGFR1 population (generates "ES" FGFR1) (Dunham-Ems *et al.*, 2009). Thus, changes in FGFR1 global chromatin kinetics during generalized gene activation may reflect FGFR1 interactions with CBP and with RSK1. By increasing the residence time of FGFR1 and its CBP, as well as the RSK1 partners at the target genes, the transcriptional reaction may be initiated and carried to the completion of RNA transcript. Based on the comparison of RNA Pol II (Darzacq *et al.*, 2007) and FGFR1 chromatin oscillation (Dunham-Ems *et al.*, 2009) rates, we postulate a model (Figure 4.6b) in which nuclear FGFR1, CBP and RSK1 collisions at the chromatin sites may drive specific transcriptional reactions,

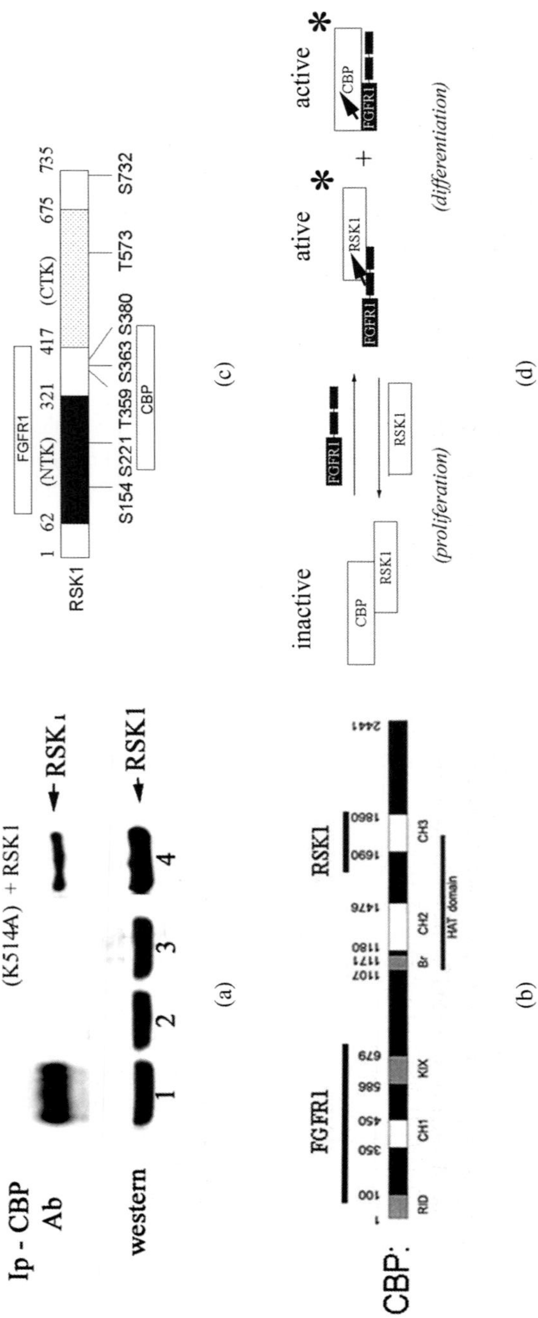

Figure 4.5. (*Continued*)

Figure 4.5. Interactions among FGFR1, CBP and RSK1 (Fang *et al.*, 2005; Hu *et al.*, 2004). (a) Endogenous RSK1–CBP complex is disrupted by wild type FGFR1 or by kinase-inactive FGFR1(K514A) and is restored by the expression of an additional RSK1. TE671 cells were transfected with FGFR1 or FGFR1(K514/A) (8 μg) and Flag/RSK1 (2 μg) or with control plasmids as indicated. Cell lysates were either immunoprecipitated (Ip) with a CBP antibody and then immunobloted with an anti-RSK1 antibody, or the lysates were directly immunobloted with the RSK1 antibody (western) (Fang *et al.*, 2005). FGFR1 and FGFR1(K514A) do not affect RSK1 expression (western) but they prevent CBP–RSK1 interaction. (b) FGFR1 and RSK1 bind to distinct CBP domains (Fang *et al.*, 2005). The binding regions in CBP were established by CBP–GST binding assays (Fang *et al.*, 2005). (c) FGFR1 and CBP bind to the overlapping regions in RSK1 (Hu *et al.*, 2004). (d) CBP, FGFR1 and RSK1 are in dynamic equilibrium, in which nuclear accumulation of FGFR1 disrupts the RSK1–CBP complex while an increase in RSK1 restores the CBP–RSK1 binding. Serum and other mitogens promote RSK1–CBP binding in proliferating cells, thereby blocking CBP transactivation of genes involved in cell differentiation. cAMP, PKC and other signaling pathways that differentiate cells, trigger nuclear accumulation of FGFR1. Nuclear FGFR1 binds to RSK1 and dissociates the inactive RSK–CBP complex, allowing CBP and RSK1 to exert their functions as a RNA Pol II activator, histone acetyltransferase (CBP) and a ssTF/histone kinase (RSK) (Cheung *et al.*, 2000a; Nakajima *et al.*, 1996; Sassone-Corsi *et al.*, 1999). The direct binding of FGFR1 to CBP and RSK1 augments their transcription-activating functions and activates genes involved in differentiation (Fang *et al.*, 2005). Figure 4.5a was reprinted from *J. Biological Chemistry* 280, pp. 28458–28456, 2005, with permission of American Society for Biochemistry and Molecular Biology Inc.

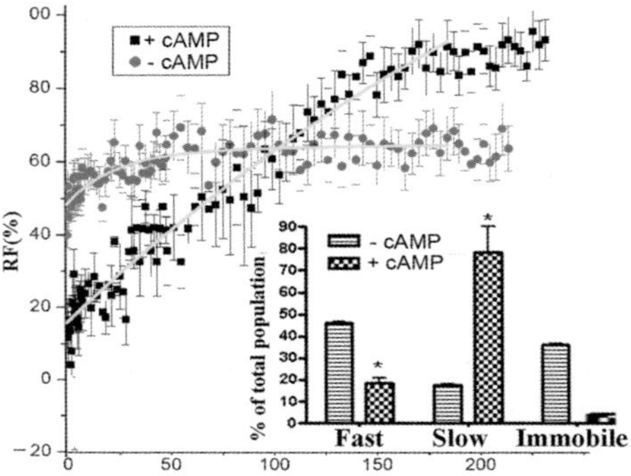

R1-EGFP population	recovery rate (s)	
hyperkinetic $(t_{1/2})_s$	0.29 +/- 0.02 "F"	> 0.20 "F"
hypokinetic $(t_{1/2})_s$	24.41 +/- 3.48 "S"	168.41 +/-12.52* "ES"
	control	+dBcAMP

(a)

Figure 4.6. (a) Transcription activation affects R1-EGFP nuclear FRAP mobility. Bimodal analyses of the R1-EGFP recovery demonstrates that nuclear FGFR1 contains hyperkinetic ("F": fast recovering $t_{1/2} < 1$ sec; present in nucleoplasm), hypokinetic ("S": Slow recovering, $t_{1/2} = 24$ sec; chromatin-associated), and non-recovering (immobile, nuclear matrix-associated) populations (bar graph). dB cAMP treatment eliminates the immobile population, reduces the hyperkinetic population and increases the hypokinetic population ($p < 0.001$). The recovery half-time of hypokinetic R1-EGFP is increased ($p < 0.001$), thereby generating an Extra-Slow ("ES"; $t_{1/2} = 168$ sec) R1-EGFP. These effects are reproduced by the transfection of R1-binding CBP and are reversed by antisense CBP or transcription inhibitors (Dunham-Ems *et al.*, 2009). Transfection of RSK1 converts the "S" R1 into "ES" without increasing the total hypokinetic population (Dunham-Ems *et al.*, 2009). (b) Kinetics of RNA Pol II (RPII) "on/off" gene oscillations at different phases of transcription are based on Darzacq *et al.* (Darzacq *et al.*, 2007). Kinetic model of INFS: R1 that is not engaged in transcription associates with nuclear matrix and is immobile. Activation of transcription by cAMP releases R1 from the matrix via R1 interaction with 23-kDa FGF2, thus generating "F" R1, which engages in rapid ($t_{1/2} < 1$ sec) "non-productive" molecular collisions and chromatin scanning. R1-CBP binding converts "F" R1 into hypokinetic R1 ($t_{1/2}$ 5–10 sec). We propose "S" R1 represents R1-CBP oscillations that drive the formation of the RNA Pol II PIC. CBP binding to DNA-associated transcription factors may extend the CBP-R1 chromatin residence, thereby promoting initiation. During transcription activation, the rate of oscillations is further reduced ("S" R1 converts into "ES", ($t_{1/2} > 50$ sec), possibly reflecting R1 and RSK1 binding events and formation of the productive elongating complexes. Figure 4.6a was reprinted from *Molec. Biol. Cell* 20, pp. 2406, 2009.
*PIC: preinitiation complex.

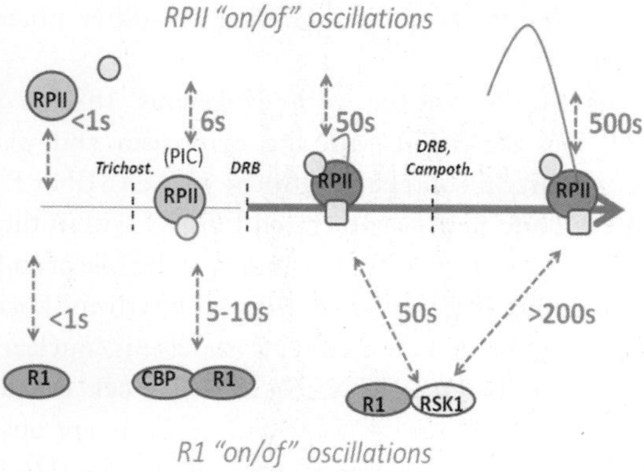

(b)

Figure 4.6b. (*Continued*)

i.e. formation of the preinitiation complex (PIC) [the on/off rate PIC chromatin oscillation (Darzacq *et al.*, 2007) is similar to the FGFR1 "S" population] and the transcript elongation (the rate of the elongating RNA Pol II oscillation is similar to the FGFR1 "ES"). The future investigations will link FGFR1, CBP and RSK1 chromatin kinetics to the transcription of specific gene loci and to the functions of the R1-CBP and R1-RSK1 transcriptional "machines".

4.4. INFS Mediates the Differentiation of Multipotent Neural Stem Cells and Pluripotent Embryonic Stem Cells

4.4.1. *Studies in human brain neural progenitor cells*

Distinct roles for cell surface and nuclear FGFR1 were demonstrated in cultured human neuronal progenitor cells (HNPC) (Stachowiak *et al.*, 2003b). HNPC are isolated from fetal human brains and serve as an established model for neural development (Piper *et al.*, 2001; Piper *et al.*, 2000). When stimulated with exogenous FGF2, HNPC proliferate and display characteristics of undifferentiated cells with short processes with an absence of voltage-dependent K^+ or Na^+ currents. However, when HNPC are treated with cAMP, forskolin, or BMP-7, there is a cell cycle exit, outgrowth of dendrites and axons and the expression of neuron-specific βIII tubulin, MAP2, NF-L, α-internexin, glutamate receptors and voltage-dependent and neurotransmitter receptor-regulated ionic currents.

Figure 4.2c shows the morphology of proliferating non-differentiated and neuronal-differentiated HNPC.

In proliferating HNPC treated with exogenous 18-kDa FGF2 or other mitogens, FGFR1 was associated with the cytoplasm and plasma membrane (Stachowiak *et al.*, 2003b). In contrast, treatment with cAMP or BMP7 induced an accumulation of FGF2 and newly synthesized FGFR1 within the nuclear interior (Figure 4.2b) (Stachowiak *et al.*, 2003a). In support of the role of endogenous nuclear FGFR1 in neuronal differentiation, transfection with tyrosine kinase (TK)-deleted dominant negative receptor mutants [either cytoplasmic/nuclear FGFR1(TK−) or exclusively nuclear FGFR1(SP−/NLS)(TK−)] block the differentiation of HNPC induced by cAMP (Figure 4.7b). Similar effects are observed in human umbilical cord blood-derived neural stem/progenitor cells (HUCB-NSC) (Fang *et al.*, 2005; Hu *et al.*, 2004). HUCB-NSC (described in Chapter 10) were isolated from the non-hematopoietic fraction of cord blood and differentiate into neurons, astrocytes or oligodenrocytes (Buzanska *et al.*, 2002). Stimulation with dBcAMP promotes neuronal differentiation, accompanied by activation of FGFR1 gene, nuclear accumulation of FGFR1 and is blocked by FGFR1 (SP−/NLS)(TK−) (Fang *et al.*, 2005; Sun *et al.*, 2005). Thus, these studies demonstrate that nuclear accumulation of FGFR1 is necessary for cAMP- or BMP7-induced differentiation of neural stem/progenitor cells.

In support of the role of nuclear FGFR1 in neuronal differentiation, FGFR1 transfection in HNPC results in FGFR1 accumulation in the cytoplasm as well as the nucleus and induction of long process outgrowth (Figure 4.7a) (Stachowiak *et al.*, 2003a). Furthermore, HNPC transfected with a mutant FGFR1 (FGFR1(SP−/NLS), in which the signal peptide (SP) was replaced with a nuclear localization signal (NLS) to direct it specifically to the nucleus, also showed the growth of long processes. In contrast, the membrane-associated non-nuclear FGFR1/R4 chimera in which the atypical FGFR1 transmembrane domain (partially hydrophilic/ β-sheet) was replaced with the typical (hydrophobic/α-helical) FGFR4 domain, fails to induce differentiation (Figure 4.7a) (Stachowiak *et al.*, 2003a). The morphological differentiation of HNPC induced by transfected cytoplasmic/nuclear FGFR1 or exclusively nuclear FGFR1(SP−/NLS) is accompanied by the induction of neuron-specific proteins including, neuron-specific enolase (NSE) and MAP-2. Thus, nuclear FGFR1 accumulation is sufficient to induce neuronal differentiation in the absence of other factors (Stachowiak *et al.*, 2003a). Neuronal differentiation by FGFR1(SP−/NLS), was abolished by the deletion of the TK domain but was still elicited by its kinase inactive FGFR1(K514A)(SP−/NLS)(TK−) mutant. Thus, like the stimulation by CBP and transcription, the TK domain but not TK activity of nuclear FGFR1 is essential for the stimulation of neuronal development.

The functional significance of the interaction between CBP and RSK1 for cell differentiation was first established by Nakajima *et al.* (1996), who showed that transfection of a RSK1 fragment that affects the CBP–RSK1 interaction influences cell differentiation. We found that the induction of neurite outgrowth by cAMP or by transfection of FGFR1(SP−/NLS) or kinase-inactive FGFR1(K541/A)(SP−/NLS) was blocked by co-transfected RSK1. Similarly, neuronal differentiation induced by transfection with CBP was also prevented by co-transfected RSK1 (Figure 4.7d) (Fang *et al.*, 2005). Thus, the dynamic equilibrium between FGFR1, CBP and RSK1 (outlined on Figure 4.5d) appears instrumental in regulating cell differentiation as well as gene activation.

4.4.2. *Neuronal development of pluripotent mouse (m)ESC is also directed by INFS*

Treatment of mESC with retinoic acid (RA) induces cells to withdraw from the proliferative/self-renewal cycle and differentiate. Many of the differentiating cells acquire the neuronal phenotype (Figure 4.8b) (Kehoe, 2008). Biochemical and immunocytochemical analyses show that RA induces nuclear accumulation of FGFR1 that precedes full differentiation. RA treatment induces an accumulation of FGFR1 in nuclear domains containing less dense chromatin (euchromatin) (Figure 4.8a), followed by an outgrowth of βIII tubulin-positive long neurites. The effects of RA are mimicked by transfection with nuclear FGFR1(SP−/NLS), which induces outgrowth of βIII tubulin-positive neurites in the absence of RA stimulation (Figure 4.8c) (Kehoe, 2008). Thus, the INFS mechanisms appear to be engaged in the differentiation of diverse types of stem-like cells.

4.4.3. *In vivo studies of INFS/FGFR1 function*

Attempts to evaluate the roles of FGFR1 in cell development turned out to be quite difficult at first glance. FGFR1 knock-out mice die during early embryonic development, indicating an essential role of FGFR1 (Deng *et al.*, 1994; Yamaguchi *et al.*, 1994). More recently, a successful CRE-lox-induced FGFR1 mutation was described; the mutation shows defects in neural development, affecting both cell proliferation and differentiation (Ohkubo *et al.*, 2004; Pirvola *et al.*, 2002). These effects of FGFR1 knock-out may likely reflect the loss of FGFR1 signaling at the cell surface and the INFS, respectively. Indeed, in SVZ neural stem/progenitor

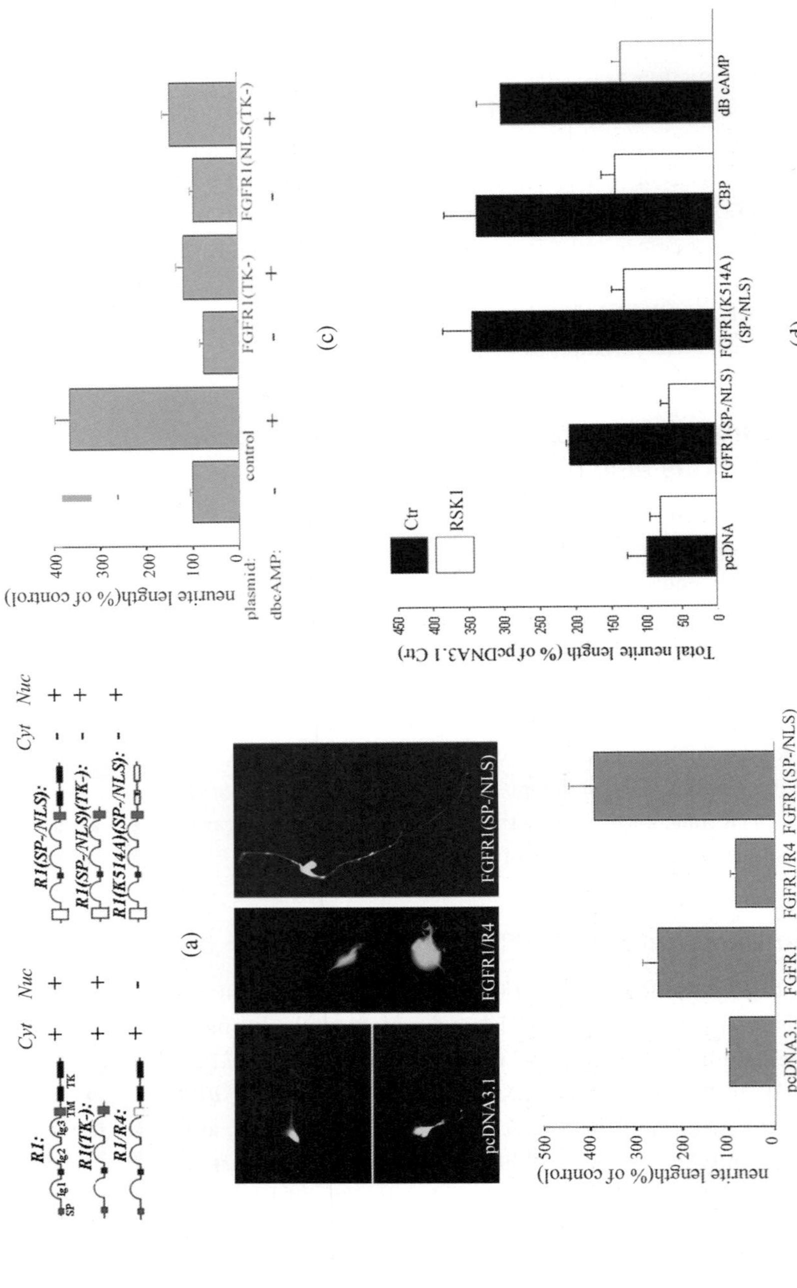

Figure 4.7. (*Continued*)

Figure 4.7. Nuclear FGFR1 stimulates the differentiation of HNPC isolated from fetal human brain. (a) FGFR1 (R1) mutants directed to different subcellular locales: Cyt (cytoplasm + membranes) and Nuc. To direct FGFR1 exclusively into the cell nucleus, we replaced the SP with a NLS. TK deletion generates dominant negative cytoplasmic and nuclear FGFR1(TK−) or nuclear FGFR1(SP−/NLS)(TK−) that blocks cAMP, retinoic acid or BMP7-induced differentiation and gene activities. Replacing the FGFR1 TM domain with the FGFR4 TM arrests chimerical FGFR1/R4 in cytoplasmic membranes. The K514/A mutation inactivates FGFR1 TK and prevents RSK1 activation but not the CBP activation. (b) Effects of transfected cytoplasmic/nuclear FGFR1, exclusively nuclear FGFR1(SP−/NLS), or membrane-associated FGFR1/R4 on neurite outgrowth in HNPC. Cells were co-transfected with pEGFP plasmid, expressing enhanced green fluorescent protein. Examples of fluorescent cells showing typical changes in morphology 4 days after transfection are shown. The majority of FGFR1 or FGFR1(SP−/NLS) expressing cells produced two long processes. Bar graph: the length of the single longest process in individual transfected cells was measured (Stachowiak *et al.*, 2003a). The average length of neurite processes in FGFR1 and FGFR1(SP−/NLS) transfected cells was significantly longer than in control pcDNA3.1 transfected HNPC ($p < 0.0001$). HNPC transfected with FGFR1/R4 show no additional neurite outgrowth. (c) Dominant negative FGFR1 blocks dB–cAMP-induced neurite outgrowth. HNPC were transfected with cytoplasmic/nuclear FGFR1(TK−) or nonmembranous, exclusively nuclear FGFR1(SP−/NLS)(TK−) (Stachowiak *et al.*, 2003a). (d) Roles of CBP, RSK1 and FGFR1 interactions in cell differentiation. Neural progenitor-like cells were transfected with CBP, FGFR1(SP−/NLS) or FGFR1(K514A)(SP−/NLS) and with RSK1 or control plasmids and with pEGFP. Receptor- or CBP-transfected cells as well as dB cAMP-treated cells produced two or more processes. Total length of cellular processes is expressed as a percentage of the dBcAMP non-treated cells transfected with control vectors. Figures 4.7a–4.7c were reprinted from *J. Neurochem.* 84, pp. 1302–1305, 2003, with permission of Wiley-Blackwell Publishing Ltd., International Society of Neurochemistry. Figure 4.7d was reprinted from *J. Biol Chem.* 280(31), pp. 28460, 2005, with permission of American Society for Biochemistry and Molecular Biology.

*HNPC: human neuronal progenitor cells, NLS: nuclear localization signal, Nuc: nucleus, SP: signal peptide, TK: tyrosine kinase, TM: transmembrane.

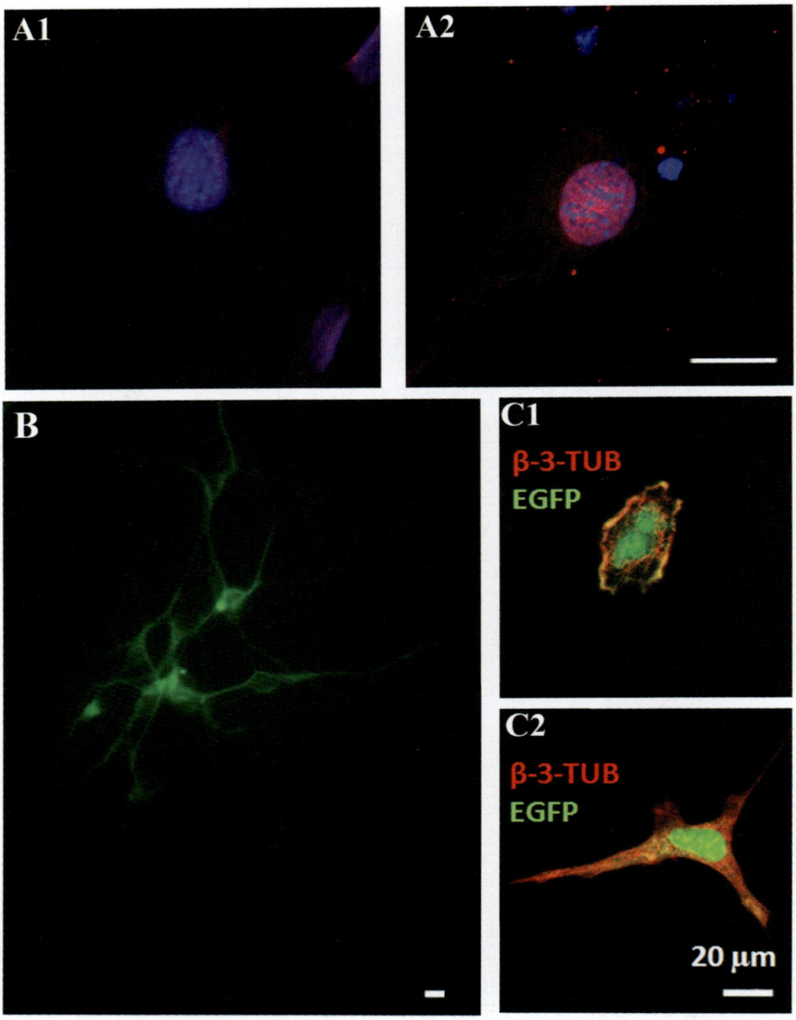

Figure 4.8. RA induces nuclear accumulation of FGFR1 and differentiation of mESC. (a) Non-differentiated mESC were maintained in LIF-containing, self-renewal promoting medium (a1) or were treated with RA (a2 – 2 days). Cells were immunostained with polyclonal FGFR1 Ab + Alexa488. DNA was stained using DAPI (Blue). After RA treatment, FGFR1 accumulates in the nuclear euchromatin-like regions with less concentrated DAPI fluorescence. (b) Morphological differentiation and expression of βIII tubulin (green) in mESC treated with RA for 7 days. (c) Transfection of nuclear R1(SP−/NLS) induces outgrowth of βIII-tubulin neurites in the absence of RA stimulation. Non-differentiated mESC cultures were transfected with control EGFP (c1) or with FGFR1(SP−/NLS) + EGFP (c2).
*mESC: mouse embryonic stem cells, RA: retinoic acid.

cells (NS/PC), FGFR1 is associated with the cytoplasm, while in differentiating cells, FGFR1 translocates to the nucleus. Nuclear accumulation of FGFR1 can also be observed in developing midbrain SNc neurons as they grow their telencephalic projections (Fang *et al.*, 2005). As this development is completed, FGFR1 localization becomes predominantly cytoplasmic (Bharali *et al.*, 2005; Fang *et al.*, 2005; Stachowiak *et al.*, 2003a). To examine the role of FGF receptor signaling in the postnatal development of dopamine-producing neurons, the kinase-deleted FGFR1(TK−) has been employed. The expression of FGFR1(TK−) was restricted to dopamine neurons and other catecholaminergic cells by fusion of FGFR1(TK−) to the 4.5-kb rat tyrosine hydroxylase (TH) gene promoter (Klejbor *et al.*, 2006). This promoter was shown to be active in the ventral aspects of mesencephalon [future substantia nigra compacta (SNc) and the ventral tegmental area (VTA)] at E11, and at later times during postmitotic development (Banerjee *et al.*, 1992; Schimmel *et al.*, 1999).

In TH–FGFR1(TK−) mice, significant reductions in the size of TH-immunoreactive neurons are observed in SNc and VTA at postnatal days 0 and 360 (Klejbor *et al.*, 2006). Newborn mice have a reduced density of DA neurons in both SN and VTA. Reduced density of DA transporter in the striatum further demonstrates an impaired development of the nigrostriatal DA projection. Behaviorally, TH–FGFR1(TK−) mice display impaired sensory gaiting, social interactions and cognitive deficits that are corrected by antipsychotic agents. These neurodevelopmental, anatomical, biochemical and behavioral alterations are similar to those observed in human schizophrenia. Indeed, in humans, changes in FGF and its receptors FGFR1 have been described in the brains of schizophrenia and bipolar patients (Ovalle *et al.*, 2001); a large association study suggests that genetic variation of the FGF receptor increases the risk of developing schizophrenia (O'Donovan *et al.*, 2009). These combined evidence suggest that changes in the FGF system contributes to schizophrenia and possibly a wide range of related psychiatric disorders (Terwisscha van Scheltinga *et al.*, 2009). The genetic links reported for schizophrenia, in addition to FGFs include cyclic AMP enzymes, MAPK signaling pathways, transcription co-activator complexes (De Luca *et al.*, 2003; Millar *et al.*, 2005; O'Donovan *et al.*, 2009; O'Neill *et al.*, 2004) and chromatin modifications. These mechanisms have been shown to influence neuronal development via interaction with the INFS. Hence, such disorders might result in part from impaired INFS mechanism in developing neurons.

Recently, direct evidence for INFS stimulation of brain neuronogenesis was obtained in mice using novel nanoparticle-mediated gene transfer into the neural stem/progenitor cells in the subventricular zone of brain lateral ventricles. Transfections of the nuclear FGFR1(SP−/NLS) or HMW–FGF2, which mobilizes endogenous nuclear FGFR1, block proliferation and stimulate differentiation as well as neuronal development of the neural stem/progenitor cells (Bharali *et al.*, 2005; Stachowiak, 2009). These studies show that *in vivo* targeting of the INFS mechanism may effectively reinstate neuronogenesis in the adult brain.

4.5. Activation of INFS in Cancer Cells

The rather unique case of neuronal differentiation of pheochromocytoma cells by growth factor activation of tyrosine kinase receptors is an interesting example of INFS signaling. The earliest work on the use of FGFs as extracellular factors for the differentiation of PC12 pheochromocytoma cell line was performed independently by the laboratories of Gordon Guroff (Togari *et al.*, 1985) and Lloyd Greene (Rydel and Greene, 1987). The former lab reported a weak transient neurite outgrowth while the latter achieved more robust and stable outgrowth in PC12 cultures treated with exogenous FGF2 peptides. Recent work appears to resemble the former results (Hausott *et al.*, 2008; Lin *et al.*, 2009). In our view, the likely explanation lies with a rather inefficient delivery of FGFR1 to the cell nucleus upon treatment with extracellular FGF. Nerve growth factor (NGF), which produces a more homogeneous neurite outgrowth response over 7 to 10 days, may be more efficient due to robust activation of INFS.

The PC12 cells continually proliferate *in vitro* in the presence of serum, however when treated with NGF, PC12 exits the cell cycle, grows neurites and differentiates into catecholamine-producing sympathetic-like neurons that express neuronal markers such as doublecortin and tyrosine hydroxylase (Stachowiak, 2009) (Figure 4.9b). This differentiation process is preceded by nuclear accumulation of FGFR1 (Figure 4.9a) and the stimulation of NGF receptors can be circumvented by direct transfection of DNA encoding nuclear FGFR1(SP−/NLS) or nuclear HMW–FGF2 (Figure 4.9b).

INFS-induced neuronal-like differentiation is observed in another cancer cell line, human neuroblastoma BE(2)C. Treatment with cAMP, causes the BE(2)C cells to seize proliferation, express nuclear FGFR1 and differentiate into neurons that express NeuN. Again, these effects can be reproduced directly by transfection of nuclear form of FGFR1 (Stachowiak, unpublished observations).

When INFS is disabled, as it appears to be in the cancer cell line TE671 medulloblastoma due to the near total absence of FGFR1, the cells proliferate aggressively and display non-differentiated features. When INFS is re-established in TE671 cells by transfection of FGFR1(SP−/NLS), evidence of cell differentiation can be observed. FGFR1(SP−/NLS) induced neuronal-like differentiation also

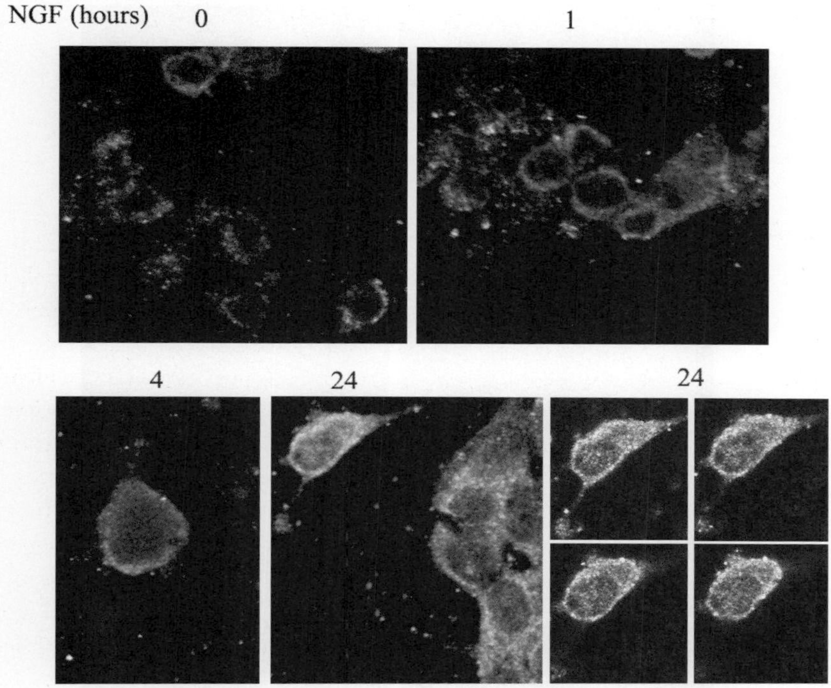

(a)

Figure 4.9. (a) NGF stimulates nuclear accumulation of FGFR1 in PC12 cells. Cells were maintained in the absence (0 hours) or with NGF (50 ng/ml) for 1–24 hours and were immunostained with monoclonal FGFR1 McAb6 + Alexa488. Confocal sections through the middle of the cell nucleus are shown. Bottom right panel: Images of the individual confocal sections (0.5 μm apart) show intranuclear speckle domains with high FGFR1 content. (b) INFS stimulates DCX and TH expression by PC12 cells. Cultured PC12 cells were transfected with pcDNA3.1 (control) or HMW–FGF2 or FGFR1(SP−/NLS) expressing plasmids (transfection of cells was verified using EGFP plasmid, not shown) and were analyzed 5 days after transfection. Some EGFP-transfected cultures were incubated with NGF, 1 day after transfection for an additional 4 days, after which the cells were fixed and stained with TH (blue) or DCX (red) antibodies. Panels: (a1,a2) control DNA, (b1,b2) control DNA + NGF, (c1,c2) HMW–FGF2, (d1,d2) FGFR1(SP−/NLS) . In control non-stimulated cells (−NGF), the TH or DCX immunoreactivity was weak. However, both were markedly increased by treatment with NGF. The effects of NGF on DCX and TH expression were mimicked by transfection of HMW–FGF2 or FGFR1(SP−/NLS). Quantitative analysis of confocal images collected in the linear range of the fluorescent signal using Image J (Peng *et al.*, 2002) confirmed that stimulation of INFS by DNA transfection or treatment with NGF increased intensity of immunofluorescence four- to tenfold (not shown). Thus, the over-expression of nuclear FGFR1 or nuclear FGF2 ligand is sufficient to induce neuronal-like differentiation in the absence of additional differentiation promoting stimulation (NGF).

*DCX: doublecortin, NGF: nerve growth factor, TH: tyrosine hydroxylase.

(b)

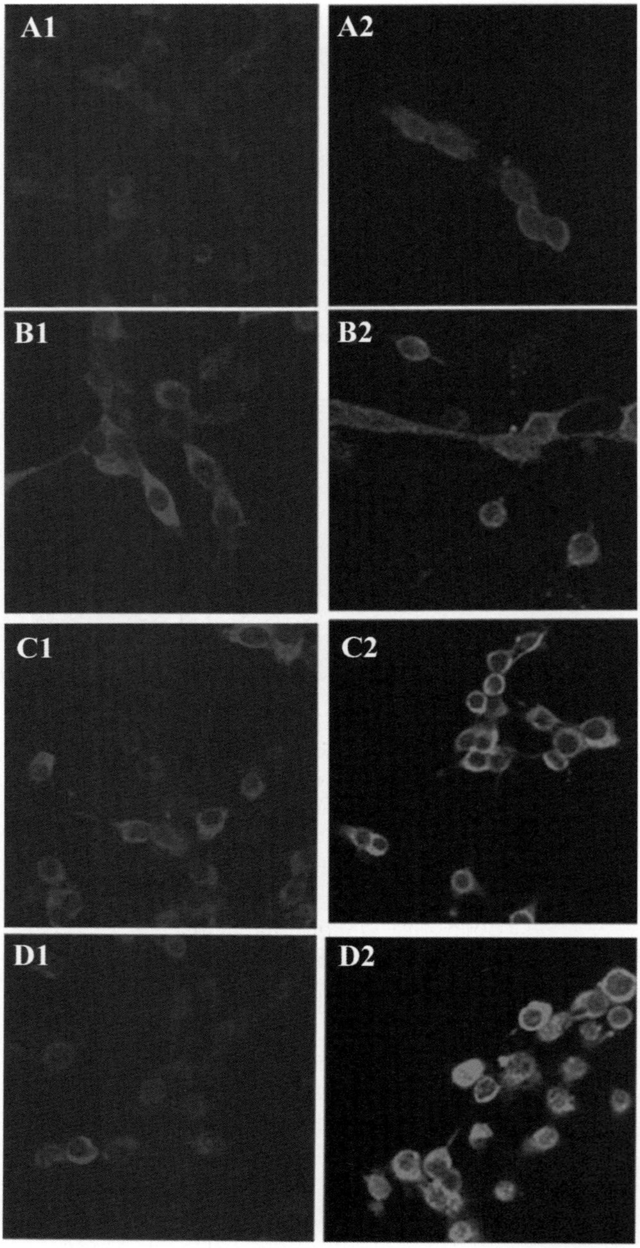

Figure 4.9. (*Continued*)

in SHSY5Y neuroblastoma cells further supporting its anti-oncogenic action (Fang *et al.*, 2005; Stachowiak *et al.*, 2003b). The TE671 and SHSY5Y cells express little of endogenous FGFR1 (Peng *et al.*, 2001; Peng *et al.*, 2002). Hence, the reconstitution of nuclear FGFR1 signaling in those cancer cells inhibits proliferation and induces differentiation (Fang *et al.*, 2005).

4.6. Conclusions

Development follows a specific genetic blueprint, regulated by stochastic extragenomic signals. These diverse signals — which include mechanical elements such as vascular, neuronal, glial and ciliary microstructures, anchorage signals from the cell surface and extracellular matrix-associated immobile molecules, diffusible growth factors, cytokines, neurotransmitters and hormones — are integrated and computed by universal network modules acting as molecular switches that direct cells toward different developmental phases. Universal, integrative network modules responsible for self-renewal and mitotic expansion have been identified. The INFS pathway appears to constitute an analogous universal module that directs cells toward postmitotic development, i.e. neuronal differentiation. The function of INFS can be described as a "feed-forward-and-gate" module (Figure 4.10) (Fang *et al.*, 2005; Stachowiak *et al.*, 2007). The INFS pathway is activated by classical signaling cascades, initiated by diverse membrane receptors that activate sequence specific transcription factors (ssTFs). The same stimuli trigger a parallel signal (nuclear FGFR1) that "feeds forward" the stimulation to the common and limiting transcriptional coactivator, CBP. Nuclear FGFR1 releases CBP from inactive complexes and the coupled activation of CBP and RSK1 by INFS and ssTFs enables the coordinated regulation of the multi-gene program involved in differentiation.

The feed-forward loops are commonly found in biological networks (Chapter 1, Figure 1.2b). Their functions may vary from pulse generators (for instance, via gate opening) or response delaying circuits. An important feature of developmental differentiation is the "permanent", often irreversible status of the achieved state. Such a state could result from positive feedback regulation, where the expression of the central genes in the module is activated by their products (Chapter 1, Figure 1.2a). For instance, in the INFS, the FGF2 ligand and its nuclear receptor FGFR1 appear to form a common two-node feedback loop, whereby their complex drives the expression of these two factors. This positive feedback loop may lead to a bi-stable "low" or a "high" INFS signal. Once activated by other factors, the INFS stays switched on even if the original activators are no longer present. Such mechanism is illustrated in Figure 4.10b.

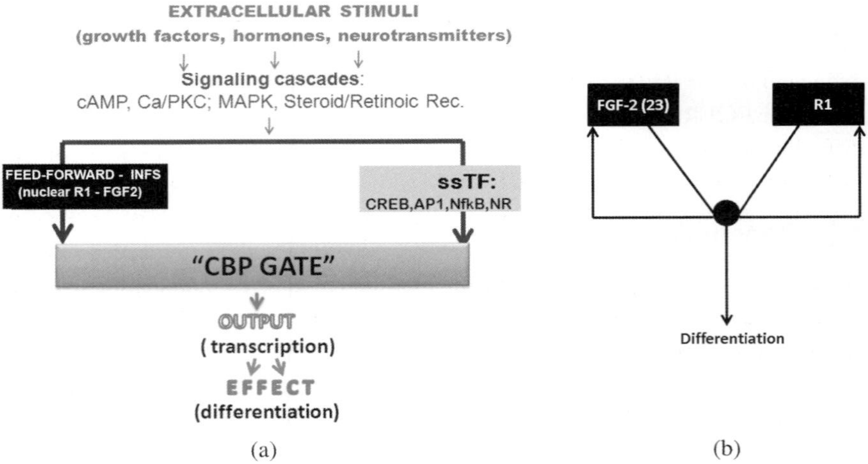

Figure 4.10. (a) "Feed-Forward-And-Gate" signaling module of INFS is initiated by diverse stimuli that activate various signaling cascades and sequence-specific transcription factors (ssTFs) (Fang *et al.*, 2005). The stimuli trigger nuclear translocation of FGFR1 (R1-INFS), a parallel signal that "feeds forward" the stimulation to CBP, the functional equivalent of the "End Gate". CBP is present in limited quantities and, when released from inactive complexes by nuclear FGFR1 (R1), enables gene activation and neuronal differentiation. Transfection of an engineered nuclear FGFR1(SP−/NLS) is sufficient to open the CBP "gate" and trigger neuronal differentiation. (b) In the INFS, the FGF2 ligand and its nuclear receptor FGFR1 (R1) appear to form a common feedback loop whereby their complex drives the expression of these two factors. Hence, once activated by other factors, the INFS generates stable cell differentiation-promoting signals. Figure 4.10a reprinted from *J. Biol Chem* 280(31), pp. 28461, 2005, with permission of American Society for Biochemistry and Molecular Biology.
*CBP: CREB-binding protein, ssTF: sequence specific transcription factors.

One of the apparent merits of the network motifs is prediction. The discovery of the pluripotency gene network suggested that the over-expression of the few "pluripotency" genes linked by positive feedback loops may suffice to generate the pluripotent state in differentiated somatic cells. Testing this prediction has led to a new successful technology for the generation of pluripotent stem-like cells, IPC (Chapter 5). The Feed-Forward-and-Gate module of the INFS predicts that direct activation of the INFS via over-expression of nuclear FGFR1 or its HMW FGF2 ligand, should suffice to convert the proliferating stem/progenitor-like cells into their differentiated progeny, thus bypassing the daunting need to identify proper combinations of the many of external and internal factors needed for such a transition. This has been shown to work not only *in vitro* in many cultured cells (as described in the present chapter), but recently also *in vivo*. As discussed in Chapter 11, the transfer of the INFS genes reinstates an active neuronogenic development by the stem cells in the mature brain.

Another general theme that emerges from this previous work is the dynamic nature of gene regulation by INFS (Figure 4.6). The mobilization of FGFR1 from the nuclear matrix is followed by rapid Brownian oscillations, allowing chromatin scanning for the FGFR1 partner proteins (Dunham-Ems *et al.*, 2009). Stochastic generations of slower oscillating FGFR1/CBP and FGFR1/RSK1 interactions with chromatin, thus create efficacious molecular complexes that support transcription reactions (Dunham-Ems *et al.*, 2009). The general hypothesis is that increasing the time of association between the colliding factors promotes specific interactions and the formation of "gene-regulating machines".

At the present time, it remains unknown whether INFS mechanisms may also be engaged in gene inhibition and, if so, whether nuclear FGFR1 may suppress the mitotic and multipotency genes or genes that promote alternative (i.e. non-neuronal) development. If so, is such an inhibition mediated directly by FGFR1 binding to repressed genes or is the mechanism indirect? Future studies will address these hypotheses and develop further practical means to control the INSF module for the sake of regenerative medicine.

Acknowledgements

We gratefully acknowledge grant support from the New York State Stem Cell Science (NYSTEM).

Bibliography

Angenstein, F., Greenough, W. T. and Weiler, I. J. 1998. Metabotropic glutamate receptor-initiated translocation of protein kinase p90rsk to polyribosomes: A possible factor regulating synaptic protein synthesis, *Proc Natl Acad Sci U S A* 95:15078–15083.

Arany, Z., Sellers, W. R., Livingston, D. M. and Eckner, R. 1994. E1A-associated p300 and CREB-associated CBP belong to a conserved family of coactivators, *Cell* 77:799–800.

Arese, M., Chen, Y., Florkiewicz, R. Z., Gualandris, A., Shen, B. and Rifkin, D. B. 1999. Nuclear activities of basic fibroblast growth factor: Potentiation of low-serum growth mediated by natural or chimeric nuclear localization signals, *Mol Biol Cell* 10:1429–1444.

Banerjee, S. A., Hoppe, P., Brilliant, M. and Chikaraishi, D. M. 1992. 5′ flanking sequences of the rat tyrosine hydroxylase gene target accurate tissue-specific, developmental, and transsynaptic expression in transgenic mice, *J Neurosci* 12:4460–4467.

Bellot, F., Crumley, G., Kaplow, J. M., Schlessinger, J., Jaye, M. and Dionne, C. A. 1991. Ligand-induced transphosphorylation between different FGF receptors, *EMBO J* 10:2849–2854.

Bharali, D. J., Klejbor, I., Stachowiak, E. K., Dutta, P., Roy, I., Kaur, N., Bergey, E. J., Prasad, P. N. and Stachowiak, M. K. 2005. Organically modified silica nanoparticles: A nonviral vector for *in vivo* gene delivery and expression in the brain, *Proc Natl Acad Sci U S A* 102:11539–11544.

Bilak, M. M., Hossain, W. A. and Morest, D. K. 2003. Intracellular fibroblast growth factor produces effects different from those of extracellular application on development of avian cochleovestibular ganglion cells *in vitro*, *J Neurosci Res* 71:629–647.

Blencowe, B. J., Nickerson, J. A., Issner, R., Penman, S. and Sharp, P. A. 1994. Association of nuclear matrix antigens with exon-containing splicing complexes, *J Cell Biol* 127:593–607.

Bouche, G., Gas, N., Prats, H., Baldin, V., Tauber, J. P., Teissie, J. and Amalric, F. 1987. Basic fibroblast growth factor enters the nucleolus and stimulates the transcription of ribosomal genes in ABAE cells undergoing G0–G1 transition, *Proc Natl Acad Sci U S A* 84:6770–6774.

Boyes, J., Byfield, P., Nakatani, Y. and Ogryzko, V. 1998. Regulation of activity of the transcription factor GATA-1 by acetylation, *Nature* 396:594–598.

Bryant, D. M. and Stow, J. L. 2005. Nuclear translocation of cell-surface receptors: Lessons from fibroblast growth factor, *Traffic* 6:947–954.

Buzanska, L., Machaj, E. K., Zablocka, B., Pojda, Z. and Domanska-Janik, K. 2002. Human cord blood-derived cells attain neuronal and glial features *in vitro*, *J Cell Sci* 115:2131–2138.

Carpenter, G. 2003. Nuclear localization and possible functions of receptor tyrosine kinases, *Curr Opin Cell Biol* 15:143–148.

Cheung, P., Allis, C. D. and Sassone-Corsi, P. 2000a. Signaling to chromatin through histone modifications, *Cell* 103:263–271.

Cheung, P., Tanner, K. G., Cheung, W. L., Sassone-Corsi, P., Denu, J. M. and Allis, C. D. 2000b. Synergistic coupling of histone H3 phosphorylation and acetylation in response to epidermal growth factor stimulation, *Mol Cell* 5:905–915.

Chrivia, J. C., Kwok, R. P., Lamb, N., Hagiwara, M., Montminy, M. R. and Goodman, R. H. 1993. Phosphorylated CREB binds specifically to the nuclear protein CBP, *Nature* 365:855–859.

Clarke, W. E., Berry, M., Smith, C., Kent, A. and Logan, A. 2001. Coordination of fibroblast growth factor receptor 1 (FGFR1) and fibroblast growth factor-2 (FGF2) trafficking to nuclei of reactive astrocytes around cerebral lesions in adult rats, *Mol Cell Neurosci* 17:17–30.

Creighton, T. E. 1984. Proteins, structures, and molecular properties. W. H. Freeman and Company, New York, pp. 87–90.

Darzacq, X., Shav-Tal, Y., de Turris, V., Brody, Y., Shenoy, S. M., Phair, R. D. and Singer, R. H. 2007. *In vivo* dynamics of RNA polymerase II transcription, *Nat Struct Mol Biol* 14:796–806.

De Luca, A., Conti, E., Grifone, N., Amati, F., Spalletta, G., Caltagirone, C., Bonaviri, G., Pasini, A., Gennarelli, M., Stefano, B., Berti, L., Mittler, G., Meisterernst, M., Dallapiccola, B. and Novelli, G. 2003. Association study between CAG trinucleotide repeats in the PCQAP gene (PC2 glutamine/Q-rich-associated protein) and schizophrenia, *Am J Med Genet B Neuropsychiatr Genet* 116B:32–35.

Deng, C. X., Wynshaw-Boris, A., Shen, M. M., Daugherty, C., Ornitz, D. M. and Leder, P. 1994. Murine FGFR-1 is required for early postimplantation growth and axial organization, *Genes Dev.* 8:3045–3057.

Dono, R. and Zeller, R. 1994. Cell-type-specific nuclear translocation of fibroblast growth factor-2 isoforms during chicken kidney and limb morphogenesis, *Dev Biol* 163:316–330.

Dunham-Ems, S.M., Lee, Y. W., Stachowiak, E. K., Pudavar, H., Claus, P., Prasad, P. N. and Stachowiak, M. K. 2009. Fibroblast growth factor receptor-1 (FGFR1) nuclear dynamics reveal a novel mechanism in transcription control, *Mol Biol Cell* 20:2401–2412.

Dunham-Ems, S. M., Pudavar, H. E., Myers, J. M., Maher, P. A., Prasad, P. N. and Stachowiak, M. K. 2006. Factors controlling fibroblast growth factor receptor-1's cytoplasmic trafficking and its regulation as revealed by FRAP analysis, *Mol Biol Cell* 17:2223–2235.

Dunham, S. M., Pudavar, H. E., Prasad, P. N. and Stachowiak, M. K. 2004. Cellular signaling and protein-protein interactions studied using fluorescence recovery after photobleaching, *J Phys Chem B* 108:10540–10546.

Ebert, A. D., Laussmann, M., Wegehingel, S., Kaderali, L., Erfle, H., Reichert, J., Lechner, J., Beer, H. D., Pepperkok, R. and Nickel, W. 2010. Tec-kinase-mediated phosphorylation of fibroblast growth factor 2 is essential for unconventional secretion, *Traffic* 6:813–826.

Eckner, R., Yao, T. P., Oldread, E. and Livingston, D. M. 1996. Interaction and functional collaboration of p300/CBP and bHLH proteins in muscle and B-cell differentiation, *Genes Dev* 10:2478–2490.

Eisenberg, D., Schwarz, E., Komaromy, M. and Wall, R. 1984. Analysis of membrane and surface protein sequences with the hydrophobic moment plot, *J Mol Biol* 179:125–142.

Fang, X., Stachowiak, E. K., Dunham-Ems, S. M., Klejbor, I. and Stachowiak, M. K. 2005. Control of CREB-binding protein signaling by nuclear fibroblast growth factor receptor-1: A novel mechanism of gene regulation, *J Biol Chem* 280:28451–28462.

Florkiewicz, R. Z., Baird, A. and Gonzalez, A. M. 1991. Multiple forms of bFGF: Differential nuclear and cell surface localization, *Growth Factors* 4:265–275.

Fu, X. D. and Maniatis, T. 1990. Factor required for mammalian spliceosome assembly is localized to discrete regions in the nucleus, *Nature* 343:437–441.

Glass, C. K., Rose, D. W. and Rosenfeld, M. G. 1997. Nuclear receptor coactivators, *Curr Opin Cell Biol* 9:222–232.

Gonzalez, A. M., Berry, M., Maher, P. A., Logan, A. and Baird, A. 1995. A comprehensive analysis of the distribution of FGF-2 and FGFR1 in the rat brain, *Brain Res* 701:201–226.

Goodman, R. H. and Smolik, S. 2000. CBP/p300 in cell growth, transformation, and development, *Genes Dev* 14:1553–1577.

Gospodarowicz, D., Lui, G. M. and Cheng, J. 1982. Purification in high yield of brain fibroblast growth factor by preparative isoelectric focusing at pH 9.6, *J Biol Chem* 257:12266–12276.

Gospodarowicz, D., Neufeld, G. and Schweigerer, L. 1987. Fibroblast growth factor: Structural and biological properties, *J Cell Physiol Suppl* Suppl 5:15–26.

Grothe, C., Zachmann, K. and Unsicker, K. 1991. Basic FGF-like immunoreactivity in the developing and adult rat brainstem, *J Comp Neurol* 305:328–336.

Hausott, B., Schlick, B., Vallant, N., Dorn, R. and Klimaschewski, L. 2008. Promotion of neurite outgrowth by fibroblast growth factor receptor 1 overexpression and lysosomal inhibition of receptor degradation in pheochromocytoma cells and adult sensory neurons, *Neuroscience* 153:461–473.

Heasley, L. E., Benedict, S., Gleavy, J. and Johnson, G. L. 1991. Requirement of the adenovirus E1A transformation domain 1 for inhibition of PC12 cell neuronal differentiation, *Cell Regul* 2:479–489.

Horbinski, C., Stachowiak, E. K., Chandrasekaran, V., Miuzukoshi, E., Higgins, D. and Stachowiak, M. K. 2002. Bone morphogenetic protein-7 stimulates initial dendritic growth in sympathetic neurons through an intracellular fibroblast growth factor signaling pathway, *J Neurochem* 80:54–63.

Horvai, A. E., Xu, L., Korzus, E., Brard, G., Kalafus, D., Mullen, T. M., Rose, D. W., Rosenfeld, M. G. and Glass, C. K. 1997. Nuclear integration of JAK/STAT and Ras/AP-1 signaling by CBP and p300, *Proc Natl Acad Sci U S A* 94:1074–1079.

Hu, Y., Fang, X., Dunham, S. M., Prada, C., Stachowiak, E. K. and Stachowiak, M. K. 2004. 90-kDa ribosomal S6 kinase is a direct target for the nuclear fibroblast growth factor receptor 1 (FGFR1): Role in FGFR1 signaling, *J Biol Chem* 279:29325–29335.

Itoh, N. and Ornitz, D. M. 2011. Fibroblast growth factors: From molecular evolution to roles in development, metabolism and disease, *J Biochem* 149:121–130.

Janknecht, R. and Hunter, T. 1996. Transcription. A growing coactivator network, *Nature* 383:22–23.

Joy, A., Moffett, J., Neary, K., Mordechai, E., Stachowiak, E. K., Coons, S., Rankin-Shapiro, J., Florkiewicz, R. Z. and Stachowiak, M. K. 1997. Nuclear accumulation of FGF-2 is associated with proliferation of human astrocytes and glioma cells, *Oncogene* 14:171–183.

Kehoe, D., Stachowiak, E. K., Stachowiak, M. K., Lee, Y.-W. and Tzanakakis, E. 2008. New mechanism in neuronogenic differentiation of embroynic stem cells, *In Society For Neuroscience Annual Meeting*, Washington, DC.

Klejbor, I., Myers, J. M., Hausknecht, K., Corso, T. D., Gambino, A. S., Morys, J., Maher, P. A., Hard, R., Richards, J., Stachowiak, E. K. and Stachowiak, M. K. 2006. Fibroblast growth factor receptor signaling affects development and function of dopamine neurons — inhibition results in a schizophrenia-like syndrome in transgenic mice, *J Neurochem* 97:1243–1258.

Klimaschewski, L., Meisinger, C. and Grothe, C. 1999. Localization and regulation of basic fibroblast growth factor (FGF-2) and FGF receptor-1 in rat superior cervical ganglion after axotomy, *J Neurobiol* 38:499–506.

Kraus, W. L., Manning, E. T. and Kadonaga, J. T. 1999. Biochemical analysis of distinct activation functions in p300 that enhance transcription initiation with chromatin templates, *Mol Cell Biol* 19:8123–8135.

Kwok, R., Lundblad, J., Chrivia, J., Richards, J., Bachinger, H., Brennan, R., Roberts, S., Green, M., Goodman, R. and Strahl, K. 1998. Histone acetylation and transcriptional regulartory mechanisms, *Genes Dev* 12:599–606.

Kwok, R. P., Lundblad, J. R., Chrivia, J. C., Richards, J. P., Bachinger, H. P., Brennan, R. G., Roberts, S. G., Green, M. R. and Goodman, R. H. 1994. Nuclear protein CBP is a coactivator for the transcription factor CREB, *Nature* 370:223–226.

Leadbeater, W. E., Gonzalez, A. M., Logaras, N., Berry, M., Turnbull, J. E. and Logan, A. 2006. Intracellular trafficking in neurones and glia of fibroblast growth factor-2, fibroblast growth factor receptor 1 and heparan sulphate proteoglycans in the injured adult rat cerebral cortex, *J Neurochem* 96:1189–1200.

Lin, W. F., Chen, C. J., Chang, Y. J., Chen, S. L., Chiu, I. M. and Chen, L. 2009. SH2B1beta enhances fibroblast growth factor 1 (FGF1)-induced neurite outgrowth through MEK-ERK1/2-STAT3-Egr1 pathway, *Cell Signal* 21:1060–1072.

Lundblad, J. R., Kwok, R. P., Laurance, M. E., Harter, M. L. and Goodman, R. H. 1995. Adenoviral E1A-associated protein p300 as a functional homologue of the transcriptional co-activator CBP, *Nature* 374:85–88.

Maher, P. A. 1996. Nuclear Translocation of fibroblast growth factor (FGF) receptors in response to FGF-2, *J Cell Biol* 134:529–536.

Marchetti, L., Sabbieti, M. G., Agas, D., Menghi, M., Materazzi, G., Menghi, G. and Hurley, M. M. 2006. PGF2alpha increases FGF-2 and FGFR2 trafficking in Py1a rat osteoblasts via clathrin independent and importin beta dependent pathway, *J Cell Biochem* 97:1379–1392.

Matsuda, S., Desaki, J., Okumura, N., Shiosaka, S., Imaoka, S. and Sakanaka, M. 1992. Basic fibroblast growth factor-like immunoreactivity in the trigeminal proprioceptive and motor systems, *Brain Res* 577:92–100.

Matsuda, S., Ii, Y., Desaki, J., Yoshimura, H., Okumura, N. and Sakanaka, M. 1994. Development of Purkinje cell bodies and processes with basic fibroblast growth factor-like immunoreactivity in the rat cerebellum, *Neuroscience* 59:651–662.

Merienne, K., Pannetier, S., Harel-Bellan, A. and Sassone-Corsi, P. 2001. Mitogen-regulated RSK2-CBP interaction controls their kinase and acetylase activities, *Mol Cell Biol* 21:7089–7096.

Millar, J. K., Pickard, B. S., Mackie, S., James, R., Christie, S., Buchanan, S. R., Malloy, M. P., Chubb, J. E., Huston, E., Baillie, G. S., Thomson, P. A., Hill, E. V., Brandon, N. J., Rain, J. C., Camargo, L. M., Whiting, P. J., Houslay, M. D., Blackwood, D. H., Muir, W. J. and Porteous, D. J. 2005. DISC1 and PDE4B are interacting genetic factors in schizophrenia that regulate cAMP signaling, *Science*, 310:1187–1191.

Misteli, T. 2001. Protein dynamics: Implications for nuclear architecture and gene expression, *Science* 291:843–847.

Moffett, J., Kratz, E., Florkiewicz, R. and Stachowiak, M. K. 1996. Promoter regions involved in density-dependent regulation of basic fibroblast growth factor gene expression in human astrocytic cells, *Proc Natl Acad Sci U S A* 93:2470–2475.

Moffett, J., Kratz, E., Myers, J., Stachowiak, E. K., Florkiewicz, R. Z. and Stachowiak, M. K. 1998a. Transcriptional regulation of fibroblast growth factor-2 expression in human astrocytes: Implications for cell plasticity, *Mol Biol Cell* 9:2269–2285.

Moffett, J., Kratz, E. and Stachowiak, M. K. 1998b. Increased tyrosine phosphorylation and novel cis-acting element mediate activation of the fibroblast growth factor-2 (FGF-2) gene by nicotinic acetylcholine receptor. New mechanism for trans-synaptic regulation of cellular development and plasticity, *Brain Res Mol Brain Res* 55:293–305.

Mohammadi, M., Dikic, I., Sorokin, A., Burgess, W. H., Jaye, M. and Schlessinger, J. 1996. Identification of six novel autophosphorylation sites on fibroblast growth factor receptor 1 and elucidation of their importance in receptor activation and signal transduction, *Mol Cell Biol* 16:977–989.

Myers, J. M., Martins, G. G., Ostrowski, J. and Stachowiak, M. K. 2003. Nuclear trafficking of FGFR1: A role for the transmembrane domain, *J Cell Biochem* 88:1273–1291.

Nakajima, T., Fukamizu, A., Takahashi, J., Gage, F. H., Fisher, T., Blenis, J. and Montminy, M. R. 1996. The signal-dependent coactivator CBP is a nuclear target for pp90RSK, *Cell* 86:465–474.

Nakanishi, Y., Kihara, K., Mizuno, K., Masamune, Y., Yoshitake, Y. and Nishikawa, K. 1992. Direct effect of basic fibroblast growth factor on gene transcription in a cell-free system, *Proc Natl Acad Sci U S A* 89:5216–5220.

Neganova, I. and Lako, M. 2008. G1 to S phase cell cycle transition in somatic and embryonic stem cells, *J Anat* 213:30–44.

Niwa, H. 2007. How is pluripotency determined and maintained? *Development* 134:635–646.

O'Donovan, M. C., Norton, N., Williams, H., Peirce, *et al.* 2009. Analysis of 10 independent samples provides evidence for association between schizophrenia and a SNP flanking fibroblast growth factor receptor 2, *Mol Psychiatry* 14:30–36.

O'Neill, M. J., Bleakman, D., Zimmerman, D. M. and Nisenbaum, E. S. 2004. AMPA receptor potentiators for the treatment of CNS disorders, *Curr Drug Targets CNS Neurol Disord* 3:181–194.

Ohkubo, Y., Uchida, A. O., Shin, D., Partanen, J. and Vaccarino, F. M. 2004. Fibroblast growth factor receptor 1 is required for the proliferation of hippocampal progenitor cells and for hippocampal growth in mouse, *J Neurosci* 24:6057–6069.

Ovalle, S., Zamanillo, D., Andreu, F., Farre, A. J. and Guitart, X. 2001. Fibroblast growth factor-2 is selectively modulated in the rat brain by E-5842, a preferential sigma-1 receptor ligand and putative atypical antipsychotic, *Eur J Neurosci* 13:909–915.

Peng, H., Moffett, J., Myers, J., Fang, X., Stachowiak, E. K., Maher, P., Kratz, E., Hines, J., Fluharty, S. J., Mizukoshi, E., Bloom, D. C. and Stachowiak, M. K. 2001. Novel nuclear signaling pathway mediates activation of fibroblast growth factor-2 gene by type 1 and type 2 angiotensin II receptors, *Mol Biol Cell* 12:449–462.

Peng, H., Myers, J., Fang, X., Stachowiak, E. K., Maher, P. A., Martins, G. G., Popescu, G., Berezney, R. and Stachowiak, M. K. 2002. Integrative nuclear FGFR1 signaling (INFS) pathway mediates activation of the tyrosine hydroxylase gene by angiotensin II, depolarization and protein kinase C, *J Neurochem* 81:506–524.

Piper, D. R., Mujtaba, T., Keyoung, H., Roy, N. S., Goldman, S. A., Rao, M. S. and Lucero, M. T. 2001. Identification and characterization of neuronal precursors and their progeny from human fetal tissue, *J Neurosci Res* 66:356–368.

Piper, D. R., Mujtaba, T., Rao, M. S. and Lucero, M. T. 2000. Immunocytochemical and physiological characterization of a population of cultured human neural precursors, *J Neurophysiol* 84:534–548.

Pirvola, U., Ylikoski, J., Trokovic, R., Hebert, J. M., McConnell, S. K. and Partanen, J. 2002. FGFR1 is required for the development of the auditory sensory epithelium, *Neuron* 35:671–680.

Powell, P. P. and Klagsbrun, M. 1991. Three forms of rat basic fibroblast growth factor are made from a single mRNA and localize to the nucleus, *J Cell Physiol* 148:202–210.

Puchacz, E., Stachowiak, E. K., Florkiewicz, R. Z., Lukas, R. J. and Stachowiak, M. K. 1993. Basic fibroblast growth factor (bFGF) regulates tyrosine hydroxylase and proenkephalin mRNA levels in adrenal chromaffin cells, *Brain Res* 610:39–52.

Reilly, J. F. and Maher, P. A. 2001. Importin beta-mediated nuclear import of fibroblast growth factor receptor: role in cell proliferation, *J Cell Biol* 152:1307–1312.

Robinson, M. L., MacMillan-Crow, L. A., Thompson, J. A. and Overbeek, P. A. 1995. Expression of a truncated FGF receptor results in defective lens development in transgenic mice, *Development* 121:3959–3967.

Rydel, R. E. and Greene, L. A. 1987. Acidic and basic fibroblast growth factors promote stable neurite outgrowth and neuronal differentiation in cultures of PC12 cells, *J Neurosci* 7:3639–3653.

Sabbieti, M. G., Marchetti, L., Gabrielli, M. G., Menghi, M., Materazzi, S., Menghi, G., Raisz, L. G. and Hurley, M. M. 2005. Prostaglandins differently regulate FGF-2 and FGF receptor expression and induce nuclear translocation in osteoblasts via MAPK kinase, *Cell Tissue Res* 319:267–278.

Sassone-Corsi, P., Mizzen, C. A., Cheung, P., Crosio, C., Monaco, L., Jacquot, S., Hanauer, A. and Allis, C. D. 1999. Requirement of Rsk-2 for epidermal growth factor-activated phosphorylation of histone H3, *Science* 285:886–891.

Schechter, J. E. 1992. Is cellular disruption the mechanism of release of basic fibroblast growth factor from anterior pituitary gonadotropes? *Tissue Cell* 24:791–802.

Schimmel, J. J., Crews, L., Roffler-Tarlov, S. and Chikaraishi, D. M. 1999. 4.5 kb of the rat tyrosine hydroxylase 5′ flanking sequence directs tissue specific expression during development and contains consensus sites for multiple transcription factors, *Brain Res Mol Brain Res* 74:1–14.

Schmahl, J., Kim, Y., Colvin, J. S., Ornitz, D. M. and Capel, B. 2004. Fgf9 induces proliferation and nuclear localization of FGFR2 in Sertoli precursors during male sex determination, *Development* 131:3627–3636.

Sheng, Z., Lewis, J. A. and Chirico, W. J. 2004. Nuclear and nucleolar localization of 18-kDa fibroblast growth factor-2 is controlled by C-terminal signals, *J Biol Chem* 279:40153–40160.

Sherman, L., Stocker, K. M., Morrison, R. and Ciment, G. 1993. Basic fibroblast growth factor (bFGF) acts intracellularly to cause the transdifferentiation of avian neural crest-derived Schwann cell precursors into melanocytes, *Development* 118:1313–1326.

Somanathan, S., Stachowiak, E. K., Siegel, A. J., Stachowiak, M. K. and Berezney, R. 2003. Nuclear matrix bound fibroblast growth factor receptor is associated with splicing factor rich and transcriptionally active nuclear speckles, *J Cell Biochem* 90:856–869.

Stachowiak, E. K., Maher, P. A., Tucholski, J., Mordechai, E., Joy, A., Moffett, J., Coons, S. and Stachowiak, M. K. 1997a. Nuclear accumulation of fibroblast growth factor receptors in human glial cells–association with cell proliferation, *Oncogene* 14:2201–2211.

Stachowiak, E. K., Roy, I., Lee, Y.-W., Capacchietti, M., Aletta, J. M., Prasad P. N. and Stachowiak, M. K. 2009. Targeting novel Integrative Nuclear FGFR1 Signaling by nanoparticle-mediated gene transfer stimulates neurogenesis in adult brain, *Integr Biol (Camb)* 1:394–403.

Stachowiak, M. K., Fang, X., Myers, J. M., Dunham, S. M., Berezney, R., Maher, P. A. and Stachowiak, E. K. 2003a. Integrative nuclear FGFR1 signaling (INFS) as a part of a universal "feed-forward-and-gate" signaling module that controls cell growth and differentiation, *J Cell Biochem* 90:662–691.

Stachowiak, M. K., Fang, X. H., Myers, J. M., Dunham, S. M., Berezney, R., Maher, P. A. and Stachowiak, E. K. 2003b. Integrative nuclear FGFR1 signaling (INFS) as a part of a universal "feed-forward-and-gate" signaling module that controls cell growth and differentiation, *J Cell Biochem* 90:662–691.

Stachowiak, M. K., Goc, A., Hong, J. S., Poisner, A., Jiang, H. K. and Stachowiak, E. K. 1994a. Regulation of tyrosine hydroxylase gene expression in depolarized non-transformed bovine adrenal medullary cells: Second messenger systems and promoter mechanisms, *Brain Res Mol Brain Res* 22:309–319.

Stachowiak, M. K., Maher, P. A., Joy, A., Mordechai, E. and Stachowiak, E. K. 1996a. Nuclear accumulation of fibroblast growth factor receptors is regulated by multiple signals in adrenal medullary cells, *Mol Biol Cell* 7:1299–1317.

Stachowiak, M. K., Maher, P. A., Joy, A., Mordechai, E. and Stachowiak, E. K. 1996b. Nuclear localization of functional FGF receptor 1 in human astrocytes suggests a novel mechanism for growth factor action, *Brain Res Mol Brain Res* 38:161–165.

Stachowiak, M. K., Maher, P. A. and Stachowiak, E. K. 2007. Integrative Nuclear Signaling in Cell Development-A Role for FGF Receptor-1, *DNA Cell Biol* 26:811–826.

Stachowiak, M. K., Moffett, J., Joy, A., Puchacz, E., Florkiewicz, R. and Stachowiak, E. K. 1994b. Regulation of bFGF gene expression and subcellular distribution of bFGF protein in adrenal medullary cells, *J Cell Biol* 127:203–223.

Stachowiak, M. K., Moffett, J., Maher, P., Tucholski, J. and Stachowiak, E. K. 1997b. Growth factor regulation of cell growth and proliferation in the nervous system. A new intracrine nuclear mechanism, *Mol Neurobiol* 15:257–283.

Sun, W., Buzanska, L., Domanska-Janik, K., Salvi, R. J. and Stachowiak, M. K. 2005. Voltage-sensitive and ligand-gated channels in differentiating neural stem-like cells derived from the nonhematopoietic fraction of human umbilical cord blood, *Stem Cells* 23:931–945.

Szebenyi, G. and Fallon, J. F. 1999. Fibroblast growth factors as multifunctional signaling factors, *Int Rev Cytol* 185:45–106.

Takahashi, K. and Yamanaka, S. 2006. Induction of pluripotent stem cells from mouse embryonic and adult fibroblast cultures by defined factors, *Cell* 126:663–676.

Terwisscha van Scheltinga, A. F., Bakker, S. C. and Kahn, R. S. 2009. Fibroblast Growth Factors in Schizophrenia, *Schizophr Bull.*

Togari, A., Dickens, G., Kuzuya, H. and Guroff, G. 1985. The effect of fibroblast growth factor on PC12 cells, *J Neurosci* 5:307–316.

Vlodavsky, I., Bar-Shavit, R., Ishai-Michaeli, R., Bashkin, P. and Fuks, Z. 1991. Extracellular sequestration and release of fibroblast growth factor: A regulatory mechanism? *Trends Biochem Sci* 16:268–271.

von Mikecz, A., Zhang, S., Montminy, M., Tan, E. M. and Hemmerich, P. 2000. CREB-binding protein (CBP)/p300 and RNA polymerase II colocalize in transcriptionally active domains in the nucleus, *J Cell Biol* 150:265–273.

Wei, X., Somanathan, S., Samarabandu, J. and Berezney, R. 1999. Three-dimensional visualization of transcription sites and their association with splicing factor-rich nuclear speckles, *J Cell Biol* 146:543–558.

Wolffe, A. P. and Pruss, D. 1996. Targeting chromatin disruption: Transcription regulators that acetylate histones, *Cell* 84:817–819.

Woodward, W. R., Nishi, R., Meshul, C. K., Williams, T. E., Coulombe, M. and Eckenstein, F. P. 1992. Nuclear and cytoplasmic localization of basic fibroblast growth factor in astrocytes and CA2 hippocampal neurons, *J Neurosci* 12:142–152.

Wu, J. and Janknecht, R. 2002. Regulation of the ETS transcription factor ER81 by the 90-kDa ribosomal S6 kinase 1 and protein kinase A, *J Biol Chem* 277:42669–42679.

Yamaguchi, T. P., Harpal, K., Henkemeyer, M. and Rossant, J. 1994. fgfr-1 is required for embryonic growth and mesodermal patterning during mouse gastrulation, *Genes Dev* 8:3032–3044.

Part II

From Mechanisms to Enabling Technologies for Stem Cells

Generation, Characterization and Propagation of Human-Induced Pluripotent Stem Cells

5

Sheena Abraham, Marion J. Riggs, Ena Xiao, Khaled Alsayegh, Venkat Gadepalli, and Raj R. Rao

Human induced pluripotent stem cells (hiPSCs) generated from reprogrammed adult cells have gained recognition as a potential cell source for regenerative biomedical therapies. This potential provides opportunities for (a) generating many differentiated cell types, (b) creating unique disease-specific *in vitro* cell models, and (c) applications in customized medicine. This chapter will focus on the generation of hiPSCs using viral and non-viral approaches, characterization of their pluripotent capabilities, their differentiation capabilities and development of culture conditions for their stable propagation.

5.1. Introduction

Genetic reprogramming of adult somatic cells to express key genes and factors necessary for maintaining self-renewal in embryonic stem cells (ESC) produces induced pluripotent stem cells (iPSCs). These iPSCs are ESC-like in their ability to renew indefinitely through symmetric cell division, while maintaining *pluripotency* (the ability to differentiate into any cell types from the three germ layers: endoderm, mesoderm, and ectoderm).

The derivation of iPSCs is a significant advancement in the field of regenerative biomedical research, where ethical issues involving the use of pluripotent stem cells from embryos can be circumvented. These iPSCs are shown to resemble ESCs in their morphology, gene expression patterns, cell surface markers, doubling time, and embryoid body formation, details of which are discussed later in this chapter. However, further characterization using bioinformatics approaches, such as global gene expression profiling, methylation pattern analysis, and computational modeling of molecular framework are needed to fully assess their relationship.

From a clinical perspective, the process of transplantation is a complex and challenging medical procedure due to the innate possibility of rejection of transplanted cells/tissues by the host. However, with well-defined induction methods producing patient-specific cell types, the potential for immunological rejections might be avoided, making cell and organ transplantation feasible. In addition, *in vitro* propagation of iPSCs will make personalized drug screening and toxicity testing possible.

The initial derivation of human ESC lines contributed to the framework for the subsequent generation of human iPSCs. Elaborate characterization of ESCs allowed for analysis of genes and pathways, highlighting pluripotency and self-renewal, which in turn identified key factors that were necessary for induction of pluripotency in normal somatic cells (Rao *et al.*, 2004). In seminal studies in 2006, Takahashi and Yamanaka over-expressed *Oct4*, *Sox2*, *c-Myc*, *Klf4* in mouse fibroblasts and induced the pluripotent state in these cells (Takahashi *et al.*, 2006). Subsequent studies in humans with the same four genes contributed to the generation of human iPSCs (Takahashi *et al.*, 2007). In parallel studies, over-expression of *Nanog*, *Lin28*, *Oct4* and *Sox2* also successfully reverted human somatic cells to the pluripotent state (Yu *et al.*, 2007). Although cells derived from the Yamanaka group failed to contribute to the germline in chimeras, later improvements in protocols (especially in the selection of gene promoters) resulted in iPSCs remarkably similar to ESCs. A transposon-based approach (*PiggyBac*) was proposed as a better alternative for viral delivery of key factors (Woltjen *et al.*, 2009; Yu *et al.*, 2009). The following section focuses on the current status of the different induction approaches that have been used for generating human iPSCs.

5.2. Induction Approaches for Generating hiPSCs

5.2.1. *Viral induction*

In these approaches, conventional retroviral or lentiviral vectors have been used to deliver the selected transcription factors into the genome of adult somatic cells. Viral induction methods involve infecting cells with (a) "Floxed" retroviral vectors, (b) adenoviral vectors, or (c) episomal vectors (Figure 5.1), and are discussed in detail below.

5.2.1.1. *Floxed viral vectors*

In this strategy, cDNA for Oct4, Sox2, Nanog, Klf4, cMyc and GFP were cloned into the retroviral vector devoid of viral genes with the Cre recombinant site flanking the loxP region separately transfected into packaging cells capable of producing envelope protein of viruses. The viral supernatant was combined with the ESC

(5) culture conditions, (6) method of identification of reprogrammed cells, and (7) method of characterization of the reprogrammed line (Maherali and Hochedlinger, 2008). Compounding this experimental layer of variability is the biological complexity driven by the heterogeneity in the cells produced during the reprogramming process, due to the low efficiency in any reprogramming protocol (Smith *et al.*, 2009). The heterogeneity in reprogramming is a significant challenge as research is predicated upon reproducibility that will require a reduction in cross-lab variability. In order to address this dilemma, there is significant impetus to develop uniform standards that are accepted by the scientific community at large. Increasingly, it is becoming apparent that pluripotency itself is a heterogeneous state with different cells demonstrating varying potentials (Enver *et al.*, 2009; Smith *et al.*, 2009). For example, some pluripotent stem cells are able to form teratomas but are unable to meet the stricter test of chimeric formation or germ-line transmission (Daley *et al.*, 2009). These cells are said to meet the criteria for functional pluripotency, but not the more stringent criteria for developmental pluripotency (Smith *et al.*, 2009). Likewise, a cell may exhibit the molecular markers for pluripotency, while not having the capability to generate embryoid bodies or well-differentiated teratomas, thus failing the functional pluripotency test. Therefore, critical to the development of uniform standards is an increased understanding of the pluripotency and distinct potentials within this class of cells. A set of industry-accepted uniform standards should take the application of the cells into consideration. For example, in the therapeutic setting, the emphasis is on a safe and functional cellular product that can be generated reproducibly in sufficient numbers. Thus, it can be argued that a cell with significant plasticity that might pass the molecular pluripotency test, but cannot form teratomas may be safer clinically (Ellis *et al.*, 2009). However, in basic developmental biology research, it may be critical to demonstrate developmental pluripotency using the most stringent tests for pluripotency (Maherali and Hochedlinger, 2008). A goal for application-appropriate standards should be an optimized set of characterization criteria balancing stringency with ease of entry for new investigator contributions (Ellis *et al.*, 2009).

5.3.2. *Classical assays for demonstrating pluripotency*

Research is demonstrating greater heterogeneity in pluripotent cells than we first thought with significant impacts on their behavior and differentiation capabilities. This may be representative of different cellular states corresponding to different developmental time points or variability of reprogramming hiPSCs. The best approach to understanding hiPSCs for the foreseeable future is to use assays developed in other cellular models such as mESCs and hESCs. Incidentally, many of the markers developed to determine pluripotency in hESCs originated in earlier work performed on embryonic carcinoma cells (Andrews *et al.*, 2005). Thus, it is conceivable that a desirable therapeutic strategy is to reprogram cells to a non-tumorogenic but plastic intermediate state, or even to directly program one somatic cell to another somatic cell type (Vierbuchen *et al.*, 2010). In any case, established methods for ascertaining pluripotency in hESCs can be employed for a thorough characterization of hiPSCs and to better understand pluripotency and the reprogramming process (Figure 5.2).

For ethical reasons, the most stringent test for pluripotency can only be performed on mice. Varying reports exist on the success of tetraploid complementation from miPSCs — the most stringent test for developmental pluripotency, in which the

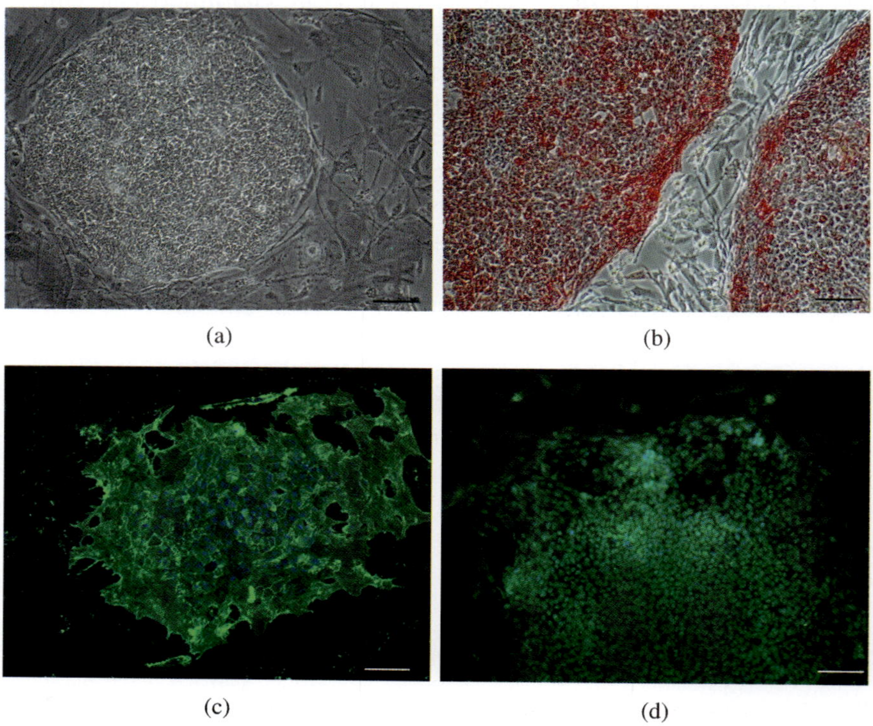

Figure 5.2. Human iPS cells propagated on inactivated mouse embryonic fibroblasts maintain (a) High nuclear to cytoplasmic ratio and tight boundaries; (b) Positive expression of alkaline phosphatase; (c) Stage specific embryonic antigen 4, SSEA4; (d) Transcription factor Oct3/4. Scale bar = 100 μm.

pluripotent cells give rise to the entire offspring. Most reports state that miPSCs have not generated live mice from tetraploid complementation and thus fail to meet the strictest definition of pluripotency (Daley *et al.*, 2009; Maherali and Hochedlinger, 2008). One published report stated the successful production of live pups through tetraploid complementation in 3 out of 37 induced lines (Zhao *et al.*, 2009). The establishment of chimeric mice from blastocyst injection of mouse iPSC and germ-line transmission has been routinely demonstrated; however, many of the derived lines that meet other less strict pluripotent qualifications, such as forming well-differentiated teratomas comprised of tissue from each of the three germ layers, fail at blastocyst chimerism or germ-line transmission (Daley *et al.*, 2009).

The strictest pluripotent assay that can be ethically conducted on hESCs is the teratoma-forming assay in immunocompromised mice. Recently, experimental evidence has suggested that hESCs are more representative of mouse epiblast stem cells (mEpiSCs) than mESCs, which represent a later developmental stage than cells from the inner cell mass (Enver *et al.*, 2009; Silva and Smith, 2008; Tesar *et al.*, 2007; Toyooka *et al.*, 2008). Interestingly, mEpiSCs can form teratomas but cannot contribute to chimerism or germ-line transmission (Daley *et al.*, 2009; Silva and Smith, 2008). Thus, even hESCs may fail some developmental pluripotency standards. HiPSCs have been shown to generate well-differentiated teratomas (Takahashi *et al.*, 2007); however, given the wide accessibility of labs to generate iPSCs and enter the field, along with the time-consuming nature of *in vivo* techniques, many labs will undoubtedly be tempted to settle for quicker *in vitro* measures to suggest pluripotent capabilities of an induced line.

5.3.3. *Systems biology of pluripotency*

HiPSCs, hESCs, and mESCs, provide powerful models for the understanding of pluripotency, development, and differentiation. The principles of reprogramming and differentiation are unified mathematically by depicting cell fates as balanced states, also known as attractors, in a dynamical system using energy landscape theory (MacArthur *et al.*, 2008). Attractor states are stable solutions to a set of equations representing states in equilibrium (Enver *et al.*, 2009). This computational model is in its infancy but experimental evidence is mounting for its justification; and there is considerable optimism for the strength of its application. The foundations for this theoretical framework first emerged in the 1940s by the physicist Max Delbruck and have been detailed extensively by Macarthur *et al.* (2009). In 1957, developmental biologist Conrad Waddington depicted development as an epigenetic landscape with an increasing bifurcation of valleys as development progressed, forming the basis for some recent studies by Vierbuchen *et al.* (2010).

The hills of the valleys, acting as a barrier between different cell types and tissue specialization during development, could be viewed as a marble rolling down a hill in between the valleys adopting a particular cell fate. A consequence of this theory is that sufficient perturbation will cause a transition from one cell type (i.e. balanced state) to another (Arias and Hayward, 2006). Another consequence of the attractor state landscape is that cells sufficiently near a particular attractor will be pulled to that state as a basin of local energy minima and that the system will exhibit multistability with many coexisting attractor states (each resistant to weak perturbations) (Chang *et al.*, 2008; Hemberger *et al.*, 2009). Reprogramming a somatic cell to a pluripotent state is remarkable evidence for the energy landscape theory of cell states, where pluripotency is one of many attractor states. The ability to use different cell types and various methods for reprogramming further builds evidence for the application and relevance of this model (Macarthur *et al.*, 2009).

SSEA3 downregulation is the first and most rapid marker of differentiation (Draper *et al.*, 2002; Fenderson *et al.*, 1987). Heterogeneity of the hESC compartment by SSEA3 expression can be theorized as follows: Two stochastically interconverting cell populations marked by changes between SSEA3+ and SSEA3− expressions represent differences in energy level (i.e. sub-states) within the same attractor state. The SSEA3− cells exist at a higher energy level and are closer to the "commitment barrier" (i.e. edge of the well) and thus have greater probability of escaping the stem cell compartment and differentiating (Enver *et al.*, 2009). Experimental evidence in the mESCs supports the view that embryonic stem cells exist in the ground state. Upon inhibition of ERK and FGF signaling, the mESCs perpetually self-renew by intrinsic mechanisms (in the absence of LIF or any other pluripotent maintenance factors). These results suggest that pluripotency and self-renewal requires prevention of differentiation (i.e. pluripotent cells from leaving the ground state), as opposed to an active process that maintains pluripotency (Ying *et al.*, 2008).

Key to applying the landscape theory is determining the factors that can cause a transition from one state to another (e.g. transcription factors and growth factors) and delineating the components that stabilize the energy well, e.g. epigenetic mechanisms (Hemberger *et al.*, 2009). Each metastable state should be associated with a transcriptional circuitry that can be depicted in terms of a gene regulatory network, as well as being defined by other global architectures, such as a signature epigenome. Pluripotent cells including hiPSCs have a characteristic epigenetic profile, consisting of bivalent chromatin marks comprised of histone 3 lysine 27 trimethylation (H3K27Me3) and histone 3 lysine 4 trimethylation (H3K4Me3). Each mark is associated with inhibition and activation of transcription, respectively. The polycomb group repressive complex 2 (PRC2) consisting of polycomb group proteins is found at key developmental genes and is associated with the transcriptional inhibitory H3K27Me3 modification (Bernstein *et al.*, 2006; Boyer *et al.*, 2006; Lee *et al.*, 2006). Increasingly, high-throughput technology for genomic sequencing, RNA profiling, proteomics, DNA methylation, and histone modifications are becoming accessible to many researchers to generate the data needed to put Waddington's Canal on quantitative grounds (MacArthur *et al.*, 2008). Energy landscape theory has the potential to shape our understanding of the optimal cells, parameters, and conditions for reprogramming to the pluripotent state or for directly programming somatic cell to somatic cell transitions.

5.4. Directed Differentiation Studies Involving hiPSCs

The attention that hiPSC research has garnered in the past few years has been based on its potential for generation of *in vitro*-based disease models as well as development of appropriate therapies using patient-specific immunologically compatible cells. The differentiation potential of hiPSCs can be demonstrated by the generation of germ layer derivatives *in vitro* (embryoid body formation; Figure 5.3) and *in vivo* (teratoma formation) (Takahashi *et al.*, 2007). However, the key to realizing the potential of hiPSCs to generate patient- and disease-specific cell lines lies in its capability for directed differentiation.

Employing established protocols for directing differentiation in hESCs have yielded similar results in hiPSCs, reinforcing the functional similarities between the two cell sources. However, comparably fewer differentiation studies have been attempted. In this section, we summarize the directed differentiation strategies

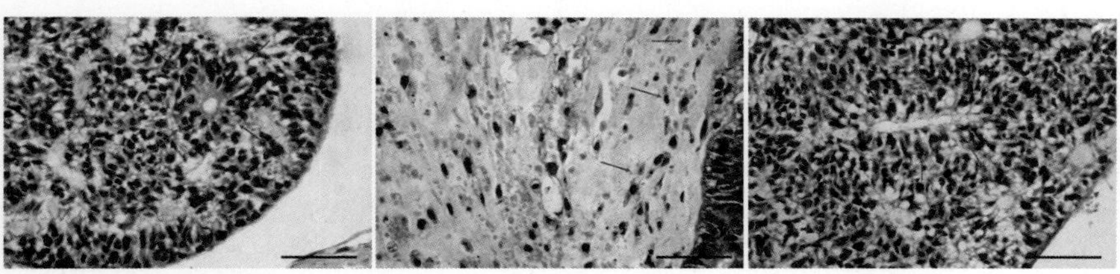

(a) (b) (c)

Figure 5.3. Tri-lineage differentiation is evident in embryoid bodies generated from hiPSC cells cultured on inactivated mouse embryonic fibroblasts. Trilineage potential is demonstrated as (a) ectodermal differentiation, neuroepithelial; (b) mesodermal differentiation, fibrous connective; and (c) endodermal differentiation; intestinal. Scale bar = 50 μm.

adopted to terminally differentiate hiPSCs into cell types from the three embryonic germ layers (ectoderm, endoderm and mesoderm).

5.4.1. *Differentiation down the ectodermal lineage*

Within the ectodermal germ layer, neuronal differentiation has been achieved mostly using the EB formation method with medium supplementation involving specific growth factors. Differentiation of hiPSCs into functionally mature motor neurons has been achieved by two methods: (1) EB formation in addition to enrichment for motor neuron differentiation using retinoic acid and the Shh pathway agonist purmorphamine, and (2) mechanical isolation of neural rosettes from hiPSC cultures further cultured on laminin-coated dishes in the presence of retinoic acid, purmorphamine and neurotrophic factors. Comparisons between motor neurons derived from hiPSCs and hESCs have shown that the two ectodermal derivatives are indistinguishable, both physiologically and molecularly (Saravanan *et al.*, 2009).

Other neuronal derivatives such as midbrain dopamine and spinal motoneurons were achieved by the dual inhibition of SMAD signaling pathway. Inhibitors such as Noggin and SB431542 are known to individually enhance neural differentiation within EBs (Elkabetz *et al.*, 2008; Smith *et al.*, 2008), and have also been shown to induce neural differentiation in adherent cultures as well (Chambers *et al.*, 2009). In another report, midbrain dopamine (mDA) neurons were derived using an improved protocol, where rosettes that formed within hiPSC cultures were isolated and expanded, followed by a suspension culture to promote neural progenitor (NP) sphere formation. Further specification to mDA pathway was achieved by adhesion culture of the spheres onto modified slides and supplementation of growth medium with dibutyryl cAMP (Cai *et al.*, 2009).

5.4.2. *Differentiation down the mesodermal lineage*

Mesodermal differentiation of hiPSCs to form hematopoietic derivates includes suspension as well as adherent cultures, whereas cardiac specification and adipogenesis required EB formation. A transient exposure of hiPSCs to MEDII medium (HepG2: human hepatocarcinoma cell line, conditioned medium) enhanced initial differentiation of the hiPSCs into mesodermal derivatives during embryoid body formation. Preferential differentiation into hematopoietic progenitors in addition to other mesodermal elements was observed with the concurrent reduction in ectodermal and endodermal derivatives (Lu *et al.*, 2009). Another study detailing the induction of hematopoietic and endothelial differentiation utilizes strategies identical to that used for hESCs. Coculturing with a murine stromal cell line was shown to be successful in hematopoietic and endothelial induction of several hiPSC lines (Kyung-Dal *et al.*, 2009).

Other mesodermal derivatives include cardiomyocyte differentiation of hiPSCs using EB formation. Following 4 days of EB formation in growth medium supplemented with FBS pretested for cardiac differentiation, the EBs were seeded onto gelatin-coated dishes and maintained for extended periods for contracting EB formation. These studies suggests that the efficiency of iPSCs to form contracting EBs was similar to human ESCs (Zhang *et al.*, 2009; Zwi *et al.*, 2009). Adipogenic differentiation of hiPSCs was obtained by initiation of EB formation with transient retinoic acid supplementation. Followed by the seeding of 12-day-old EBs onto poly-L-ornithine- and fibronectin-coated plates and further maintained for 10 days in adipogenesis-inducing growth medium (Taura *et al.*, 2009).

5.4.3. *Differentiation down the endodermal lineage*

Few studies have reported the differentiation of hiPSCs to cells derived from the endoderm and include hepatic- or hepatocyte-like cells. In the first study, Song and colleagues induced differentiation of hiPSCs into hepatic cells through a stepwise protocol that involved (1) endoderm induction, (2) hepatic specification, (3) hepatoblast expansion, (4) hepatic maturation, leading up to (5) matured hepatocytes, modified from an earlier protocol used to induce hepatic differentiation of hESCs (Song *et al.*, 2009). Human fibroblasts were initially reprogrammed using *Oct4, Sox2, Klf4*, and *Utf1* genes and further differentiated down endodermal lineages. Subsequent testing of the resulting differentiated cells exhibited liver-related functions and demonstrated many characteristics of hESC-derived hepatocytes. In another similar study, Sullivan *et al.* (2010) reported the derivation of human hepatic endoderm cells from iPSCs derived from male and female sexes and two varying ethnic groups. A modified protocol based on the differentiation of hESCs into hepatocyte-like cells was used. The differentiated cells exhibited hepatic morphology and characteristics similar to that reported by Song and colleagues. However, the use of *c-Myc* in the second study was stated to be a possible interference with the differentiation of iPSCs since *c-Myc* is essential for the self-renewal and differentiation of ESCs.

Thus, the use of reprogrammed somatic cells for basic and translational research might provide a viable alternative to hESCs. As seen in the various aforementioned studies, given that hiPSCs follow normal developmental progression, the prospects of developing disease models as well as patient-specific lines seem to be within reach.

5.5. Propagation and Culture Systems

Given the similarities between hiPSCs and hESCs, the propagation and culture methodologies employed for hiPSCs are similar to those that have been developed for hESCs. Unlike routine mammalian cell and tissue cultures, the maintenance and propagation of hESCs and hiPSCs is tedious and expensive. Additionally, it is critical to propagate these cell lines while not only maintaining their self-renewal and pluripotent characteristics but also retaining their normal karyotype. An important hurdle in the propagation of hiPSCs is the possibility of xenogenic contamination due to the use of non-human derivatives within the growth medium and murine fibroblast-based substrates. It is, thus important to develop culture protocols that avoid xenogenic contaminants and promote the use of defined media components for their maintenance.

5.5.1. *Culture medium composition*

The culture medium for human pluripotent cell propagation is traditionally supplemented with additives that regulate self-renewal and pluripotency. Attempts are being made to avoid animal feeder cells (viz. MEFs) or non-human–derived components and encourage the use of soluble factors that directly or indirectly regulate specific pathways to maintain the combined characteristics of self-renewal and pluripotency. HiPSC were so far maintained in the traditional hESC medium, which comprises of DMEM/F12 (Invitrogen), knockout serum replacement (KSR; Invitrogen), non-essential amino acids, L-glutamine, penicillin/streptomycin, β-mercaptoethanol, and basic fibroblast growth factor (bFGF) (Abraham *et al.*, 2010; Park and Daley, 2009; Takahashi *et al.*, 2007). Apart from the use of traditional hESC medium for hiPSC culture, other medium formulations developed for hESC culture (Xu *et al.*, 2001; Ludwig *et al.*, 2006) are also expected to be easily adapted for hiPSC propagation, thus presenting unique opportunities for improving the culture environment of hiPSCs.

5.5.2. *Passaging methods*

One of the primary identifying features of pluripotent cells is the cell and colony morphology that they exhibit: high nuclear to cytoplasmic ratio, prominent nucleoli and distinct colony-boundary (Figure 5.2). Successful reprogramming of adult human cells into hiPSCs can be confirmed initially by observing the above mentioned morphology. Furthermore, selection methods based on reporter markers such as antibiotics (puromycin) and lentiviral vectors that specifically express in pluripotent cells (e.g. early transposon promoter (EOS lentiviral vector), Oct4 and Sox2 enhancers) (Hotta *et al.*, 2009) help in the isolation and sub-culturing of reprogrammed adult cells more efficiently. Two basic techniques have been employed for serial sub-culturing of hiPSCs and are described below.

5.5.2.1. *Mechanical technique*

In this method, optimal IPS colonies based on size and density are dissected into nearly equal-sized sections using a fire-pulled pasteur pipette. These dissected colonies are then transferred onto a fresh feeder layer for continued passaging.

5.5.2.2. *Enzymatic/Non-enzymatic technique*

Alternatively referred to as the bulk-passaging method, this method involves dispersal either by an enzymatic treatment (Collagenase/Trypsin) or by an EDTA-based cell dissociation buffer into single cells or smaller cell clusters. The cells are subsequently collected as a pellet by centrifugation, resuspended in fresh growth medium and seeded onto a fresh feeder layer for continued passaging.

These conventional passaging techniques have certain advantages and disadvantages. The mechanical method of passaging is relatively time consuming and requires highly trained personnel when compared with the bulk-passaging methods. However, it is important to note that many studies have reported that some hESCs attain chromosomal aberrations after prolonged enzymatic passaging-based *in vitro* culture (Mitalipova *et al.*, 2005). The molecular mechanisms that lead to chromosomal instability in pluripotent stem cell cultures are unclear still, but a few technical aspects have been postulated as being responsible for the formation of cells with abnormal karyotypes. Scientists have questioned the vulnerability of certain hESC lines, culture systems, and culture adaptation. Concerns over enzymatic passaging techniques were raised as the cells experience greater stress as compared to mechanical techniques (enzymes and centrifugation). Furthermore, given that single cell suspension or smaller clusters are obtained through bulk-passaging methods, disruption of cell–cell contact within the hESC colony has been considered as a precursor to production of cells with abnormal karyotypes. The manual passaging technique, on the other hand, permits the propagation of chromosomally stable hESC lines for a longer period of time (Buzzard *et al.*, 2004; Draper *et al.*, 2004; Mitalipova *et al.*, 2005). It is useful to note a recent study that devised a culture environment allowing defined enzymatic propagation of manually passaged hESCs, while maintaining the cell lines in an undifferentiated, pluripotent, and normal state. However, it is important to note that this study relied extensively on high levels of a human recombinant basic fibroblast growth factor (hrbFGF)

(Invitrogen®), to maintain the cells in their undifferentiated state (Ellerstrom *et al.*, 2010).

5.5.3. *Culture systems*

Unlike other normal mammalian cell types, both hESCs and hiPSCs are anchorage dependent and hence need support for cell attachment to maintain their pluripotent characteristics. Scientists have advanced much in the design and development of culture systems with a specific aim of devising systems that are conducive for primarily large-scale production and secondary-controlled differentiation of these pluripotent cells into desired progenitors and/or differentiated cell types.

5.5.3.1. *Feeder cell culture systems*

These systems involved inactivated cells which are capable of maintaining the pluripotent cells in an undifferentiated state based on the provision of chemical and physical cues. Initially, hESCs were propagated on inactivated mouse embryonic fibroblast (MEF) using serum-containing medium (Reubinoff *et al.*, 2000; Thomson *et al.*, 1998). In order to avoid these xenogeneic compounds (e.g. MEF, animal-derived serum etc.) various studies have reported the successful propagation of hESCs on human-derived feeder layers such as foreskin cells (Amit *et al.*, 2003; Choo *et al.*, 2004; Hovatta *et al.*, 2003), embryonic fibroblasts (Richards *et al.*, 2002), adult skin cells (Richards *et al.*, 2003) and differentiated cells derived from hESC (Wang *et al.*, 2005). These feeder-based systems have been used in the derivation and propagation of hiPSCs and have included mouse fibroblast (Meissner *et al.*, 2007), and human fetal, neonatal, and adult skin fibroblasts (Hanna *et al.*, 2008; Lowry *et al.*, 2008; Nakagawa *et al.*, 2008; Yu *et al.*, 2007).

5.5.3.2. *Feeder-free culture systems*

The feeder-free culture system for hESCs basically comprises the use of structural proteins that mostly involve extracellular matrix proteins such as laminin, collagen, fibronectin, and hyaluronic acid. Matrigel^{TM} is the most widely used feeder-free substrate but still contains components of non-human origin and undefined components, thus limiting its possible use in culturing pluripotent cells for clinical purposes. Successful studies have been carried out to establish feeder-free systems in hiPSCs by using conditioned media (MEF-CM), and high concentrations of growth factors (bFGF) (Totonchi *et al.*, 2009). Recent studies from our group have demonstrated the propensity of decellularized substrates from human fibroblasts in the stable propagation of hESCs and hiPSCs (Abraham *et al.*, 2010). Another recent study reported that tissue culture plates coated with a recombinant substrate, a fusion protein consisting of E-cadherin and the IgG Fc domain, in the presence of mTeSR1 medium supported the growth of hESCs and hiPSCs (Nagaoka *et al.*, 2010). Various biomaterial-based scaffolds have been successfully studied for undifferentiated propagation of hESCs. Examples include hydrogels (Ferreira *et al.*, 2007; Gerecht *et al.*, 2007) and calcium alginate-based microcapsule (Chayosumrit *et al.*, 2010; Siti-Ismail *et al.*, 2008) for undifferentiated propagation. From a differentiation perspective, agarose hydrogels (Dang *et al.*, 2004) and poly-L-lactic acid scaffold (Levenberg *et al.*, 2003) have found limited applications. These biomaterial-based culture systems not only have the ability to provide a three-dimensional (3D) microenvironment but can also bridge the variability of culture systems during the differentiation process in traditional culture methods. This enables the generation of uniform culture systems for undifferentiated maintenance and directed differentiation of pluripotent cells (Chayosumrit *et al.*, 2010; Gerecht *et al.*, 2007). Biomaterial-based scaffolds and feeder-free approaches have enormous potential in hiPSC cultures, propagation and differentiation and may initiate development of novel technologies.

5.5.3.3. *Bioreactors*

Propagation of pluripotent cells in bioreactors have various advantages such as maintaining specific culture conditions, large-scale production of cells, and permitting the cells to recreate their own *in vivo* milieu. The design and development of bioreactor technologies to propagate hESCs and hiPSCs is still in its infancy. As pluripotent cells are anchorage dependent, substrate-based cultures or microcarrier/hydrogel-based bioreactors have been employed. Commercially available micro carriers such as Cytodex 1 (Sigma), Cytodex 3 (Sigma), Microhex (Nunc), Cultisphere (Hyclone), and several Solohill microcarriers [plastic plus, pronectin, collagen, FACT, and Hillex II, (Solohill Engineering, Inc)] have found applications in pluripotent stem cell cultures (Nie *et al.*, 2009; Phillips *et al.*, 2008). In these studies, these microcarriers were coated with MEFs, Matrigel^{TM}, or other proteins to increase the attachment of the pluripotent stem cells. Researchers have successfully employed some of these coated microcarriers to culture hESC and trigger differentiation in stirred-suspension bioreactors (SSB) (Lock and Tzanakakis, 2009) and spinner flasks (Fernandes *et al.*, 2009; Oh *et al.*, 2009). A similar study reported the culturing of hiPSCs in spinner flasks using Matrigel^{TM}-coated microcarrier beads and observed the cell viability at different seeding densities (Kehoe *et al.*, 2009). Even though these studies reported an increase in cell numbers when compared to static culture systems, some of the concerns were based on the fact that substrates of non-human origin such as Matrigel™ or MEF coating were more conducive to cell attachment as compared to the other tested substrates. Some of the other issues involved the lack of effective subcloning method, decreased viability of cells in differentiation conditions, and the possibility of abnormal karyotypes as a result of extensive mitosis (Kehoe *et al.*, 2009). A more recent study also reported the use of an SSB system that enabled the expansion of hESCs in the absence of feeder layers of matrices (Krawetz *et al.*, 2009). The study employed additional protein supplements such as mammalian target of rapamycin (mTOR), rho kinase (ROCK), and transforming growth factor-β (TFG-β) within a SSB culture system. However, extensive studies on the interactions between these supplements still need to be carried out. When compared with other culture systems, bioreactor-based technologies offer a controlled and reproducible culture system, capable of generating clinically relevant cell numbers with minimal labor.

5.6. Summary & Future Possibilities

The promise and potential of regenerative medicine has definitely received an enormous boost through recent discoveries and research involving human-induced pluripotent stem cells. The possibility for genetically and epigenetically reprogramming somatic cells to a pluripotent state through the exogenous expression of transcription factors associated with pluripotency has yielded a new cell type with the promise for treating many human diseases. Reprogramming also provides opportunities for creating patient-specific cell lines as *in vitro* models of human diseases. However, for hiPSCs to be used extensively in research and applications, the efficiency of reprogramming needs to be drastically improved. Furthermore, procedures to control the genetic and epigenetic stability of hiPSCs need to be standardized. For eventual applications in human therapies, concerted efforts between scientists, clinicians, and engineers are needed in the development of novel culture systems that employ good manufacturing process (GMP) standards in propagation and directed differentiation paradigms.

Acknowledgements

This work was supported in part by start-up funds provided to Dr. Rao by the Department of Chemical and Life Science Engineering at Virginia Commonwealth University and funding provided by NSF CAREER Award #074556 (Rao). Mention of vendor names is for information only and does not imply endorsement.

Bibliography

Abraham, S., Sheridan, S. D., Laurent, L. C., Albert, K., Stubban, C., Ulitsky, I., Miller, B., Loring, J. F. and Rao, R. R. 2010. Propagation of human embryonic and induced pluripotent stem cells in an indirect co-culture system, *Biochem Biophys Res Commun* 393(2):211–216.

Abraham, S., Sheridan, S. D., Miller, B. and Rao, R. R. 2010. Stable propagation of human embryonic and induced pluripotent stem cells on decellularized human substrates, *Biotechnol Prog* 26(4):1126–1134.

Amit, M., Margulets, V., Segev, H., Shariki, K., Laevsky, I., Coleman, R. and Itskovitz-Eldor, J. 2003. Human feeder layers for human embryonic stem cells, *Biol Reprod* 68:2150–2156.

Andrews, P. W., Matin, M. M., Bahrami, A. R., Damjanov, I., Gokhale, P. and Draper, J. S. 2005. Embryonic stem (ES) cells and embryonal carcinoma (EC) cells: Opposite sides of the same coin, *Biochem Soc Trans* 33:1526–1530.

Arias, A. M. and Hayward, P. 2006. Filtering transcriptional noise during development: Concepts and mechanisms, *Nat Rev Genet* 7:34–44.

Bernstein, B. E., Mikkelsen, T. S., Xie, X., Kamal, M., Huebert, D. J., Cuff, J., Fry, B., Meissner, A., Wernig, M., Plath, K., Jaenisch, R., Wagschal, A., Feil, R., Schreiber, S. L. and Lander, E. S. 2006. A bivalent chromatin structure marks key developmental genes in embryonic stem cells, *Cell* 125:315–326.

Boyer, L. A., Plath, K., Zeitlinger, J., Brambrink, T., Medeiros, L. A., Lee, T. I., Levine, S. S., Wernig, M., Tajonar, A., Ray, M. K., Bell, G. W., Otte, A. P., Vidal, M., Gifford, D. K., Young, R. A. and Jaenisch, R. 2006. Polycomb complexes repress developmental regulators in murine embryonic stem cells, *Nature* 441:349–353.

Buzzard, J. J., Gough, N. M., Crook, J. M. and Colman, A. 2004. Karyotype of human ES cells during extended culture, *Nature Biotechnol* 22:381–2; author reply 382.

Cai, J., Yang, M., Poremsky, E., Kidd, S., Schneider, J. S. and Iacovitti, L. 2009. Dopaminergic Neurons Derived from Human Induced Pluripotent Stem Cells Survive and Integrate into 6-OHDA Lesioned Rats, *Stem Cells Dev* 19:1017–1023.

Chambers, S. M., Fasano, C. A., Papapetrou, E. P., Tomishima, M., Sadelain, M. and Studer, L. 2009. Highly efficient neural conversion of human ES and iPS cells by dual inhibition of SMAD signaling, *Nature Biotechnol* 27:275–280.

Chang, H. H., Hemberg, M., Barahona, M., Ingber, D. E. and Huang, S. 2008. Transcriptome-wide noise controls lineage choice in mammalian progenitor cells, *Nature* 453:544–547.

Chayosumrit, M., Tuch, B. and Sidhu, K. 2010. Alginate microcapsule for propagation and directed differentiation of hESCs to definitive endoderm, *Biomaterials* 31:505–514.

Choo, A. B., Padmanabhan, J., Chin, A. C. and Oh, S. K. 2004. Expansion of pluripotent human embryonic stem cells on human feeders, *Biotechnol Bioeng* 88:321–331.

Daley, G. Q., Lensch, M. W., Jaenisch, R., Meissner, A., Plath, K. and Yamanaka, S. 2009. Broader implications of defining standards for the pluripotency of iPSCs, *Cell Stem Cell* 4:200–1; author reply 202.

Dang, S. M., Gerecht-Nir, S., Chen, J., Itskovitz-Eldor, J. and Zandstra, P. W. 2004. Controlled, scalable embryonic stem cell differentiation culture, *Stem Cells (Dayton, Ohio)* 22:275–282.

Draper, J. S., Pigott, C., Thomson, J. A. and Andrews, P. W. 2002. Surface antigens of human embryonic stem cells: Changes upon differentiation in culture, *J Anat* 200:249–258.

Draper, J. S., Smith, K., Gokhale, P., Moore, H. D., Maltby, E., Johnson, J., Meisner, L., Zwaka, T. P., Thomson, J. A. and Andrews, P. W. 2004. Recurrent gain of chromosomes 17q and 12 in cultured human embryonic stem cells, *Nature Biotechnol* 22:53–54.

Elkabetz, Y., Panagiotakos, G., Al Shamy, G., Socci, N. D., Tabar, V. and Studer, L. 2008. Human ES cell-derived neural rosettes reveal a functionally distinct early neural stem cell stage, *Genes Dev* 22:152–165.

Ellerström, C., Hyllner, J. and Strehl, R. 2010. Single cell enzymatic dissociation of human embryonic stem cells: A straightforward, robust, and standardized culture method, *Methods Mol Biol* 584:121–134.

Ellis, J., Bruneau, B. G., Keller, G., Lemischka, I. R., Nagy, A., Rossant, J., Srivastava, D., Zandstra, P. W. and Stanford, W. L. 2009. Alternative induced pluripotent stem cell characterization criteria for *in vitro* applications, *Cell Stem Cell* 4:198–199; author reply 202.

Enver, T., Pera, M., Peterson, C. and Andrews, P. W. 2009. Stem cell states, fates, and the rules of attraction, *Cell Stem Cell* 4(5):387–397.

Fenderson, B. A., Andrews, P. W., Nudelman, E., Clausen, H. and Hakomori, S. 1987. Glycolipid core structure switching from globo- to lacto- and ganglio-series during retinoic acid-induced differentiation of TERA-2-derived human embryonal carcinoma cells, *Dev Biol* 122:21–34.

Fernandes, A. M., Marinho, P. A., Sartore, R. C., Paulsen, B. S., Mariante, R. M., Castilho, L. R. and Rehen, S. K. 2009. Successful scale-up of human embryonic stem cell production in a stirred microcarrier culture system, *Braz J Med Biol Res* 42: 515–522.

Ferreira, L. S., Gerecht, S., Fuller, J., Shieh, H. F., Vunjak-Novakovic, G. and Langer, R. 2007. Bioactive hydrogel scaffolds for controllable vascular differentiation of human embryonic stem cells, *Biomaterials* 28:2706–2717.

Fraser, M. J., Ciszczon, T., Elick, T. and Bauser, C. 1996. Precise excision of TTAA-specific lepidopteran transposons piggyBac (IFP2) and tagalong (TFP3) from the baculovirus genome in cell lines from two species of Lepidoptera, *Insect Mol Biol* 5:141–151.

Gerecht, S., Burdick, J. A., Ferreira, L. S., Townsend, S. A., Langer, R. and Vunjak-Novakovic, G. 2007. Hyaluronic acid hydrogel for controlled self-renewal and differentiation of human embryonic stem cells, *Proc Natl Acad Sci U S A* 104(27): 11298–11303.

Hanna, J., Markoulaki, S., Schorderet, P., Carey, B. W., Beard, C., Wernig, M., Creyghton, M. P., Steine, E. J., Cassady, J. P., Foreman, R., Lengner, C. J., Dausman, J. A. and Jaenisch, R. 2008. Direct reprogramming of terminally differentiated mature B lymphocytes to pluripotency, *Cell* 133(2):250–264.

Hemberger, M., Dean, W. and Reik, W. 2009. Epigenetic dynamics of stem cells and cell lineage commitment: Digging Waddington's canal, *Nat Rev Mol Cell Biol* 10(8): 526–537.

Hotta, A., Cheung, A. Y., Farra, N., Vijayaragavan, K., Seguin, C. A., Draper, J. S., Pasceri, P., Maksakova, I. A., Mager, D. L., Rossant, J., Bhatia, M. and Ellis, J. 2009. Isolation of human iPS cells using EOS lentiviral vectors to select for pluripotency, *Nat Methods* 6(5):370–376.

Houbaviy, H. B., Murray, M. F. and Sharp, P. A. 2003. Embryonic stem cell-specific microRNAs, *Dev Cell* 5(2):351–358.

Hovatta, O., Mikkola, M., Gertow, K., Stromberg, A. M., Inzunza, J., Hreinsson, J., Rozell, B., Blennow, E., Andang, M. and Ahrlund-Richter, L. 2003. A culture system using human foreskin fibroblasts as feeder cells allows production of human embryonic stem cells, *Hum Reprod* 18(7):1404–1409.

Kehoe, D. E., Jing, D., Lock, L. T. and Tzanakakis, E. M. 2009. Scalable stirred-suspension bioreactor culture of human pluripotent stem cells, *Tissue Eng Part A* 16(2):405–421.

Krawetz, R, Taiani, J. T., Liu, S., Meng, G., Li, X., Kallos, M. S. and Rancourt, D. 2009. Large-scale expansion of pluripotent human embryonic stem cells in stirred suspension bioreactors, *Tissue Eng Part C Methods* Sep 8. Epub ahead of print.

Kyung-Dal, C., Junying, Y., Kim, S. O., Giorgia, S., William, R., Maxim, V., James, T. and Igor, S. 2009. Hematopoietic and endothelial differentiation of human induced pluripotent stem cells, *Stem Cells* 27(3):559–567.

Lee, T. I., Jenner, R. G., Boyer, L. A., *et al.* 2006. Control of developmental regulators by polycomb in human embryonic stem cells, *Cell* 125(2): 301–313.

Levenberg, S., Huang, N. F., Lavik, E., Rogers, A. B., Itskovitz-Eldor, J. and Langer, R. 2003. Differentiation of human embryonic stem cells on three-dimensional polymer scaffolds, *Proc Natl Acad Sci U S A* 100(22):12741–12746.

Li, W., Wei, W., Zhu, S., Zhu, J., Shi, Y., Lin, T., Hao, E., Hayek, A., Deng, H. and Ding, S. 2009. Generation of rat and human induced pluripotent stem cells by combining genetic reprogramming and chemical inhibitors, *Cell Stem Cell* 4(1):16–19.

Lock, L. T. and Tzanakakis, E. S. 2009. Expansion and differentiation of human embryonic stem cells to endoderm progeny in a microcarrier stirred-suspension culture, *Tissue Eng Part A* 15(8):2051–2063.

Lowry, W. E., Richter, L., Yachechko, R., Pyle, A. D., Tchieu, J., Sridharan, R., Clark, A. T. and Plath, K. 2008. Generation of human induced pluripotent stem cells from dermal fibroblasts, *Proc Natl Acad Sci U S A* 105(8):2883–2888.

Lu, M., Kardel, M. D., O'Connor, M. D. and Eaves, C. J. 2009. Enhanced generation of hematopoietic cells from human hepatocarcinoma cell-stimulated human embryonic and induced pluripotent stem cells, *Exp Hematol* 37(8):924–936.e4.

Ludwig, T. E., Levenstein, M. E., Jones, J. M., Berggren, W. T., Mitchen, E. R., Frane, J. L., Crandall, L. J., Daigh, L. J., Conard, K. R., Piekarczyk, M. S., Llanas, R. A., Thomson, J. A. 2006. Derivation of human embryonic stem cells in defined conditions, *Nat Biotechnol* 24(2):185–187.

MacArthur, B. D., Ma'ayan, A. and Lemischka, I. R. 2008. Toward stem cell systems biology: From molecules to networks and landscapes, *Cold Spring Harb Symp Quant Biol* 73, 211–215.

MacArthur, B. D., Ma'ayan, A. and Lemischka, I. R. 2009. Systems biology of stem cell fate and cellular reprogramming, *Nat Rev Mol Cell Biol* 10(10):672–681.

Maherali, N. and Hochedlinger, K. 2008. Guidelines and techniques for the generation of induced pluripotent stem cells, *Cell Stem Cell* 3(6):595–605.

Marson, A., Levine, S. S., Cole, M. F., Frampton, G. M., Brambrink, T., Johnstone, S., Guenther, M. G., Johnston, W. K., Wernig, M., Newman, J., Calabrese, J. M., Dennis, L. M., Volkert, T. L., Gupta, S., Love, J., Hannett, N., Sharp, P. A., Bartel,

D. P., Jaenisch, R. and Young, R. A. 2008. Connecting microRNA genes to the core transcriptional regulatory circuitry of embryonic stem cells, *Cell* 134(3):521–533.

Meissner, A., Wernig, M. and Jaenisch, R. 2007. Direct reprogramming of genetically unmodified fibroblasts into pluripotent stem cells, *Nat Biotechnol* 25(10):1177–1181.

Mitalipova, M. M., Rao, R. R., Hoyer, D. M., Johnson, J. A., Meisner, L. F., Jones, K. L., Dalton, S. and Stice, S. L. 2005. Preserving the genetic integrity of human embryonic stem cells, *Nat Biotechnol* 23(1):19–20.

Nakagawa, M., Koyanagi, M., Tanabe, K., Takahashi, K., Ichisaka, T., Aoi, T., Okita, K., Mochiduki, Y., Takizawa, N. and Yamanaka, S. 2008. Generation of induced pluripotent stem cells without Myc from mouse and human fibroblasts. *Nat Biotechnol* 26(1): 101–106.

Nagaoka, M., Si-Tayeb, K., Akaike, T. and Duncan, S. A. 2010. Culture of human pluripotent stem cells using completely defined conditions on a recombinant E-cadherin substratum, *BMC Dev Biol* 10(1):60

Nie, Y., Bergendahl, V., Hei, D. J., Jones, J. M. and Palecek, S. P. 2009. Scalable culture and cryopreservation of human embryonic stem cells on microcarriers, *Biotechnol Prog* 25(1):20–31.

Oh, S. K., Chen, A. K., Mok, Y., Chen, X., Lim, U. M., Chin, A., Choo, A. B. and Reuveny, S. 2009. Long-term microcarrier suspension cultures of human embryonic stem cells, *Stem Cell Res* Mar 4. Epub ahead of print.

Okita, K., Nakagawa, M., Hyenjong, H., Ichisaka, T. and Yamanaka, S. 2008. Generation of mouse induced pluripotent stem cells without viral vectors, *Science* (3225903): 949–953.

Park, I. H. and Daley, G. Q. 2009. Human iPS cell derivation/reprogramming, *Curr Protocols in Stem Cell Biol* Chapter 4, Unit 4A.1.

Phillips, B. W., Horne, R., Lay, T. S., Rust, W. L., Teck, T. T. and Crook, J. M. 2008. Attachment and growth of human embryonic stem cells on microcarriers, *J Biotechnol* 138(1–2):24–32.

Rao, R. R., Calhoun, J. D., Qin, X., Rekaya, R., Clark, J. K. and Stice, S. L. 2004. Comparative transcriptional profiling of two human embryonic stem cell lines, *Biotechnol Bioeng* 88(3):273–286.

Reubinoff, B. E., Pera, M. F., Fong, C. Y., Trounson, A. and Bongso, A. 2000. Embryonic stem cell lines from human blastocysts: Somatic differentiation *in vitro*, *Nat Biotechnol* 18(4):399–404.

Richards, M., Fong, C. Y., Chan, W. K., Wong, P. C. and Bongso, A. 2002. Human feeders support prolonged undifferentiated growth of human inner cell masses and embryonic stem cells, *Nat Biotechnol* 20(9):933–936.

Richards, M., Tan, S., Fong, C. Y., Biswas, A., Chan, W. K. and Bongso, A. 2003. Comparative evaluation of various human feeders for prolonged undifferentiated growth of human embryonic stem cells, *Stem Cells* 21(5):546–556.

Roberts, R. M., Telugu, B. P. and Ezashi, T. 2009. Induced pluripotent stem cells from swine (Sus scrofa): Why they may prove to be important, *Cell Cycle* 8(19):3078–3081.

Saravanan, K., Bennett, G. N., Michaela, P., Joy, A. U., Laura, R., Anne, L., Anne, E. C., Amander, T. C., Steve, A. G., Kathrin, P., Martina, W. P., Harley, I. K. and William, E. L. 2009. Directed differentiation of human-induced pluripotent stem cells generates active motor neurons, *Stem Cells* 27(4):806–811.

Silva, J., Barrandon, O., Nichols, J., Kawaguchi, J., Theunissen, T. W. and Smith, A. 2008. Promotion of reprogramming to ground state pluripotency by signal inhibition, *PLoS Biol* 6(10):e253.

Silva, J. and Smith, A. 2008. Capturing pluripotency, *Cell* 132(4):532–536.

Siti-Ismail, N., Bishop, A. E., Polak, J. M. and Mantalaris, A. 2008. The benefit of human embryonic stem cell encapsulation for prolonged feeder-free maintenance, *Biomaterials* 29(29):3946–3952.

Smith, J. R., Vallier, L., Lupo, G., Alexander, M., Harris, W. A. and Pedersen, R. A. 2008. Inhibition of Activin/Nodal signaling promotes specification of human embryonic stem cells into neuroectoderm, *Dev Biol* 313(1):107–117.

Smith, K. P., Luong, M. X. and Stein, G. S. 2009. Pluripotency: Toward a gold standard for human ES and iPS cells, *J Cell Physiol* 220(1):21–29.

Song, Z., Cai, J., Liu, Y., Zhao, D., Yong, J., Duo, S., Song, X., Guo, Y., Zhao, Y., Qin, H., Yin, X., Wu, C., Che, J., Lu, S., Ding, M. and Deng, H. 2009. Efficient generation of hepatocyte-like cells from human induced pluripotent stem cells, *Cell Res* 19(11): 1233–1242.

Stadtfeld, M., Nagaya, M., Utikal, J., Weir, G. and Hochedlinger, K. 2008. Induced pluripotent stem cells generated without viral integration, *Science* 322(5903):945–949.

Stadtfeld, M. and Hochedlinger, K. 2009. Without a trace? PiggyBac-ing toward pluripotency, *Nat Methods* 6(5):329–330.

Sullivan, J, G., Hay, D. C., Park, I. H., Fletcher, J., Hannoun, Z., Payne, C. M., Dalgetty, D., Black, J. R., Ross, J. A., Samuel, K., Wang, G., Daley, G. Q., Lee, J. H., Church, G. M., Forbes, S. J., Iredale, J. P. and Wilmut, I. 2010. Generation of functional human hepatic endoderm from human induced pluripotent stem cells, *Hepatology* 51(1): 329–35.

Takahashi, K., Tanabe, K., Ohnuki, M., Narita, M., Ichisaka, T., Tomoda, K. and Yamanaka, S. 2007. Induction of pluripotent stem cells from adult human fibroblasts by defined factors, *Cell* 131(5):861–872.

Taura, D., Noguchi, M., Sone, M., Hosoda, K., Mori, E., Okada, Y., Takahashi, K., Homma, K., Oyamada, N., Inuzuka, M., Sonoyama, T., Ebihara, K., Tamura, N., Itoh, H., Suemori, H., Nakatsuji, N., Okano, H., Yamanaka, S. and Nakao, K. 2009. Adipogenic differentiation of human induced pluripotent stem cells: Comparison with that of human embryonic stem cells, *FEBS Letters* 583(6): 1029–1033.

Tesar, P. J., Chenoweth, J. G., Brook, F. A., Davies, T. J., Evans, E. P., Mack, D. L., Gardner, R. L. and McKay, R. D. 2007. New cell lines from mouse epiblast share defining features with human embryonic stem cells, *Nature* 448(7150):196–199.

Thomson, A. J., Itskovitz-Eldor, J., Shapiro, S. S., Waknitz, M. A., Swiergiel, J. J., Marshall, V. S. and Jones, J. M. 1998. Embryonic stem cell lines derived from human blastocysts, *Science* 282(5391):1145–1147.

Totonchi, M., Taei, A., Seifinejad, A., Tabebordbar, M., Rassouli, H., Farrokhi, A., Gourabi, H., Aghdami, N., Hosseini-Salekdeh, G. and Baharvand, H. 2009. Feeder- and serum-free establishment and expansion of human induced pluripotent stem cells, *The Int J Dev Biol* 54(5):877–886.

Toyooka, Y., Shimosato, D., Murakami, K., Takahashi, K. and Niwa, H. 2008. Identification and characterization of subpopulations in undifferentiated ES cell culture, *Development* 135(5):909–918.

Vierbuchen, T., Ostermeier, A., Pang, Z. P., Kokubu, Y., Sudhof, T. C. and Wernig, M. 2010. Direct conversion of fibroblasts to functional neurons by defined factors, *Nature* 463(7284):1035–1041.

Wang, Q., Fang, Z. F., Jin, F., Lu, Y., Gai, H. and Sheng, H. Z. 2005. Derivation and growing human embryonic stem cells on feeders derived from themselves, *Stem Cells* 23(9): 1221–1227.

Woltjen, K., Michael, I. P., Mohseni, P., Desai, R., Mileikovsky, M., Hamalainen, R., Cowling, R., Wang, W., Liu, P., Gertsenstein, M., Kaji, K., Sung, H. K. and Nagy, A. 2009. piggyBac transposition reprograms fibroblasts to induced pluripotent stem cells, *Nature* 458(7239):766–770.

Xu, C., Inokuma, M. S., Denham, J., Golds, K., Kundu, P., Gold, J. D. and Carpenter, M. K. 2001. Feeder-free growth of undifferentiated human embryonic stem cells, *Nat Biotechnol* 19(10):971–974.

Ying, Q. L., Wray, J., Nichols, J., Batlle-Morera, L., Doble, B., Woodgett, J., Cohen, P. and Smith, A. 2008. The ground state of embryonic stem cell self-renewal, *Nature* 453(7194):519–523.

Yu, J., Vodyanik, M. A., Smuga-Otto, K., Antosiewicz-Bourget, J., Frane, J. L., Tian, S., Nie, J., Jonsdottir, G. A., Ruotti, V., Stewart, R., Slukvin, I. I. and Thomson, J. A. 2007. Induced pluripotent stem cell lines derived from human somatic cells, *Science* 318(5858):1917–1920.

Yu, J., Hu, K., Smuga-Otto, K., Tian, S., Stewart, R., Slukvin, I. I. and Thomson, J. A. 2009. Human induced pluripotent stem cells free of vector and transgene sequences, *Science* 324(5928):797–801.

Zhang, J., Wilson, G. F., Soerens, A. G., Koonce, C. H., Yu, J., Palecek, S. P., Thomson, J. A. and Kamp, T. J. 2009. Functional cardiomyocytes derived from human induced pluripotent stem cells, *Circ Res* 104(4):e30–e41.

Zhao, X. Y., Li, W., Lv, Z., Liu, L., Tong, M., Hai, T., Hao, J., Guo, C. L., Ma, Q. W., Wang, L., Zeng, F. and Zhou, Q. 2009. iPS cells produce viable mice through tetraploid complementation, *Nature* 461(7260):86–90.

Zwi, L., Caspi, O., Arbel, G., Huber, I., Gepstein, A., Park, I. H. and Gepstein, L. 2009. Cardiomyocyte differentiation of human induced pluripotent stem cells, *Circulation* 120(15):1513–1523.

The Importance of Physiologically Inspired Physicochemical Parameters on Hematopoietic Stem-Cell Maintenance and Lineage-Specific Differentiation in *Ex Vivo* Cultures

6

Stephan Lindsey and Eleftherios T. Papoutsakis

6.1. Introduction

6.1.1. *The thesis for the chapter*

Stem cell localization reflects the needs of these cells for maintenance, renewal and need-driven differentiation. The stem cell niche is dictated not only by cells that co-occupy the body locale of the stem cells, but also by the physicochemical parameters or nutrient availability of that locale, including the pH, pO_2, physical forces due to flow (or lack thereof), and the amount (i.e. abundance, lack or limited supply) of nutrient and growth factors. These issues have largely remained out of the radar screen of modern stem cell biology and bioengineering, but they now emerge as potent, if not indispensable, modulators of stem cell fate. Here we focus on issues associated with *ex vivo* expansion of hematopoietic stem and progenitor cells (HSPCs).

6.1.2. *Hematopoietic culture applications and challenges, and the importance of the physicochemical parameters of the stem cell niche*

Stem and progenitor cells from the bone marrow (BM), mobilized peripheral blood (mPB), or cord blood (CB) can be used to restore hematopoietic activity in patients whose BM has been ablated by chemotherapy (Bertolini *et al.*, 1997; Brugger *et al.*,

1995; Decaudin *et al.*, 2004; Scheding *et al.*, 2004; Shirvaikar *et al.*, 2008; Srour *et al.*, 1999; Tura *et al.*, 2007; Verfaillie, 2002; Williams *et al.*, 1996). However, current protocols may be limited by donor cell availability and the long time (1.5–4+ weeks) needed to restore neutrophil and platelet counts to normal levels. By reducing the required harvest size, *ex vivo* expansion of hematopoietic progenitors has the potential to allow the following: (1) patients with low BM cellularity to serve as their own donors, (2) replacement of invasive BM harvests (1–2 liters) with an aliquot drawn in a doctor's office, (3) use of a single PB harvest, and (4) use of CB cells for adult transplants. Of equal or greater importance, *ex vivo* expansion has the potential to enhance the transplant outcome. For example, increased doses of granulocytic progenitors have been shown to decrease the variability in patient engraftment and increase the survival rate (Bacigalupo *et al.*, 1995; Perez-Simon *et al.*, 2003). In addition, by producing large numbers of post-progenitors with the potential to rapidly generate mature neutrophils and platelets, it may be possible to completely prevent cytopenias associated with high-dose chemotherapy (Timmins *et al.*, 2009). Finally, *ex vivo* expansion offers the potential for repeated cycles of chemotherapy and cell transplants from a single BM or PB harvest (that is aliquotted and frozen for subsequent culture). Other applications include production of cycling stem and progenitor cells for gene therapy, the expansion of dendritic cells for immunotherapy (Cui *et al.*, 2002; Di Nicola *et al.*, 1998; Saraya and Reid, 1996; Siena *et al.*, 1995), and production of red blood cells and platelets for transfusions (McAdams *et al.*, 1996b; Timmins and Nielsen, 2009). Clinical trials (Bertolini *et al.*, 1997; Brugger *et al.*, 1995; Williams *et al.*, 1996) have demonstrated that *ex vivo*-expanded hematopoietic cells can be safely re-infused to immunocompromised patients who have undergone intensive chemotherapy for cancer.

Hematopoietic cultures remain among the most challenging culture systems because of the intrinsically heterogeneous and highly variable nature of the cultured cells. Heterogeneity and variability derive from both patient-to-patient variability and the complexity of hematopoietic differentiation (Figure 6.1). While some potential applications can be realized using standard hematopoietic culture techniques, effective control of the culture environment will be required to consistently produce the cell populations needed for the different applications noted above. There has been extensive research regarding the effects of hematopoietic growth factors (HGFs) on cell expansion and differentiation patterns [reviewed by Kelly *et al.* (2009), McAdams *et al.* (1996a), Sauvageau *et al.* (2004), and Timmins and Nielsen (2009)]. Beyond cytokines, a fast-accumulating body of literature demonstrates that physiological parameters such as culture pH, pO_2 and flow-induced shear effects can also have dramatic effects, and this is the focus of the chapter. Bioengineering advances have not kept pace with the rapid developments in

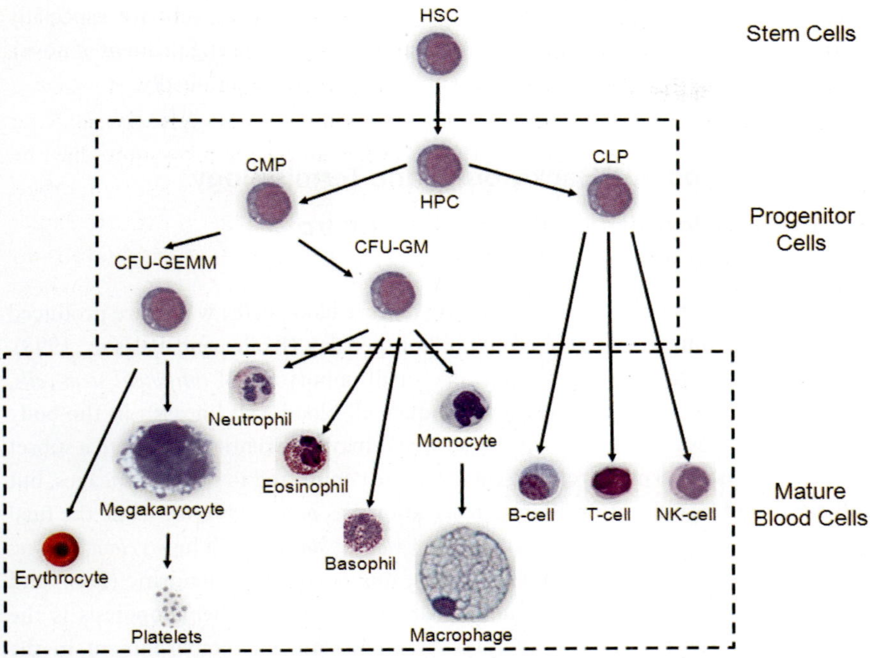

Figure 6.1. The hematopoietic hierarchy showing a few of the differentiation steps for how hematopoietic stem cells in the bone marrow are committed and expand in numbers to generate all myeloid and lymphoid blood cells in the circulation.

hematopoietic cell biology. The true potential of *ex vivo* expansion of hematopoietic cells will only be realized when culture systems or bioreactors of sufficient simplicity, flexibility, and economic efficacy have been developed. Expansion protocols on the horizon are likely to be used in a number of clinical trials in stem, progenitor and mature/differentiated hematopoietic cell transplantation therapies. One challenge for the future will be to further increase the productivity and reliability of the *ex vivo* expansion protocols. In this context, the goal of stem cell bioengineering is to define culture conditions (beyond media and cytokines) to improve the degree of control over the proliferation and differentiation of specific progenitors and blood lineages. Such goals will require increased understanding of the synergism between cytokines and other culture parameters (pH, pO_2, flow effects), and the engagement of synthetic or semisynthetic matrices to present cytokines to cells, thus providing a suitable "chemical" and "elastic" or protective microenvironment for the cells. Such developments should be largely driven by understanding the natural milieu or "niche" of the stem, progenitor and more differentiated cells, as well as the evolutionary forces that have shaped the "niche". These aspects are especially emphasized in this chapter. One, of course, cannot exclude the engagement of novel, completely synthetic niches, but those will be hard to define rationally.

6.2. Hematopoietic Background and Terminology

6.2.1. *Hematopoiesis, stroma and in vitro colony assays*

Hematopoiesis is the process of generating mature blood cells, which are produced in the average human at a rate of 400 billion per day (Koller and Palsson, 1993). Hematopoietic cells originate from a very small population of *totipotent stem cells*, which proliferate and differentiate to produce all blood–cell lineages in the body (Figure 6.1). Less primitive *pluripotent stem cells* may differentiate only into a subset of lineages. The intermediate stage cells that are committed to specific lineages, but still capable of significant proliferation are known as *progenitor cells*, while the final non-proliferative functional cells are termed *mature blood cells*. The *myeloid lineages* include the erythroid (E), granulocytic (G), monocytic (M), dendritic (DC), and megakaryocytic (Mk) lineages. The primary site of *in vivo* hematopoiesis is the red bone marrow (BM). The complex hematopoietic microenvironment in the BM cavities between blood vessels is potentiated by a variety of (mostly) non-hematopoietic supportive cells, collectively known as *stroma*. Stromal cells consist primarily of macrophages, endothelial and adventitial reticular cells, adipocytes, osteoblasts and osteoclasts, and differentiated non-hematopoietic stem cells known as mesenchymal stem cells (MSCs) (Caplan, 2009; Ho *et al.*, 2008; Morrison and Spradling, 2008) that provide both soluble and membrane-bound growth factors to hematopoietic cells. Stromal cells also produce an extracellular matrix that affects proliferation and differentiation. Colony-forming cell (CFC) or colony-forming unit (CFU) assays are *in vitro* assays that are widely used to assess the frequency and number of hematopoietic stem and progenitor cells (Coulombel, 2004). A small number of sample cells is placed in a semisolid medium containing a suitable mixture of growth factors, and incubated for 2 weeks. Cultures are then scored for the presence of colonies of 50 or more cells that originated from a single cell 2 weeks earlier. Colony types are assessed by color, with colonies containing: white cells termed colony-forming unit GM (CFU-GM), and red cells termed burst-forming-unit E (BFU-E). The long-term culture-initiating cell (LTC-IC) assay is the most common assay for more primitive cells. For this assay, sample cells are placed on a supportive stromal cell layer and maintained in culture for 5–8 weeks. After that, cells are placed into a colony assay to measure CFCs. During the 5–8-week culture period,

CFCs differentiate into mature cells, and the more primitive LTC-ICs (believed to be a subset of the stem cells) differentiate into CFCs, which are then detected by the colony assay. A variant of this assay, known as long-term hematopoietic culture-initiating cell (LTHC-IC) (Traycoff *et al.*, 1995) eliminates the use of preformed stroma and offers substantial time savings (4 vs. 8 weeks total time) and improved reproducibility. *In vivo* assays (Gordon, 1993) and employing animal models provide more a sophisticated, clinically relevant assessment of HSC potency and potential.

6.2.2. *Sources of hematopoietic stem and progenitor cells (HSPCs)*

There are three main sources of hematopoietic stem and progenitor cells (Korbling and Anderlini, 2001). BM is collected under general anesthesia via multiple needle aspirations to the sternum and/or pelvis. Cord blood (CB) is obtained non-invasively from the umbilical cord of newborn infants. Mobilized peripheral blood (mPB) cells are generally obtained after a stem cell mobilization, which is achieved by the administration of one of several chemotherapeutic drugs and/or hematopoietic growth factors (HGFs) to the patient. Nucleated cells are then collected from the patients using a blood-processing machine (PB apheresis). The expansion potential of different sources of hematopoietic cells, and even of different samples of the same type of cells, is subject to large variations. CB has the greatest expansion potential (Cairo and Wagner, 1997; Lansdorp *et al.*, 1993), while PB has the greatest variation due to the wide variety of mobilization regimens and disease state of the patients (Janssen, 1995). CD34 is a surface glycoprotein of unknown function that is found on primitive cells from the quiescent stem cells to the highly proliferative progenitors [ca. 1% of collected mononuclear cells (MNCs) (Krause *et al.*, 1996)]. Methods have been developed for the selection of $CD34^+$ cells (de Wynter *et al.*, 1995). When compared to those for MNCs, cultures initiated with $CD34^+$ cells have much greater expansion potential. However, $CD34^+$ selection is expensive and often results in significant cell loss. A more recently explored source of hematopoietic cells is the placenta (Dzierzak and Robin, 2010), which is an available, otherwise wasted tissue like the umbilical cord. Finally, in the last few years, embryonic stem cells (ESCs) and induced pluripotent cells (iPCs) are explored as a longer-term, more accessible and stable source of HSPCs (Carpenter *et al.*, 2003; Choi *et al.*, 2009; Lengerke and Daley, 2010; Wang *et al.*, 2005). Like other applications of ESCs and iPCs, exploration of these cells as a practical source of HSPCs is still in its infancy, but is the focus of an ever-increasing research activity.

6.2.3. *Hematopoietic growth factors (HGFs) and cytokines*

While stroma-containing cultures produce their own cytokines, stroma-free (commonly called "suspension") cultures require the addition of exogenous HGFs (Audet *et al.*, 2002; Kelly *et al.*, 2009; Panuganti *et al.*, 2010; Sauvageau *et al.*, 2004). Cytokines essential for cell survival and proliferation are interleukin 3 (IL-3), stem cell factor (SCF), and/or flt3 ligand (FL). IL-3 prevents programmed cell death (apoptosis) of committed progenitor cells, and SCF (Brandt *et al.*, 1994) and FL (Takahira *et al.*, 1996) prevent apoptosis of early progenitor cells. FL is structurally and functionally similar to SCF except that it does not act on the E lineage (McKenna *et al.*, 1995). IL-6 acts in concert with other cytokines to induce cycling of additional stem and progenitor cells (Leary *et al.*, 1988). Thus, SCF/IL-3/IL-6 has become a common cytokine combination for broad expansion across multiple hematopoietic lineages. Granulocyte colony-stimulating factor (G-CSF) and IL-11 have also been shown to bring additional dormant primitive hematopoietic cells into the cell cycle (Leary *et al.*, 1992). GM-CSF is a broad-acting growth factor that induces proliferation of G, M, DC and E progenitors (Clark and Kamen, 1987). Erythropoietin (Epo) (Fisher, 2003; Kaushansky, 2006), thrombopoietin (Tpo) (de Sauvage *et al.*, 1994; Kaushansky, 2006) and G-CSF (Clark and Kamen, 1987; Kaushansky, 2006) are powerful proliferation and maturation factors for the E, Mk, and G lineages, respectively. Yet, in addition to the G-CSF mentioned above, Tpo and Epo have been found to promote the expansion of primitive hematopoietic stem cells, and thus, are frequently used in that context as well, not merely for promoting the expansion of their respective lineages. There are also cytokines which inhibit or enhance hematopoietic cell proliferation or differentiation (depending on the target cells and other cytokines used), including TGF-β (Hatzfeld *et al.*, 1991) and TNF-α (Rusten *et al.*, 1994).

6.3. The Physiology Inspires: Bone Marrow Organization, pO_2 and pH Levels, and the Impact of Flow-Induced Shear Flow in Hematopoiesis

6.3.1. *Physiological measurements and observations*

The organization of BM has been studied using primarily long bones (Figure 6.2) from rodents and rabbits (Chow *et al.*, 2001a; Chow *et al.*, 2001b; Kaplan *et al.*, 2007; Lichtman, 1984; Mayani *et al.*, 1992; Tavassoli, 1974). The vasculature is dominated by the venous system, which is made up of vascular sinuses draining

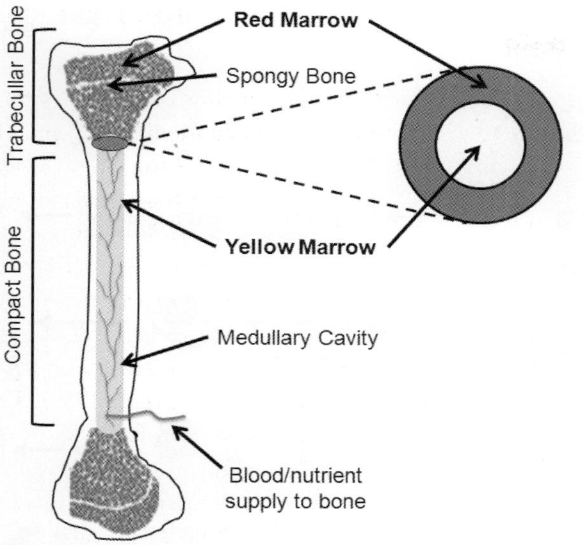

Figure 6.2. The long bone is the site of the red and yellow (fatty) bone marrow, the predominant site of adult hematopoiesis. The most primitive and thus, hematopoietic potent stem cells are assumed to be located mostly at the long bone endosteum, mostly in the red marrow. Copyright of the two parts of the figure is public: From Wikipedia.

into the great central vein. In adults, the area around the central artery and central veins is occupied by the fatty yellow marrow (found mostly in the compact part area of the long bone), while the hematopoietically active red marrow resides near the bone endosteum and (especially) in the trabecular part of the long bone (Morrison and Spradling, 2008; ter Huurne *et al.*, 2010; Xie *et al.*, 2009) (Figure 6.2). Hematopoiesis takes place in the extravascular spaces between the sinuses. The venous system supports the passage of maturing hematopoietic cells, which are being released from the BM hematopoietic compartment (BMHC). The thin sinus wall is composed of endothelial cells, which line the sinus lumen, and adventitial reticular cells, which form an incomplete abluminal outer coat. Hematopoietic cells must cross the sinus wall to enter the circulatory blood. The most primitive stem and progenitor cells reside in the extravascular (frequently referred to as perivascular) areas, the furthest away from the sinuses and next to the endosteum. As cells differentiate, they move closer to the sinus lining (Figure 6.3). Specifically, the (motile) granulocytes (G cells) typically develop away from the sinus wall in G loci and as they mature, they move near the wall where they develop microvilli that facilitate their crossing into circulation. The (immotile) erythroid (E) cells develop

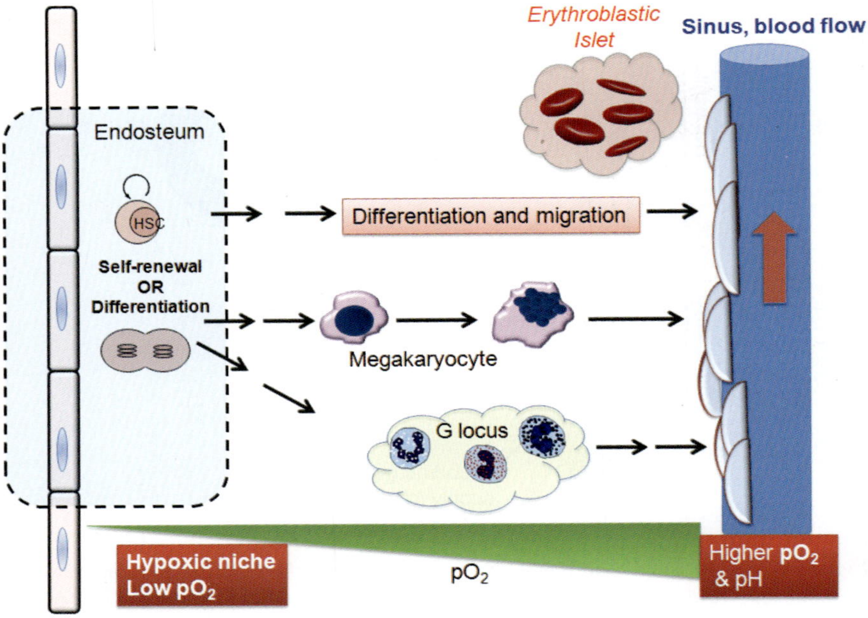

Figure 6.3. Schematic representation of the differentiation and migration of hematopoietic stem cells from the primitive endosteal niche (low, hypoxic pO_2) to a bone marrow sinus (higher pO_2 and pH) for release of differentiated cells into the blood circulation.

in erythroblastic islets (one to two circular layers of E cells surrounding one or two macrophages; diameter of ca. 50–80 mm) near the sinus wall, with mature E cells bordering the vascular wall and immature cells (i.e, basophilic erythroblasts) away from the wall. Megakaryocytes (Mks) differentiate in the BM vasculature: they undergo a variation of the normal cell cycle, termed endomitosis, to form polyploid cells ($\geq$8N DNA content) (Vitrat *et al.*, 1998). Committed Mks migrate from the hematopoietic BM compartment towards the endothelial lining of the BM sinusoids, where they mature and extend long, branched cytoplasmic protrusions (termed proplatelets) through gaps of the endothelium into the vasculature (Hattori *et al.*, 2003; Schulze and Shivdasani, 2005) (Figures 6.3 and 6.4). Mk cells encounter mechanical stresses as they deform to penetrate the gaps of the sinusoid walls, and shear forces by the exposure to blood flow. The pulmonary circulation is another important site of Mk maturation and platelet biogenesis (Dunois-Larde *et al.*, 2009): Mks may enter the BM circulation and reach the lungs where they shed proplatelets. As such, Mk cells encounter shear forces in circulation as well as mechanical strain in the lung vasculature. Such stresses/forces were recently shown to be essential

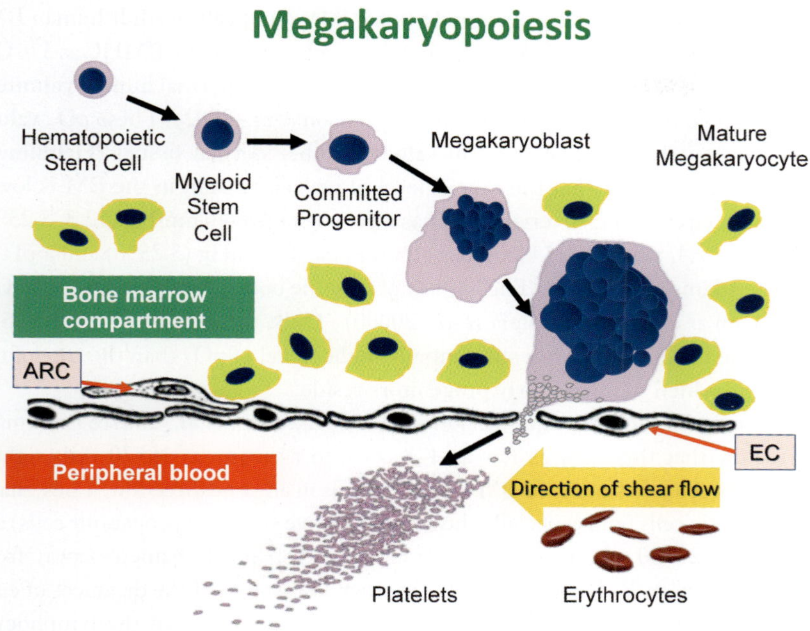

Figure 6.4. Bone marrow megakaryocytes migrate to and mature near the bone marrow sinuses, where they penetrate the endothelial sinus wall and release proplatelets into circulation under shear flow. A similar process takes places in the lung vasculature from less mature Mks that enter circulation and complete the Mk maturation process in the lungs.

for Mk maturation and proplatelet/platelet biogenesis (Dunois-Larde *et al.*, 2009; Junt *et al.*, 2007). Additionally, Mk maturation is enhanced by oxidative stress (as demonstrated by the positive impact of high O_2 tensions on Mk maturation) as has been shown by us (Mostafa *et al.*, 2000a) and, most recently, by others (Motohashi *et al.*, 2010). Thus, Mk maturation and platelet release appear to be stress-induced processes.

There is a wide variation in pO_2 levels in different tissues of the human/animal body and even within a single tissue due to frequent significant spatial heterogeneities in tissue organization (Kietzmann *et al.*, 1999; Thomas *et al.*, 1994). In normal mouse subcutaneous tissue, the mean measured pO_2 was 30 mmHg and values as low as 12 mmHg were recorded (Helmlinger *et al.*, 1997). Similarly, low pO_2 values have been widely reported for other normal tissues [e.g. Vaupel (1996)]. The lowest pO_2 value in a tissue is found furthest from the vessel wall (Helmlinger *et al.*, 1997). There are few direct pO_2 and pH measurements in human BM. Ishikawa and Ito (1988) reported a mean pO_2 of 52 ± 15 mmHg (4.85–8.8%) in healthy adult human BM. Pennathur-Das and Levitt (1987) reported the average pO_2 in the BMHC as 5% O_2, while other measurements of bone marrow aspirates from normal human volunteer donors yielded mean pO_2 values of 6.6% (Harrison *et al.*, 2002). These pO_2 values are approximately the same as mean values of other normal tissues (Helmlinger *et al.*, 1997), and show that the overall level of oxygen tension in the BM is lower than that in circulation. Arterial blood has a pO_2 of ca. 95–40 mmHg (12.5–5.25%) and a pH of 7.4, while venal blood has a pO_2 of ca. 40 mmHg (5.25%) and a pH of 7.35 (McAdams *et al.*, 1997). The highest pO_2 in the body is in the lungs (16% O_2) (Kietzmann *et al.*, 1999; Mostafa *et al.*, 2000b). Thus, the sites of Mk maturation and platelet production have a substantially higher level of pO_2 than the core of the BMHC, in which stem and early progenitors reside.

Marrow microphotographs (Chow *et al.*, 2001a; Chow *et al.*, 2001b; Lichtman, 1981) show that there are as few as 4–8 cells, to as many as 16–20 cells tightly packed between two sinuses or BM arterial vessels in any one direction. Thus, many hematopoietic cells (and especially the most primitive stem and progenitor cells) are 2–8 cells (or 20–80 μm, assuming a conservative 10-μm cell diameter) away from the closest sinus or BM arterial vessels. Another assessment of the distances of cells from sinuses or arteries is the size (80–1200 μm in diameter) of the lymphocyte nodules (these nodules are between sinuses), in which lymphocytes are organized in the hematopoietic compartment of the human BM (Chow *et al.*, 2001a; Chow *et al.*, 2001b). The non-uniform vasculature of BM suggests spatial variations in pH and pO_2 (Chow *et al.*, 2001a; Chow *et al.*, 2001b; McAdams *et al.*, 1997). It has been demonstrated experimentally that pH drops from 7.35 to ca. 7.1 within about 25 μm from a blood vessel in normal subcutaneous tissue (rabbit) (Martin and Jain, 1994). Finally, it should be noted that the lowest pH and pO_2values are observed the furthest away from the vessel wall (Helmlinger *et al.*, 1997). In view of the fact that hematopoietic cells are metabolically very active (Chow *et al.*, 2001a; Chow *et al.*, 2001b; Koller and Palsson, 1993), substantial pH and pO_2 (and possibly other nutrients such as glucose or cytokines) gradients exist in the BM hematopoietic compartment (Figure 6.3). This is further supported by several indirect studies, where it was noted that the most primitive stem cells that give rise to BM regeneration are relatively resistant to ionizing radiation (Allalunis *et al.*, 1983) [this is similar to the radioresistance of hypoxic cells in solid tumors (Helmlinger *et al.*, 1997)]. It was thus suggested several years ago (Allalunis *et al.*, 1983) that "a hypoxic component of hematopoietic cells could be important in the regenerative process within the marrow after such myelotoxic trauma". The most recent studies discussed in detail below have now established this hypothesis as a fact. In any case, it is important to note that the prevailing *mean BM* (i.e. blood in the arterial and venus system, and the hematopoietic tissue) pO_2 is about one third of the normal atmospheric pO_2 (i.e. 5–7% vs. 21% O_2). Thus, the standard practice of culturing hematopoietic cells in 20% O_2 incubators creates a hyperoxic environment, which may be detrimental to the cells. We note that most "low" pO_2 studies (see below) on hematopoietic cells have been carried out under 5% O_2, a value which better approximates the likely pO_2 value in the BM hematopoietic compartment near the sinus wall.

In summary, experimental evidence suggests that the pH, pO_2, and nutrient/cytokine concentration conditions in the BMHC are quite different from the conditions prevailing in the circulating blood and from those used in cell culture experiments and processes. We had proposed several years ago (Chow *et al.*, 2001a; Chow *et al.*, 2001b; Koller *et al.*, 1992c; LaIuppa *et al.*, 1998; McAdams *et al.*, 1997; McAdams *et al.*, 1998; Yang *et al.*, 2002) that pH and pO_2 gradients play an important role in cell differentiation and proliferation of hematopoietic cells in synergy with HGFs and cytokines, and had hypothesized that these effects can be exploited for practical *ex vivo* expansion applications. Both hypotheses have been now extensively supported by recent work in our laboratory and in many others, as detailed below.

6.3.2. *Modeling to estimate pO_2 and pH levels and gradients in the BM hematopoietic compartment (BMHC)*

6.3.2.1. *Oxygen*

As discussed above, the BM is a tissue of complex architectural organization, which includes granulopoietic loci, erythroblastic islets, and lymphocytic nodules. Chow *et al.* (2001a; 2001b) developed homogeneous and heterogeneous/composite Kroghian models to estimate pO_2 distributions in the BM hematopoietic compartment (BMHC) and to conservatively simulate pO_2-limited cellular architectures. These models were built based on biophysical and metabolic data of hematopoietic cells and characteristics of the BM physiology. For the homogeneous Krogh models, they constructed a tissue cylinder solely occupied by various BM cell types. Krogh cylinders occupied by granulocytic progenitors (the most metabolically active stage of the most abundant cell type) provide a physiologically relevant limiting case of the smallest cylinder leading to oxygen limitations, while adipocyte-occupied cylinders represent the other extreme of the thickest cylinders leading to oxygen limitation. Although the number of possible cellular architectures is large, all simulated pO_2 profiles fall between those two extreme cases. For cellular architectures of physiological significance, more complex biophysical transport Kroghian models were considered, such as multilayer Kroghian models and two-dimensional Kroghian models. These models are built to simulate complex cellular architectures in which erythroid clusters or individual cells (megakaryocytes or adipocytes) are located in the BMHC, predominantly occupied by mature granulocytes. pO_2 distributions in colony-type cellular arrangements (erythroblastic islets, granulopoietic loci, and lymphocytic nodules) were examined by multilayer Kroghian models. Simulation results based on these models show that most hematopoietic progenitors experience low pO_2 values, and that the most primitive stem cells, which are located even further away from BM sinuses, are likely located in a very low pO_2 environment, commonly referred to as hypoxic.

6.3.2.2. *pH*

Although pH gradients have been modeled using a similar effective diffusion and reaction model (Martin and Jain, 1994), this is not likely to prove effective in modeling pH gradients in tissues in view of the complexities associated with the bicarbonate buffer, and CO_2 and lactate production and transport. Since the hematopoietic progenitors are very glycolytic and are metabolically almost as active as hybridoma and tumor cells (Chow *et al.*, 2001a), and since lactate formation is the main reason (Carlsson and Acker, 1988) for the generation of pH gradients in spheroids, it is possible to estimate upper and lower limits for pH gradients from analogous profiles in primary tissues (Helmlinger *et al.*, 1997) and tumors (Gorlach and Acker, 1994; Helmlinger *et al.*, 1997). There is a reasonable correlation between pO_2 and pH in primary subcutaneous tissue (Helmlinger *et al.*, 1997), and there is a good correlation between mean pO_2 and pH profiles in tumors *in vivo* (Helmlinger *et al.*, 1997) or *ex vivo* (Gorlach and Acker, 1994). As would be expected, the slope $\Delta pH/\Delta pO_2$ appears to depend on cell metabolism. If this is accepted as generally valid, it is possible to use the metabolic characteristics of hematopoietic cells (Chow *et al.*, 2001a) and the estimated pO_2 gradients (Chow *et al.*, 2001a; Chow *et al.*, 2001b) to estimate pH gradients. One would surely conclude that areas of large ΔpO_2 gradients are also areas of large ΔpH gradients, and that pH levels and ΔpH gradients are likely to play a significant role in regulating stem- and progenitor-cell differentiation.

6.4. The Stem-Cell Niche as Related to Low Oxygen Tension

6.4.1. *The importance of the hypoxic endosteum as a site for stem cell renewal and maintenance*

It is now well accepted that HSCs reside in discrete microenvironmental domains within the bone marrow that are collectively termed the "stem cell niche" (Schofield, 1978). The quiescent state is now accepted as a characteristic and indispensable property for the maintenance of HSCs and likely all primitive stem cells (Arai *et al.*, 2004). Several recent studies suggest that the true and most primitive, and thus potent, HSCs reside next to the endosteal surface of the bone in proximity to stromal cells and osteoblasts, and that hematopoietic differentiation proceeds radially towards the longitudinal axis of the marrow (Nilsson *et al.*, 2001; Schofield, 1978; Winkler *et al.*, 2010; Xie *et al.*, 2009). Levesque *et al.* (Levesque *et al.*, 2007) used confocal laser scanning microscopy to show that, in the murine system, the endosteum at the bone BM interface is hypoxic, with constitutive expression of hypoxia-inducible transcription factor-1 alpha (HIF-1α), and that HSPC mobilization induces expansion of the hypoxic areas in the BM due to the proliferation of myeloid progenitors that occurs during HSPC mobilization. Genetic studies have confirmed that the endosteal region of the bone marrow is the principal site of HSC niche and hematopoiesis (Calvi *et al.*, 2003; Zhang *et al.*, 2003). The unique microenvironment of the endosteum not only represents the anatomical location where the HSCs reside, but also potentiates signaling cascades that maintain definitive stem cell properties and regulates the stem cell pool, presumably by interacting with stroma cells that regulate stem cell function (Dao *et al.*, 2007; Jones and Wagers, 2008; Morrison and Spradling, 2008; ter Huurne *et al.*, 2010; Zhang *et al.*, 2003). In addition, residence of adult HSCs in this BM niches is essential for their quiescence and long-term functions (Schofield, 1978; Wilson and Trumpp, 2006; Morrison and Spradling, 2008; Nilsson *et al.*, 2001; ter Huurne *et al.*, 2010). Arai *et al.* (2004) have shown that that HSCs expressing receptor tyrosine kinase Tie2 are quiescent and anti-apoptotic, and adhere to osteoblasts in the endosteum BM niche. *In vitro*, the interaction of Tie2 with its ligand angiopoietin-1 (Ang-1) induced cobblestone formation, while *in vivo*, it maintained the long-term repopulating activity of HSCs. Ang-1 enhanced the ability of HSCs to become quiescent and induced HSC adhesion to the bone.

As recently discussed (Winkler *et al.*, 2010), it is hypothesized that vascular niches, which are abundantly supplied with nutrients and oxygen by sinusoidal blood, may represent "proliferative niches," whereas endosteal niches, which are generally inaccessible to blood nutrients and maintain a low oxygen concentration, could represent more "quiescent niches" for HSCs (Figure 6.3). Building on the seminal work of Parmar *et al.* (2007), and in order to investigate the functional attributes of these niches, Winkler *et al.* (2010) perfused murine BM with the vital DNA dye Hoechst 33342 *in vivo* before BM cell collection. These experiments aimed to assess the position of HSCs and multi-potent myeloid progenitors relative to blood flow. Independently, Xie *et al.* (2009) used real-time imaging to identify the location and frequency of HSCs in murine long bones. These findings suggest that the most potent HSC niches are low in oxygen concentration and supported by locally produced growth factors due to poor access by blood flow. In summary, the endosteal niches are now firmly established as the hypoxic environment for HSC maintenance and possibly renewal.

6.4.2. *Low oxygen tension or mild hypoxia is an essential element of the stem-cell niche and not only for hematopoietic stem cells*

As discussed above, HSCs and the supporting cells of the stem cell niche found within the endosteum are predominantly located at the lowest end of an oxygen gradient in the BM with the implication that regionally defined hypoxia plays a fundamental role in regulating stem cell function (Mazumdar *et al.*, 2010; Parmar *et al.*, 2007; Winkler *et al.*, 2010). Although hypoxia is not generally encountered in most healthy tissues, as discussed above, BM represents a unique tissue type given its complicated hierarchical organization emanating from a small population of resting stem cells, all in the endosteum area and mostly in the trabecular part of the long bones (Xie *et al.*, 2009). The data discussed above seem to strongly support the hypothesis that stem cell metabolism is regulated by very low oxygen levels and oxygen gradients within the stem-cell niche (Mohyeldin *et al.*, 2010). Supporting this hypothesis, the very primitive $Lin^-Sca^+Kit^+$ (LSK) cells express high levels of glycolytic and oxidative repair proteins, indicating that primitive HSCs cells are adapted to anaerobic metabolism (Unwin *et al.*, 2006). Exposure to high levels of oxygen damages the functions of HSCs (Cipolleschi *et al.*, 1993; Ivanovic *et al.*, 2000a; Parmar *et al.*, 2007), explaining further why HSCs prefer to reside in areas of low pO_2. Yet, the significance of hypoxia signaling in HSCs has been only recently explored. Takubo *et al.* (2010) and Simsek *et al.* (2010) have shown that hypoxia regulates the metabolic state of HSCs and protects their integrity by controlling HIF-1α. Mazumdar *et al.* (2010) have shown that oxygen regulates stem cell maintenance through HIF-1α-controlled Wnt/β-catenin signaling. Tothova *et al.* (2007) have shown that the FoxO family of transcriptional regulators (a subgroup of Forkhead transcription factors which modulate responses to oxidative stress, DNA damage, cell cycle arrest and general homeostasis) are necessary for protecting the HSC compartment from oxidative stress, and that *in vivo* treatment of mice, with defective FoxO gene expression, with the antioxidant N-acetyl-S-cysteine reversed the FoxO gene defects. Taken together, this body of recent work provides molecular understanding for studies published several years ago whereby *ex vivo* maintenance and expansion of primitive HSCs was shown to be greatly improved by maintaining the cell cultures at very low oxygen levels (1–5% O_2) (Cipolleschi *et al.*, 1993; Danet *et al.*, 2003; Ivanovic *et al.*, 2000a; Ivanovic *et al.*, 2000b; Koller *et al.*, 1992a; Koller *et al.*, 1992b; Koller *et al.*, 1992c).

The impact and (likely) necessity for low oxygen tension and/or hypoxia for stem cell maintenance and expansion is now becoming more broadly accepted as being applicable to other stem cells such as the BM mesenchymal stem cells (MSCs)(Grayson *et al.*, 2007; Grayson *et al.*, 2006a; Hirao *et al.*, 2006; Rosova *et al.*, 2008), neuronal stem cells (Mazumdar *et al.*, 2010; Shingo *et al.*, 2001), as well as embryonic stem cells (ESCs) and induced pluripotent cells (iPCs) [recently reviewed by Mohyeldin *et al.* (2010)]. These findings suggest that the stem cell niche of these various tissues is likely anoxic due to mass transfer resistances (diffusion and lack of perfusion) similar to that established in the BM tissue (Chow *et al.*, 2001a; Chow *et al.*, 2001b), as already discussed.

6.5. Beyond the Stem Cell Compartment: Low Oxygen Tension Impacts Proliferation and Differentiation Potential of Various Hematological Lineages, and More Broadly Stem Cell Differentiation

6.5.1. *Phenomenological studies*

Beyond stem cells, oxygen tension can also greatly influence both cell proliferation and differentiation of hematopoietic lineages. The role of pO_2 with respect to desirable outcomes from hematopoietic cultures has been long recognized, starting with hematopoietic colony assays on semisolid media. The size and number of hematopoietic colonies in semisolid medium is significantly enhanced under low pO_2 (Bradley *et al.*, 1978; Broxmeyer *et al.*, 1989; Broxmeyer *et al.*, 1990; Katahira and Mizoguchi, 1987; Maeda *et al.*, 1986; Rich, 1986; Rich and Kubanek, 1982). In early studies in liquid cultures, pO_2 effects have been quite variable, with optimal production reported at 20% O_2 for murine BM cultures (Cipolleschi *et al.*, 1993) and 5% O_2 for CB MNC cultures (Koller *et al.*, 1992a; Koller *et al.*, 1992c). Subsequent, more detailed and better controlled studies have established that the effects of thrombopoietin (Tpo), Epo and G-CSF on megakaryocytes, erythrocytes and granulocytes, respectively, are dependent on pO_2 (Hevehan *et al.*, 2000; LaIuppa *et al.*, 1998; Mostafa *et al.*, 2000b; Mostafa *et al.*, 2001). As discussed, in the case of megakaryocytes (Mks), relatively undifferentiated progenitor cells mature adjacent to BM sinus walls, extend protrusions (proplatelets) into blood vessels, and subsequently release platelets within the sinusoidal space or in lung capillaries (Junt *et al.*, 2007). Platelet release can be viewed as the culmination of terminal Mk differentiation, and because the sites for platelet release have higher levels of pO_2 than the core of the BM, where stem and progenitor cells reside, this example highlights the "oxygen paradigm" as it relates to hematopoietic differentiation. HSCs reside in a low oxygen state and move to a more oxygenated region to differentiate. In support of this hypothesis, Mks generated from primary human $CD34^+$ cells collected from G-CSF mobilized peripheral blood show increased markers of differentiation, such as greater expansion, increased polyploidization, and increased proplatelet formation (Mostafa *et al.*, 2000a). Moreover, pO_2 modulates the expression of cytokine receptors, transcription factors, and lineage-specific markers in cultured human megakaryocytes (Mostafa *et al.*, 2001). During cultures of cord blood and bone marrow-derived mononuclear cells, reduced pO_2 resulted in a substantial increase in both the number and frequency of granulocyte-macrophage colony-forming units (CFU-GM), erythroid burst-forming units (BFU-E), and granulocyte erythrocyte macrophage megakaryocyte colony-forming units (CFU-GEMM) (Koller *et al.*, 1992a). Low pO_2 also stimulates the activation or differentiation of cells generated from bone marrow precursors, including cells of the monocyte-macrophage lineage (Bradley *et al.*, 1978; Broxmeyer *et al.*, 1990; Koller *et al.*, 1992a).

Furthermore, low pO_2 is an integral component of the native BM microenvironment of the human MSCs. Decreased (3%) pO_2 is also a key parameter that influences the *in vitro* proliferative lifespan and differentiation potential of BM-derived MSCs (Grayson *et al.*, 2006b). In these experiments, culturing cells at reduced pO_2 resulted in increased proliferation, but reduced the differentiation capacity of MSCs into adipogenic progeny. Similarly, hypoxia blocks the growth and differentiation of osteoblasts (and thus bone formation), while strongly stimulating osteoclast formation (and thus bone resorption). pO_2 also affects the differentiation of many other stem and progenitor cells, including neuronal (Studer *et al.*, 2000) and cytotrophoblast stem cells, (Genbacev *et al.*, 1996), and ESCs (Ezashi *et al.*, 2005) among others.

6.5.2. *Reactive-oxygen species (ROS) and antioxidants*

Several mechanisms may be responsible for the effects of pO_2 on hematopoietic cultures. The recent work on the importance of mild hypoxia on the maintenance and, possibly, renewal of the most primitive stem cell compartment was discussed above (Cipolleschi *et al.*, 1993; Ivanovic *et al.*, 2000a; Parmar *et al.*, 2007; Takubo *et al.*, 2010; Mazumdar *et al.*, 2010; Simsek *et al.*, 2010). It is clear that ROS play a significant role in terms of both damage and signaling in the stem cell compartment, and apparently also at the more differentiated states of progenitors and beyond (Ito *et al.*, 2006; Nordberg and Arner, 2001; Sauer *et al.*, 2001; Tothova *et al.*, 2007). The literature on molecular signaling of the impact of O_2 or ROS on cells (stem cells and other) has grown enormously and is beyond the scope of this chapter, but a recent review (Naka *et al.*, 2008) provides an up-to-date summary on the impact of ROS on HSCs. In this section, our aim is to review the somewhat older literature on practical issues in the context of HSPC cultures related to O_2 effects and ROS toxicity, as it remains largely unknown and is not covered in the aforementioned 2008 review.

Decreased toxicity at low pO_2 was suggested several years ago by studies that have shown increased numbers of colonies in assays supplemented with the reducing agents 2-mercaptoethanol and α-thioglycerol (Bradley *et al.*, 1978; Ishikawa and Ito, 1988; Katahira and Mizoguchi, 1987; Rich and Kubanek, 1982). Increased colony formation due to α-thioglycerol is obtained both at low and high pO_2, but enhancement is generally greater at 20% O_2. Meagher *et al.* (Meagher *et al.*, 1988) measured increasing levels of superoxide and H_2O_2 with time in BM cultures. Addition of catalase (which scavenges H_2O_2) or mannitol (which scavenges hydroxyl radicals) increased BFU-E production by 100% and 50%, and CFU-GM production by 50% and 25%, respectively. Superoxide dismutase had only a small effect, even though it decreased superoxide to undetectable levels. These results and the effects of the three additives on H_2O_2 levels suggest that H_2O_2 inhibits maintenance of hematopoietic stem and progenitor cells. The effects of pO_2 may also be related to specific effects on individual cell types. Macrophages produce cytokines in an O_2-dependent manner (Rich, 1986), while T cells and macrophages cocultured under 5% O_2 combine to produce a factor that stimulates BFU-E proliferation (Pennathur-Das and Levitt, 1987). In our own work, we have observed increased IL-6 production under 5% O_2 in cultures of CB MNC, both in suspension and on irradiated stroma (Koller *et al.*, 1992c).

Oxygen and ROS levels are widely known (Fiegl *et al.*, 2009; Ito *et al.*, 2006; Nakata *et al.*, 2004; Nordberg and Arner, 2001; Schwarz *et al.*, 2009) to modulate gene expression of transcription factors, cytokines and growth factors, and, as such one would anticipate, a large interaction network whereby changes in O_2 or ROS concentrations will alter the locally produced growth factors, cytokines, and chemokines in the complex BM environment. Indeed, there are substantial literature as to how O_2 levels impact both cytokine production and regulation of cytokine receptor levels in hematopoietic and related cells (Fiegl *et al.*, 2009; Rogers *et al.*, 2008; Schwarz *et al.*, 2009; Tenen *et al.*, 1997). For example, transcription of the IL-6 and TNF-α genes is activated under low O_2 conditions due to upregulation in DNA binding of certain members of the C/EBP family of proteins (Yan *et al.*, 1995). The presence of C/EBP binding sites, which could potentially act as O_2-responsive elements in the M-CSFR, GM-CSFR, and G-CSFR promoters, suggests that receptor expression may change with pO_2. The importance of C/EBPα was demonstrated in C/EBP$\alpha^{-/-}$ animals in which G-CSFR mRNA levels were reduced (Tenen *et al.*, 1997). The evident analogy between hypoxia-inducible factor-1 (HIF-1) binding to the Epo gene, which results in increased Epo production during O_2 deprivation (Fandrey, 1995; Wang *et al.*, 1995; Wang and Semenza, 1993a; Wang and Semenza, 1993b) and C/EBP binding to the genes mentioned above, implicates changes in DNA-binding activity as the link between pO_2 and the perturbation of cytokine networks. This is an area of great practical interest in the context of stem cell bioengineering, aiming to achieve precise phenotypic outcomes.

6.6. The Largely Underappreciated Impact of Culture pH on Hematopoietic Stem and Progenitor Expansion

Based on the pH value of arterial and venal blood, most hematopoietic cultures have been performed at a pH of 7.4. However, as noted above, it is likely that pH varies within the BM. pH significantly affects the properties and differentiation of hematopoietic cells (McAdams *et al.*, 1997; McAdams *et al.*, 1998; Yang *et al.*, 2002). A rise in intracellular pH has been associated with the activation of platelets, lymphocytes and neutrophils (Ozaki, 1992). pH can also be a powerful regulator of cell proliferation and cytokine secretion. Rich (1988) has shown that the optimal proliferation of murine macrophages occurs between pH 7.6 and 8.0, and that Epo secretion by macrophages varies with external pH. It has also been reported that pH affects colony formation from murine BM cells (Zipori and Sasson, 1981). The myeloid lineage has demonstrated (Fischkoff and Rossi, 1990) that HL-60 (promyelocytic leukemia) cells, which normally differentiate into monocytes when cultured at pH 7.2, will instead differentiate into eosinophils when cultured at pH 7.6. In another study employing the HL-60 cell line, decreasing culture pH (7.4, 7.2, and 7.0) increased the CD13 surface protein (a human aminopeptidase N/metalloprotease which is one of the earliest markers of commitment to the myeloid lineage, and involved in several key function of myeloid cells including viral infections) surface concentration without affecting CD13 mRNA levels. Decreasing culture pH also decreased the average HL60 cell size, as well as glucose consumption and lactate production rates. Endo *et al.* (1994) have investigated the effects of pH on the human erythroid cell lines KU-812 and K562 and found that a significantly greater proportion of cells acquire a mature erythroid phenotype when cultured at pH 7.6 than when cultured at pH 7.4. The pH in static hematopoietic cultures has been reported to decline over time. The higher the cell density, the greater the pH decline, which results from metabolic acid production (Schultz *et al.*, 1992; Zipori and Sasson, 1981).

In cultures of primary HSPCs, which routinely experience pH variations of as much as 0.5 units, pH was found to have substantial effects both on progenitor cloning efficiency (as measured in liquid cultures) and on progenitor cell differentiation (as measured in methylcellulose cultures) (McAdams *et al.*, 1997; McAdams *et al.*, 1998). Cloning efficiencies of PB BPU-E in erythroid (E) cultures were ninefold higher at low pH (7.1) compared to pH (7.6). A small pH increase of 0.2 units over physiological values consistently produced significant reductions (42–85%) in cloning efficiencies for all cell types and cytokine combinations tested. Methylcellulose cultures of CB mononuclear (MNC) and PB MNC cells determined that differentiation of CFU-GM into progeny was optimal between pH 7.2 and 7.4. The differentiation of erythroid progenitors (BFU-E) progressively increased as pH was increased from 6.95 to 7.6. Flow cytometric analyses of key E surface markers CD71 and CD45RA demonstrated that E differentiation proceeds faster at high pH and is blocked at an intermediate stage by low pH. Methylcellulose cultures of PB $CD34^+$ cells exhibited similar patterns to the MNC cultures. In human primary Mk cultures from peripheral blood $CD34^+$ cells, higher pH promotes Mk cell differentiation, maturation, and apoptosis. Specifically, higher pH resulted in an earlier and higher polyploidization of $CD41^+$ Mk cells and an earlier onset of Mk cell apoptosis. Faster depletion of $CD34^+$cells and an earlier peak in the fraction of larger colony-forming Mk cells (BFU-Mks) were also observed at higher pH. To sum up, even small variations in pH substantially affected the performance of human hematopoietic cultures, with the E and Mk lineages being particularly sensitive, while PB progenitors are more sensitive to pH variations than CB progenitors.

6.7. Apoptosis, its Impact on Hematopoiesis and its Relation to pO_2 and pH

Apoptosis or programmed cell death is an essential process involved in removing superfluous or damaged cells (Kroemer *et al.*, 1995). With blood tissue being the most proliferative organ *in vivo*, apoptosis plays a major role in regulating hematopoiesis (Koury, 1992). The role of ROS in the apoptotic death of HSCs has been recently reviewed (Naka *et al.*, 2008). Beyond HSCs, it has been suggested that suppressing apoptosis (rather than inducing proliferation) may be an important regulatory role of lineage-specific cytokines such as Epo, Tpo and G-CSF (Lacronique *et al.*, 1997; Philpott *et al.*, 1997; Zauli *et al.*, 1997). In addition, other cytokines (TNF-α gnd TGF-β1) have been shown to potently induce apoptosis (Koury, 1992; Kroemer *et al.*, 1995; Sachs, 1996). Apoptosis is an important regulator of the function of mature granulocytic cells at sites of inflammation. Changes in surface marker expression allow apoptotic neutrophils to be phagocytosed intact, preventing the release of proinflammatory mediators (Dransfield *et al.*, 1995). Apoptosis may also play a significant role in regulating the proliferation and maturation of granulocytic progenitors, as indicated by G-CSF–induced suppression of apoptosis in PB $CD34^+$ progenitors (Philpott *et al.*, 1997) and the increased susceptibility to apoptosis of granulocytes in G-CSFR–deficient mice (Liu *et al.*, 1996). Similarly, studies of the Mk lineage have provided evidence that the natural fate of Mk cells is apoptotic death, which may be a trigger for platelet release (Zauli *et al.*, 1997). The role of Tpo appears to be the suppression of apoptosis in the Mk lineage long enough to allow significant differentiation and maturation to the platelet-forming stage. The features of cells stressed by oxidative damage and those undergoing apoptosis are remarkably similar (Payne *et al.*, 1995). For example, when a cell contains unrepaired DNA damage, p53 protein expression increases and induces the apoptotic pathways (Polyak *et al.*, 1997). The importance of pO_2 and ROS in apoptosis is also indicated by the ability of antioxidants and free radical scavengers to inhibit apoptosis *in vitro* (Hockenbery *et al.*, 1993; Ito *et al.*, 2006; Kroemer *et al.*, 1995; Sakagami *et al.*, 1996). Beyond apoptosis, however, the production of ROS is an important part of signaling during hematopoietic differentiation. Notably, it has been shown that ROS produced in response to stimulation of hematopoietic cells by lineage-specific HGFs (GM-CSF, IL-3, SCF, and Tpo) are necessary for lineage-specific activation of hematopoietic cells predominantly through the tyrosine phosphorylation (Sattler *et al.*, 1999).

Culture or tissue pH affects the intracellular pH by a variety of mechanisms. Decreased intracellular pH (pH_i) and/or alteration in the set point of Na^+/H^+ antiporters as an initial step in apoptosis has been established in many cell types

(Morana *et al.*, 1996; Reynolds *et al.*, 1996). Intracellular pH is an important parameter in the activation of apoptotic cascades (Hansen *et al.*, 2006) and notably of caspase-independent cascades (Kitanaka and Kuchino, 1999). Intracellular pH is also an important parameter affecting the cellular redox state, which determines how cells respond to apoptotic signals due to oxidative stress (Schafer and Buettner, 2001). It was reported that the occurrence of DNA fragmentation correlated with intracellular acidification of up to 1 pH unit in the model human hematopoietic (myeloid cell) line HL-60 (Barry and Eastman, 1992; Barry *et al.*, 1993). A pH-sensitive DNase II (active below pH 6.8) was identified as the mediator of DNA digestion in this system, in the myeloid cell line ML-1 (human myeloid leukemia) and in an IL-2–dependent T cell line (Li and Eastman, 1995; Morana *et al.*, 1996). In neutrophils, this endonuclease is the only detectable DNase (Gottlieb *et al.*, 1995). In some studies, artificial reduction of pH_i has resulted in the induction of apoptosis (Perez-Sala *et al.*, 1995). It has been also shown that intracellular alkalinization can block apoptosis (Rajotte *et al.*, 1992). The ability of G-CSF to protect neutrophils from apoptosis depends on activation of a vacuolar H^+ATPase that increases pH_i (Gottlieb *et al.*, 1995). Other studies have shown the anti-apoptotic properties of SCF and GM-CSF are mediated through a similar mechanism (Caceres-Cortes *et al.*, 1994). Given that extracellular pH often affects pH_i (Novikova *et al.*, 1993), these findings suggest that apoptotic regulation may be one of the mechanisms behind the effects of pH on hematopoietic cells, as discussed above.

6.8. Shear Forces in Stem Cell and Hematopoietic Differentiation

6.8.1. *Biomechanical forces in stem-cell biology*

The original perception that the stem cell niche (see above) is physically a nearly static one (i.e. it lacks any appreciable fluid motion, whereby only low Reynolds number interstitial flow may be present), would lead to the conclusion that shear and other mechanical forces do not affect stem cell biology and progenitor differentiation. This perception is changing fast over the last few years for all stem cell types, including MSCs (Park *et al.*, 2010; Yamamoto *et al.*, 2005), ESCs (Adamo *et al.*, 2009; Ahsan and Nerem, 2010; Masumura *et al.*, 2009; Saha *et al.*, 2006), and HSCs (Collins *et al.*, 1998a; Dunois-Larde *et al.*, 2009). However, when it comes to the differentiation of hematopoietic stem and progenitor cells, shear forces are of physiological significance for most lineages, and notably for the Mk lineage, which we discuss in some detail below. Both mature and immature hematopoietic cells are released into circulation and subjected to a spectrum of shear and biomechanical forces; in this context, the expectation that such forces would affect the differentiation and maturation of hematopoietic cells is a logical one.

6.8.2. *The importance of biomechanical forces in megakaryocytic differentiation*

Mk cells are subjected to mechanical forces (strain) as Mk cells squeeze through gaps in the endothelial lining of the BM sinusoids (Figures 6.3 and 6.4), as well as shear forces due to blood flow in the BM microvasculature or when Mk cells enter circulation and mature in the pulmonary vasculature (Dunois-Larde *et al.*, 2009). Such forces result in enhanced Mk maturation, (pro)platelet shedding and enhanced platelet biogenesis both *in vivo* (Junt *et al.*, 2007) and *in vitro* (Dunois-Larde *et al.*, 2009). The data in these recent studies suggest that shear/mechanical forces engage specific signaling (which is so far unexplored) that enhances platelet biogenesis. Although shear effects on endothelial cell biology is a well-established field, they remain mechanistically unexplored in Mk differentiation. Activation of p53 signaling in endothelial cells in response to flow-induced shear stress has been established (Zeng *et al.*, 2003). Our recent work (Fuhrken *et al.*, 2008; Fuhrken *et al.*, 2007) supports the hypothesis that p53 is a regulator of Mk maturation, modulating Mk-cell polyploidization in a specific manner, whereby p53 senses the cellular stress associated with polyploidization and regulates the process by controlling endomitosis vs. apoptosis. Furthermore, preliminary data from our laboratory suggest that shear flow alters p53 acetylation and thus, p53 activation. It appears that p53, as a known regulator of stress response, is also involved in integrating mechanical stresses during Mk maturation and proplatelet release.

6.9. Learning from Nature: Case Studies of How to Integrate the Effects of pO_2, pH, Fluid-forces, and Medium Perfusion in Processes to Generate Desirable Progenitor and Mature Hematopoietic Cells

In this section, the goal is to show how these biophysical parameters (pO_2, pH, perfusion, biomechanical forces) have been, or can be, explored to bring about desirable culture outcomes in the context of stem- and progenitor-cell cultures and bioengineering. We discuss three cases where one or more of these parameters were integrated in achieving a desirable outcome.

6.9.1. *Perfusion and low pO_2 levels for expanding the primitive HSPC compartment*

Since the most primitive compartment of hematopoiesis in the BM takes place under homeostatic perfusion (albeit at low rates), and low pO_2 conditions, such conditions would be expected to provide superior performance of the expansion of HSPCs. The development of perfusion bioreactors for stroma cocultures with control of pO_2 at about 5% demonstrated superior performance compared to any other published culture system, and notably so for the expansion of the most primitive LTC-IC cells (Koller *et al.*, 1993). Similarly, the beneficial effect of perfusion on LTC-IC output in the absence of a stromal layer was also demonstrated (Sandstrom *et al.*, 1996; Sandstrom *et al.*, 1995).

6.9.2. *Stirred hematopoietic bioreactors with pH and pO_2 control and perfusion operation for expanding myeloid cells*

A stirred bioreactor would have several advantages over static-culture systems and flat-bed bioreactors: inherent homogeneity, ability to sample more easily for culture characterization, ability to implement process–parameter control, and ease of scale-up to clinical size. On the other hand, successful growth of hematopoietic cells in stirred reactors may be problematic due to the shear sensitivity of the cells and the lack of characteristics present in static systems (e.g. cell-to-cell contact and a local microenvironment around the cells). Such bioreactors would be suitable for the expansion of mononuclear cells into granulocytes/ neutrophils, as these cells apparently grow as suspension cultures in defined media and with defined cytokine combinations (Collins *et al.*, 1996; Hevehan *et al.*, 2000; Timmins and Nielsen, 2009). PB MNCs, CB MNCs, and PB $CD34_+$ cells grew well in spinner flasks when compared to control T-flask (static) cultures (Collins *et al.*, 1998a). Culture proliferation in spinner flasks was dependent on both agitator design and RPM, and when optimized, provided better culture outcomes compared to static controls. When combined with a cell dilution feeding protocol, cultures of PB and CB MNCs in stirred bioreactor with pH and pO_2 control provided better outcomes in the expansion of total cells and CFU-GM progenitor cells (Collins *et al.*, 1998b).

6.9.3. *A three-stage process for the production of megakaryocytic cells and platelets from human CD34$^+$ cells, and the importance of biomechanical forces in a scaled-up process*

As discussed, megakaryopoiesis requires HSPC commitment (Figure 6.1) to the megakaryocytic cell (Mk) lineage. Because the degree of Mk ploidy directly correlates with the number of platelets produced (Mattia *et al.*, 2002), it is necessary to increase the ploidy of cultured Mks similar to what is observed *in vivo*, thus making it possible to obtain several thousand platelets per Mk (Mattia *et al.*, 2002). After commitment of HSPCs to Mk differentiation, the road to platelet production requires expansion of Mk progenitors and committed Mks, endomitosis accompanied polyploidization, maturation, and Mk apoptosis accompanied proplatelet release (Figure 6.4). This is a set of sequential steps, each dependent on the previous one, while amplifying the capacity of the system in terms of cell numbers. HSPCs move from the endosteum to the BM sinuses as they commit to become Mks (Figure 6.3) and penetrate the endothelial walls of the BM sinuses to release proplatelets (Figure 6.4). As discussed, a set of committed Mks can directly enter circulation through the BM sinuses to reach maturation and release proplatelets in the lung vasculature. pH and pO_2 levels increase from the endosteum to sinuses (Figure 6.3), and different sets of cytokines are important for various stages of differentiation (Panuganti *et al.*, 2010). Moreover, pO_2 and pH levels in blood circulation and the lung vasculature are considerably higher than in the endosteum region of the BM and, for that matter, than in most of the BM hematopoietic compartments between sinuses. Thus, Mk maturation requires higher pO_2 and pH settings, while HSPC and Mk progenitor settings require lower pO_2 and pH settings and different cytokine combinations. We thus argued that *ex vivo* Mk or platelet production would be better achieved by dividing the process into three separate phases (Panuganti *et al.*, 2010). Environmental conditions and growth factors must be optimized separately for Mk commitment and progenitor expansion, mature Mk production and Mk polyploidization; we set out to optimize such process in terms of cytokine combinations, and the pO_2 and pH settings (Panuganti *et al.*, 2010).

In the actual scaled-up process, while the early phase of HSPC expansion and early Mk commitment can be carried in a suspension reactor as discussed above, the process of Mk maturation, involving cell size enlargement due to polyploidization and controlled cell death for proplatelet release, will require a "protective" environment such as a packed-bed bioreactor (Sullenbarger *et al.*, 2009). As discussed, shear/mechanical stresses play a crucial role, both *in vivo* (Junt *et al.*, 2007) and *in vitro* (Dunois-Larde *et al.*, 2009), in the biogenesis of functional platelets. Thus, the ability to control Mk maturation under flow conditions is critical for providing tissue engineering solutions for designing processes to achieve large-scale production of platelets (Sullenbarger *et al.*, 2009). Incorporating shear flow in the design of processes aiming to optimize platelet release from *in vitro*-generated human Mks is a necessary step for achieving *ex vivo* platelet production, which is a goal of great practical significance in transfusion medicine (Reems *et al.*, 2010).

6.10. Summary

The stem cell niche and the microenvironment, in which progenitor cells differentiate to produce mature hematopoietic cells to be release into circulation, is characterized by low O_2 tension and O_2 gradients, pH gradients, cytokine gradients, variable BM perfusion rates (due to variable vascularization of the BM; low near the endosteum, high away from the endosteum; Figure 6.3), and in some cases, exposure to flow-induced shear forces or biomechanical forces due to cell penetration through endothelial gaps. We have reviewed the physiological, anatomical and, where available, molecular evidence for how such parameters and gradients affect the maintenance and differentiation of stem cells, significantly the expansion, differentiation and phenotypic characteristics of progenitor and differentiated cells. We have also discussed applications and integration of these parameters in process to generate stem, progenitor and differentiated hematopoietic cells for applications in regenerative and transfusion medicine.

Bibliography

Adamo, L., Naveiras, O., Wenzel, P. L., McKinney-Freeman, S., Mack, P. J., Gracia-Sancho, J., Suchy-Dicey, A., Yoshimoto, M., Lensch, M. W., Yoder, M. C., Garcia-Cardena, G. and Daley, G. Q. 2009. Biomechanical forces promote embryonic haematopoiesis, *Nature* 459:1131–U1120.

Ahsan, T. and Nerem, R. M. 2010. Fluid shear stress promotes an endothelial-like phenotype during the early differentiation of embryonic stem cells, *Tissue Eng Part A* 16:3547–3553.

Allalunis, M. J., Chapman, J. D. and Turner, A. R. 1983. Identification of a hypoxic population of bone marrow cells, *Int J Radiat Oncol Biol Phys* 9:227–232.

Arai, F., Hirao, A., Ohmura, M., Sato, H., Matsuoka, S., Takubo, K., Ito, K., Koh, G. Y. and Suda, T. 2004. Tie2/angiopoietin-1 signaling regulates hematopoietic stem cell quiescence in the bone marrow niche, *Cell* 118:149–161.

Audet, J., Miller, C. L., Eaves, C. J. and Piret, J. M. 2002. Common and distinct features of cytokine effects on hematopoietic stem and progenitor cells revealed by dose-response surface analysis, *Biotechnol Bioeng* 80:393–404.

Bacigalupo, A., Piaggio, G., Podesta, M., Figari, O., Benvenuto, F., Sogno, G., Tedone, E., Raffo, M. R., Grassia, L., Ferrero, R. and *et al.* 1995. Influence of marrow CFU-GM content on engraftment and survival after allogeneic bone marrow transplantation, *Bone Marrow Transplant* 15:221–226.

Barry, M. A. and Eastman, A. 1992. Endonuclease activation during apoptosis: The role of cytosolic Ca2+ and pH, *Biochem Biophys Res Commun* 186:782–789.

Barry, M. A., Reynolds, J. E. and Eastman, A. 1993. Etoposide-induced apoptosis in human HL-60 cells is associated with intracellular acidification, *Cancer Res* 53:2349–2357.

Bertolini, F., Battaglia, M., Pedrazzoli, P., Da Prada, G. A., Lanza, A., Soligo, D., Caneva, L., Sarina, B., Murphy, S., Thomas, T. and della Cuna, G. R. 1997. Megakaryocytic progenitors can be generated *ex vivo* and safely administered to autologous peripheral blood progenitor cell transplant recipients, *Blood* 89:2679–2688.

Bradley, T. R., Hodgson, G. S. and Rosendaal, M. 1978. The effect of oxygen tension on haemopoietic and fibroblast cell proliferation *in vitro*, *J Cell Physiol* 97:517–522.

Brandt, J. E., Bhalla, K. and Hoffman, R. 1994. Effects of interleukin-3 and c-kit ligand on the survival of various classes of human hematopoietic progenitor cells, *Blood* 83:1507–1514.

Broxmeyer, H. E., Cooper, S. and Gabig, T. 1989. The effects of oxidizing species derived from molecular oxygen on the proliferation *in vitro* of human granulocyte-macrophage progenitor cells, *Ann N Y Acad Sci* 554:177–184.

Broxmeyer, H. E., Cooper, S., Lu, L., Miller, M. E., Langefeld, C. D. and Ralph, P. 1990. Enhanced stimulation of human bone marrow macrophage colony formation *in vitro* by recombinant human macrophage colony-stimulating factor in agarose medium and at low oxygen tension, *Blood* 76:323–329.

Brugger, W., Heimfeld, S., Berenson, R. J., Mertelsmann, R. and Kanz, L. 1995. Reconstitution of hematopoiesis after high-dose chemotherapy by autologous progenitor cells generated *ex vivo*, *N Engl J Med* 333:283–287.

Caceres-Cortes, J., Rajotte, D., Dumouchel, J., Haddad, P. and Hoang, T. 1994. Product of the steel locus suppresses apoptosis in hemopoietic cells. Comparison with pathways activated by granulocyte macrophage colony-stimulating factor, *J Biol Chem* 269:12084–12091.

Cairo, M. S. and Wagner, J. E. 1997. Placental and/or umbilical cord blood: An alternative source of hematopoietic stem cells for transplantation, *Blood* 90:4665–4678.

Calvi, L. M., Adams, G. B., Weibrecht, K. W., Weber, J. M., Olson, D. P., Knight, M. C., Martin, R. P., Schipani, E., Divieti, P., Bringhurst, F. R., Milner, L. A., Kronenberg, H. M. and Scadden, D. T. 2003. Osteoblastic cells regulate the haematopoietic stem cell niche, *Nature* 425:841–846.

Caplan, A. 2009. Why are MSCs therapeutic? New data: New insight, *J Pathol* 217:318–324.

Carlsson, J. and Acker, H. 1988. Relations between pH, oxygen partial pressure and growth in cultured cell spheroids, *Int J Cancer* 42:715–720.

Carpenter, M. K., Rosler, E. and Rao, M. S. 2003. Characterization and differentiation of human embryonic stem cells, *Cloning Stem Cells* 5:79–88.

Choi, K. D., Vodyanik, M. A. and Slukvin, II. 2009. Generation of mature human myelomonocytic cells through expansion and differentiation of pluripotent stem cell-derived lin(−)CD34(+)CD43(+)CD45(+) progenitors, *J Clin Invest* 119: 2818–2829.

Chow, D. C., Wenning, L. A., Miller, W. M. and Papoutsakis, E. T. 2001a. Modeling pO(2) distributions in the bone marrow hematopoietic compartment. I. Krogh's model, *Biophys J* 81:675–684.

Chow, D. C., Wenning, L. A., Miller, W. M. and Papoutsakis, E. T. 2001b. Modeling pO(2) distributions in the bone marrow hematopoietic compartment. II. Modified Kroghian models, *Biophys J* 81:685–696.

Cipolleschi, M. G., Dello Sbarba P. and Olivotto, M. 1993. The role of hypoxia in the maintenance of hematopoietic stem cells, *Blood* 82:2031–2037.

Clark, S. C. and Kamen, R. 1987. The human hematopoietic colony-stimulating factors, *Science* 236:1229–1237.

Collins, P. C., Miller, W. M. and Papoutsakis, E. T. 1998a. Stirred culture of peripheral and cord blood hematopoietic cells offers advantages over traditional static systems for clinically relevant applications, *Biotechnol Bioeng* 59:534–543.

Collins, P. C., Nielsen, L. K., Patel, S. D., Papoutsakis, E. T. and Miller, W. M. 1998b. Characterization of hematopoietic cell expansion, oxygen uptake and glycolysis in a controlled, stirred-tank bioreactor system, *Biotechnol Prog* 14:466–472.

Collins, P. C., Papoutsakis, E. T. and Miller, W. M. 1996, *Ex vivo* culture systems for hematopoietic cells, *Curr Opin Biotechnol* 7:223–230.

Coulombel, L. 2004. Identification of hematopoietic stem/progenitor cells: Strength and drawbacks of functional assays, *Oncogene* 23:7210–7222.

Cui, Y., Golob, J., Kelleher, E., Ye, Z. H., Pardoll, D. and Cheng, L. Z. 2002. Targeting transgene expression to antigen-presenting cells derived from lentivirus-transduced engrafting human hematopoietic stem/progenitor cells, *Blood* 99:399–408.

Danet, G. H., Pan, Y., Luongo, J. L., Bonnet, D. A. and Simon, M. C. 2003. Expansion of human SCID-repopulating cells under hypoxic conditions, *J Clin Invest* 112:126–135.

Dao, M. A., Creer, M. H., Nolta, J. A. and Verfaillie, C. M. 2007. Biology of umbilical cord blood progenitors in bone marrow niches, *Blood* 110:74–81.

de Sauvage, F. J., Hass, P. E., Spencer, S. D., Malloy, B. E., Gurney, A. L., Spencer, S. A., Darbonne, W. C., Henzel, W. J., Wong, S. C., Kuang, W. J. and *et al.* 1994. Stimulation of megakaryocytopoiesis and thrombopoiesis by the c-Mpl ligand, *Nature* 369:533–538.

de Wynter, E. A., Coutinho, L. H., Pei, X., Marsh, J. C., Hows, J., Luft, T. and Testa, N. G. 1995. Comparison of purity and enrichment of CD34+ cells from bone marrow, umbilical cord and peripheral blood (primed for apheresis) using five separation systems, *Stem Cells* 13:524–532.

Decaudin, D., Vantelon, J. M., Bourhis, J. H., Farace, F., Bonnet, M. L., Guillier, M., Greissenger, N., Marracho, M. C., Assari, S., Bennaceur, A. L., Nemati, F., Michon, J., Turhan, A. G. and Boccaccio, C. 2004, *Ex vivo* expansion of megakaryocyte precursor cells in autologous stem cell transplantation for relapsed malignant lymphoma, *Bone Marrow Transplant* 34:1089–1093.

Di Nicola, M., Siena, S., Bregni, M., Longoni, P., Magni, M., Milanesi, M., Matteucci, P., Mortarini, R., Anichini, A., Parmiani, G., Drexler, I., Erfle, V., Sutter, G. and Gianni, A. M. 1998. Gene transfer into human dendritic antigen-presenting cells by vaccinia virus and adenovirus vectors, *Cancer Gene Ther* 5:350–356.

Dransfield, I., Stocks, S. C. and Haslett, C. 1995. Regulation of cell adhesion molecule expression and function associated with neutrophil apoptosis, *Blood* 85:3264–3273.

Dunois-Larde, C., Capron, C., Fichelson, S., Bauer, T., Cramer-Borde, E. and Baruch, D. 2009. Exposure of human megakaryocytes to high shear rates accelerates platelet production, *Blood* 114:1875–1883.

Dzierzak, E. and Robin, C. 2010. Placenta as a source of hematopoietic stem cells, *Trends Mol Med* 16:361–367.

Endo, T., Ishibashi, Y., Okana, H. and Fukumaki, Y. 1994. Significance of pH on differentiation of human erythroid cell lines, *Leuk Res* 18:49–54.

Ezashi, T., Das, P. and Roberts, R. M. 2005. Low O-2 tensions and the prevention of differentiation of hES cells, *Proc Natl Acad Sci U S A* 102:4783–4788.

Fandrey, J. 1995. Hypoxia-inducible gene expression, *Respir Physiol* 101:1–10.

Fiegl, M., Samudio, I., Clise-Dwyer, K., Burks, J. K., Mnjoyan, Z. and Andreeff, M. 2009. CXCR4 expression and biologic activity in acute myeloid leukemia are dependent on oxygen partial pressure, *Blood* 113:1504–1512.

Fischkoff, S. A. and Rossi, R. M. 1990. Lineage directed HL-60 cell sublines as a model system for the study of early events in lineage determination of myeloid cells, *Leuk Res* 14:979–988.

Fisher, J. W. 2003. Erythropoietin: Physiology and pharmacology update, *Exp Biol Med* 228:1–14.

Fuhrken, P. G., Apostolidis, P. A., Lindsey, S., Miller, W. M. and Papoutsakis, E. T. 2008. Tumor suppressor protein p53 regulates megakaryocytic polyploidization and apoptosis, *J Biol Chem* 283:15589–15600.

Fuhrken, P. G., Chen, C., Miller, W. M. and Papoutsakis, E. T. 2007. Comparative, genome-scale transcriptional analysis of CHRF-288–11 and primary human megakaryocytic cell cultures provides novel insights into lineage-specific differentiation, *Exp Hematol* 35:476–489.

Genbacev, O., Joslin, R., Damsky, C. H., Polliotti, B. M. and Fisher, S. J. 1996. Hypoxia alters early gestation human cytotrophoblast differentiation invasion *in vitro* and models the placental defects that occur in preeclampsia, *J Clin Invest* 97:540–550.

Gordon, M. Y. 1993. Human hematopoietic stem-cell assays, *Blood Rev* 7:190–197.

Gorlach, A. and Acker, H. 1994. The relationship of radiation sensitivity and microenvironment of human tumor cells in multicellular spheroid tissue culture, in P. Vaupel and D. F. Bruley (eds.), *Oxygen Transport to Tissue. Vol. XV*. Plenum Press, New York. 343–350.

Gottlieb, R. A., Giesing, H. A., Zhu, J. Y., Engler, R. L. and Babior, B. M. 1995. Cell acidification in apoptosis: Granulocyte colony-stimulating factor delays programmed cell death in neutrophils by up-regulating the vacuolar H(+)-ATPase, *Proc Natl Acad Sci USA* 92:5965–5968.

Grayson, W. L., Zhao, F., Bunnell, B. and Ma, T. 2007. Hypoxia enhances proliferation and tissue formation of human mesenchymal stem cells, *Biochem Biophys Res Commun* 358:948–953.

Grayson, W. L., Zhao, F., Izadpanah, R., Bunnell, B. and Ma, T. 2006a. Effects of hypoxia on human mesenchymal stem cell expansion and plasticity in 3D constructs, *J Cell Physiol* 207:331–339.

Grayson, W. L., Zhao, F., Izadpanah, R., Bunnell, B. and Ma, T. 2006b. Effects of hypoxia on human mesenchymal stem cell expansion and plasticity in 3D constructs, *J Cell Physiol* 207:331–339.

Hansen, J. M., Go, Y. M. and Jones, D. P. 2006. Nuclear and mitochondrial compartmentation of oxidative stress and redox signaling, *Annu Rev Pharmacol Toxicol* 46:215–234.

Harrison, J. S., Rameshwar, P., Chang, V. and Bandari, P. 2002. Oxygen saturation in the bone marrow of healthy volunteers, *Blood* 99:394.

Hattori, K., Heissig, B. and Rafii, S. 2003. The regulation of hematopoietic stem cell and progenitor mobilization by chemokine SDF-1, *Leuk Lymphoma* 44:575–582.

Hatzfeld, J., Li, M. L., Brown, E. L., Sookdeo, H., Levesque, J. P., O'Toole, T., Gurney, C., Clark, S. C. and Hatzfeld, A. 1991. Release of early human hematopoietic progenitors from quiescence by antisense transforming growth factor beta 1 or Rb oligonucleotides, *J Exp Med* 174:925–929.

Helmlinger, G., Yuan, F., Dellian, M. and Jain, R. K. 1997. Interstitial pH and pO2 gradients in solid tumors *in vivo*: High-resolution measurements reveal a lack of correlation, *Nat Med* 3:177–182.

Hevehan, D. L., Papoutsakis, E. T. and Miller, W. M. 2000. Physiologically significant effects of pH and oxygen tension on granulopoiesis, *Exp Hematol* 28:267–275.

Hirao, M., Tamai, N., Tsumaki, N., Yoshikawa, H. and Myoui, A. 2006. Oxygen tension regulates chondrocyte differentiation and function during endochondral ossification, *J Biol Chem* 281:31079–31092.

Ho, A. D., Wagner, W. and Franke, W. 2008. Heterogeneity of mesenchymal stromal cell preparations, *Cytotherapy* 10:320–330.

Hockenbery, D. M., Oltvai, Z. N., Yin, X. M., Milliman, C. L. and Korsmeyer, S. J. 1993. Bcl-2 functions in an antioxidant pathway to prevent apoptosis, *Cell* 75:241–251.

Ishikawa, Y. and Ito, T. 1988. Kinetics of hemopoietic stem cells in a hypoxic culture, *Eur J Haematol* 40:126–129.

Ito, K., Hirao, A., Arai, F., Takubo, K., Matsuoka, S., Miyamoto, K., Ohmura, M., Naka, K., Hosokawa, K., Ikeda, Y. and Suda, T. 2006. Reactive oxygen species act through p38 MAPK to limit the lifespan of hematopoietic stem cells, *Nat Med* 12: 446–451.

Ivanovic, Z., Bartolozzi, B., Bernabei, P. A., Cipolleschi, M. G., Rovida, E., Milenkovic, P., Praloran, V. and Dello Sbarba, P. 2000a. Incubation of murine bone marrow cells in hypoxia ensures the maintenance of marrow-repopulating ability together with the expansion of committed progenitors, *Br J Haematol* 108:424–429.

Ivanovic, Z., Dello Sbarba, P., Trimoreau, F., Faucher, J. L. and Praloran, V. 2000b. Primitive human HPCs are better maintained and expanded *in vitro* at 1 percent oxygen than at 20 percent, *Transfusion* 40:1482–1488.

Janssen, W. E. 1995. Mobilization of peripheral blood stem cells for autologous transplantation. Methods, mechanisms and role in accelerating hematopoietic recovery, *Ann N Y Acad Sci* 770:116–129.

Jones, D. L. and Wagers, A. J. 2008. No place like home: Anatomy and function of the stem cell niche, *Nat Rev Mol Cell Biol* 9:11–21.

Junt, T., Schulze, H., Chen, Z., Massberg, S., Goerge, T., Krueger, A., Wagner, D. D., Graf, T., Italiano, J. E., Jr., Shivdasani, R. A. and von Andrian, U. H. 2007. Dynamic visualization of thrombopoiesis within bone marrow, *Science* 317:1767–1770.

Kaplan, R. N., Psaila, B. and Lyden, D. 2007. Niche-to-niche migration of bone-marrow-derived cells, *Trends Mol Med* 13:72–81.

Katahira, J. and Mizoguchi, H. 1987. Improvement of culture conditions for human megakaryocytic and pluripotent progenitor cells by low oxygen tension, *Int J Cell Cloning* 5:412–420.

Kaushansky, K. 2006. Mechanisms of disease: Lineage-specific hematopoietic growth factors, *N Engl J Med* 354:2034–2045.

Kelly, S. S., Sola, C., B. S., de Lima, M. and Shpall, E. 2009. *Ex vivo* expansion of cord blood, *Bone Marrow Transplant* 44:673–681.

Kietzmann, T., Hirsch-Ernst, K. I., Kahl, G. F. and Jungermann, K. 1999. Mimicry in primary rat hepatocyte cultures of the *in vivo* perivenous induction by phenobarbital of cytochrome P-4502B1 mRNA: Role of epidermal growth factor and perivenous oxygen tension, *Mol Pharmacol* 56:46–53.

Kitanaka, C. and Kuchino, Y. 1999. Caspase-independent programmed cell death with necrotic morphology, *Cell Death Differ* 6:508–515.

Koller, M. R., Bender, J. G., Miller, W. M. and Papoutsakis, E. T. 1992a. Reduced oxygen tension increases hematopoiesis in long-term culture of human stem and progenitor cells from cord blood and bone marrow, *Exp Hematol* 20:264–270.

Koller, M. R., Bender, J. G., Miller, W. M. and Papoutsakis, E. T. 1993. Expansion of primitive human hematopoietic progenitors in a perfusion bioreactor system with IL-3, IL-6 and stem cell factor, *Biotechnology (NY)* 11:358–363.

Koller, M. R., Bender, J. G., Papoutsakis, E. T. and Miller, W. M. 1992b. Beneficial effects of reduced oxygen tension and perfusion in long-term hematopoietic cultures, *Ann N Y Acad Sci* 665:105–116.

Koller, M. R., Bender, J. G., Papoutsakis, E. T. and Miller, W. M. 1992c. Effects of synergistic cytokine combinations, low oxygen and irradiated stroma on the expansion of human cord blood progenitors, *Blood* 80:403–411.

Koller, M. R. and Palsson, B. O. 1993. Review: Tissue engineering: Reconstitution of human hematopoiesis *ex vivo*, *Biotechnol Bioeng* 42:909–930.

Korbling, M. and Anderlini, P. 2001. Peripheral blood stem cell versus bone marrow allotransplantation: Does the source of hematopoietic stem cells matter? *Blood* 98:2900–2908.

Koury, M. J. 1992. Programmed cell death (apoptosis) in hematopoiesis, *Exp Hematol* 20:391–394.

Krause, D. S., Fackler, M. J., Civin, C. I. and May, W. S. 1996. CD34: Structure, biology, and clinical utility, *Blood* 87:1–13.

Kroemer, G., Petit, P., Zamzami, N., Vayssiere, J. L. and Mignotte, B. 1995. The biochemistry of programmed cell death, *Faseb J* 9:1277–1287.

Lacronique, V., Varlet, P., Mayeux, P., Porteu, A., Gisselbrecht, S., Kahn, A. and Lacombe, C. 1997. Bcl-2 targeted overexpression into the erythroid lineage of transgenic mice delays but does not prevent the apoptosis of erythropoietin-deprived erythroid progenitors, *Blood* 90:3050–3056.

LaIuppa, J. A., Papoutsakis, E. T. and Miller, W. M. 1998. Oxygen tension alters the effects of cytokines on the megakaryocyte, erythrocyte, and granulocyte lineages, *Exp Hematol* 26:835–843.

Lansdorp, P. M., Dragowska, W. and Mayani, H. 1993. Ontogeny-related changes in proliferative potential of human hematopoietic cells, *J Exp Med* 178:787–791.

Leary, A. G., Ikebuchi, K., Hirai, Y., Wong, G. G., Yang, Y. C., Clark, S. C. and Ogawa, M. 1988. Synergism between interleukin-6 and interleukin-3 in supporting proliferation of human hematopoietic stem cells: Comparison with interleukin-1 alpha, *Blood* 71:1759–1763.

Leary, A. G., Zeng, H. Q., Clark, S. C. and Ogawa, M. 1992. Growth factor requirements for survival in G0 and entry into the cell cycle of primitive human hemopoietic progenitors, *Proc Natl Acad Sci U S A* 89:4013–4017.

Lengerke, C. and Daley, G. Q. 2010. Autologous blood cell therapies from pluripotent stem cells, *Blood Rev* 24:27–37.

Levesque, J. P., Winkler, I. G., Hendy, J., Williams, B., Helwani, F., Barbier, V., Nowlan, B. and Nilsson, S. K. 2007. Hematopoietic progenitor cell mobilization results in hypoxia with increased hypoxia-inducible transcription factor-1 alpha and vascular endothelial growth factor A in bone marrow, *Stem Cells* 25:1954–1965.

Li, J. and Eastman, A. 1995. Apoptosis in an interleukin-2-dependent cytotoxic T lymphocyte cell line is associated with intracellular acidification. Role of the Na(+)/H(+)-antiport, *J Biol Chem* 270:3203–3211.

Lichtman, M. A. 1981. The ultrastructure of the hemopoietic environment of the marrow: A review, *Exp Hematol* 9:391–410.

Lichtman, M. A. 1984. The relationship of stromal cells to hemopoietic cells in marrow, *Kroc Found Ser* 18:3–29.

Liu, F., Wu, H. Y., Wesselschmidt, R., Kornaga, T. and Link, D. C. 1996. Impaired production and increased apoptosis of neutrophils in granulocyte colony-stimulating factor receptor-deficient mice, *Immunity* 5:491–501.

Maeda, H., Hotta, T. and Yamada, H. 1986. Enhanced colony formation of human hemopoietic stem cells in reduced oxygen tension, *Exp Hematol* 14:930–934.

Martin, G. R. and Jain, R. K. 1994. Noninvasive measurement of interstitial pH profiles in normal and neoplastic tissue using fluorescence ratio imaging microscopy, *Cancer Res* 54:5670–5674.

Masumura, T., Yamamoto, K., Shimizu, N., Obi, S. and Ando, J. 2009. Shear stress increases expression of the arterial endothelial marker ephrinB2 in Mmrine ES cells via the VEGF-notch signaling pathways, *Arterioscler Thromb Vasc Biol* 29: 2125–2131.

Mattia, G., Vulcano, F., Milazzo, L., Barca, A., Macioce, G., Giampaolo, A. and Hassan, H. J. 2002. Different ploidy levels of megakaryocytes generated from peripheral or cord blood CD34+ cells are correlated with different levels of platelet release, *Blood* 99:888–897.

Mayani, H., Guilbert, L. J. and Janowskawieczorek, A. 1992. Biology of the hematopoietic microenvironment, *Eur J Haematol* 49:225–233.

Mazumdar, J., O'Brien, W. T., Johnson, R. S., LaManna, J. C., Chavez, J. C., Klein, P. S. and Simon, M. C. 2010. O-2 regulates stem cells through Wnt/beta-catenin signalling, *Nat Cell Biol* 12:1007–1013.

McAdams, T. A., Miller, W. M. and Papoutsakis, E. T. 1996a. Hematopoietic cell culture therapies (Part I): Cell culture considerations, *Trends Biotechnol* 14:341–349.

McAdams, T. A., Miller, W. M. and Papoutsakis, E. T. 1997. Variations in culture pH affect the cloning efficiency and differentiation of progenitor cells in *ex vivo* haemopoiesis, *Br J Haematol* 97:889–895.

McAdams, T. A., Miller, W. M. and Papoutsakis, E. T. 1998. pH is a potent modulator of erythroid differentiation, *Br J Haematol* 103:317–325.

McAdams, T. A., Winter, J. N., Miller, W. M. and Papoutsakis, E. T. 1996b. Hematopoietic cell culture therapies (Part II): Clinical aspects and applications, *Trends Biotechnol* 14:388–396.

McKenna, H. J., de Vries, P., Brasel, K., Lyman, S. D. and Williams, D. E. 1995. Effect of flt3 ligand on the *ex vivo* expansion of human CD34+ hematopoietic progenitor cells, *Blood* 86:3413–3420.

Meagher, R. C., Salvado, A. J. and Wright, D. G. 1988. An analysis of the multilineage production of human hematopoietic progenitors in long-term bone marrow culture: evidence that reactive oxygen intermediates derived from mature phagocytic cells have a role in limiting progenitor cell self-renewal, *Blood* 72:273–281.

Mohyeldin, A., Garzon-Muvdi, T. and Quinones-Hinojosa, A. 2010. Oxygen in stem cell biology: A critical component of the stem cell niche, *Cell Stem Cell* 7:150–161.

Morana, S. J., Wolf, C. M., Li, J., Reynolds, J. E., Brown, M. K. and Eastman, A. 1996. The involvement of protein phosphatases in the activation of ICE/CED-3 protease, intracellular acidification, DNA digestion and apoptosis, *J Biol Chem* 271:18263–18271.

Morrison, S. J. and Spradling, A. C. 2008. Stem cells and niches: Mechanisms that promote stem cell maintenance throughout life, *Cell* 132:598–611.

Mostafa, S. S., Miller, W. M. and Papoutsakis, E. T. 2000a. Oxygen tension influences the differentiation, maturation and apoptosis of human megakaryocytes, *Br J Haematol* 111:879–889.

Mostafa, S. S., Miller, W. M. and Papoutsakis, E. T. 2000b. Oxygen tension influences the differentiation, maturation and apoptosis of human megakaryocytes, *Br J Haematol* 111:879–889.

Mostafa, S. S., Papoutsakis, E. T. and Miller, W. M. 2001. Oxygen tension modulates the expression of cytokine receptors, transcription factors, and lineage-specific markers in cultured human megakaryocytes, *Exp Hematol* 29:873–883.

Motohashi, H., Kimura, M., Fujita, R., Inoue, A., Pan, X. Q., Takayama, M., Katsuoka, F., Aburatani, H., Bresnick, E. H. and Yamamoto, M. 2010. NF-E2 domination over Nrf2 promotes ROS accumulation and megakaryocytic maturation, *Blood* 115:677–686.

Naka, K., Muraguchi, T., Hoshii, T. and Hirao, A. 2008. Regulation of reactive oxygen species and genomic stability in hematopoietic stem cells, *Antioxid Redox Signal* 10:1883–1894.

Nakata, S., Matsumura, I., Tanaka, H., Ezoe, S., Satoh, Y., Ishikawa, J., Era, T. and Kanakura, Y. 2004. NF-kappa B family proteins participate in multiple steps of hematopoiesis through elimination of reactive oxygen species, *J Biol Chem* 279:55578–55586.

Nilsson, S. K., Johnston, H. M. and Coverdale, J. A. 2001. Spatial localization of transplanted hemopoietic stem cells: Inferences for the localization of stem cell niches, *Blood* 97:2293–2299.

Nordberg, J. and Arner, E. S. J. 2001. Reactive oxygen species, antioxidants, and the mammalian thioredoxin system, *Free Radic Biol Med* 31:1287–1312.

Novikova, I., Muravyeva, O. V., Cragoe, E. J. and Margolis, L. B. 1993. Study of fibroblast spreading: pH dependence, involvement of the Na+/H(+)-antiporter and PKC, *Biochim Biophys Acta* 1178:267–272.

Ozaki, Y. 1992. Effect of pH on blood cells, *Nippon Rinsho* 50:2112–2118.

Panuganti, S., Papoutsakis, E. T. and Miller, W. M. 2010. Bone marrow niche-inspired, multiphase expansion of megakaryocytic progenitors with high polyploidization potential, *Cytotherapy* 12:767–782.

Park, J. Y., Yoo, S. J., Patel, L. and Lee, S. H. 2010. Cell morphological response to low shear stress in a two-dimensional culture microsystem with magnitudes comparable to interstitial shear stress, *Biorheology* 47:165–178.

Parmar, K., Mauch, P., Vergilio, J. A., Sackstein, R. and Down, J. D. 2007. Distribution of hematopoietic stem cells in the bone marrow according to regional hypoxia, *Proc Natl Acad Sci USA* 104:5431–5436.

Payne, C. M., Bernstein, C. and Bernstein, H. 1995. Apoptosis overview emphasizing the role of oxidative stress, DNA damage and signal-transduction pathways, *Leuk Lymphoma* 19:43–93.

Pennathur-Das, R. and Levitt, L. 1987. Augmentation of in vitro human marrow erythropoiesis under physiological oxygen tensions is mediated by monocytes and T lymphocytes, *Blood* 69:899–907.

Perez-Sala, D., Collado-Escobar, D. and Mollinedo, F. 1995. Intracellular alkalinization suppresses lovastatin-induced apoptosis in HL-60 cells through the inactivation of a pH-dependent endonuclease, *J Biol Chem* 270:6235–6242.

Perez-Simon, J. A., Diez-Campelo, M., Martino, R., Sureda, A., Caballero, D., Canizo, C., Brunet, S., Altes, A., Vazquez, L., Sierra, J. and San Miguel, J. F. 2003. Impact of CD34(+) cell dose on the outcome of patients undergoing reduced-intensity-conditioning allogeneic peripheral blood stem cell transplantation, *Blood* 102: 1108–1113.

Philpott, N. J., Prue, R. L., Marsh, J. C., Gordon-Smith, E. C. and Gibson, F. M. 1997. G-CSF-mobilized CD34 peripheral blood stem cells are significantly less apoptotic than unstimulated peripheral blood CD34 cells: Role of G-CSF as survival factor, *Br J Haematol* 97:146–152.

Polyak, K., Xia, Y., Zweier, J. L., Kinzler, K. W. and Vogelstein, B. 1997. A model for p53-induced apoptosis, *Nature* 389:300–305.

Rajotte, D., Haddad, P., Haman, A., Cragoe, E. J., Jr. and Hoang, T. 1992. Role of protein kinase C and the Na+/H+ antiporter in suppression of apoptosis by granulocyte macrophage colony-stimulating factor and interleukin-3, *J Biol Chem* 267:9980–9987.

Reems, J. A., Pineault, N. and Sun, S. J. 2010, *In vitro* megakaryocyte production and platelet biogenesis: State of the art, *Transf Med Rev* 24:33–43.

Reynolds, J. E., Li, J. and Eastman, A. 1996. Detection of apoptosis by flow cytometry of cells simultaneously stained for intracellular pH (carboxy SNARF-1) and membrane permeability (Hoechst 33342), *Cytometry* 25:349–357.

Rich, I. N. 1986. A role for the macrophage in normal hemopoiesis. II. Effect of varying physiological oxygen tensions on the release of hemopoietic growth factors from bone-marrow-derived macrophages *in vitro*, *Exp Hematol* 14:746–751.

Rich, I. N. 1988. The macrophage as a production site for hematopoietic regulator molecules: Sensing and responding to normal and pathophysiological signals, *Anticancer Res* 8:1015–1040.

Rich, I. N. and Kubanek, B. 1982. The effect of reduced oxygen tension on colony formation of erythropoietic cells *in vitro*, *Br J Haematol* 52:579–588.

Rogers, H. M., Yu, X. B., Wen, J., Smith, R., Fibach, E. and Noguchi, C. T. 2008. Hypoxia alters progression of the erythroid program, *Exp Hematol* 36:17–27.

Rosova, I., Dao, M., Capoccia, B., Link, D. and Nolta, J. A. 2008. Hypoxic preconditioning results in increased motility and improved therapeutic potential of human mesenchymal stem cells, *Stem Cells* 26:2173–2182.

Rusten, L. S., Jacobsen, F. W., Lesslauer, W., Loetscher, H., Smeland, E. B. and Jacobsen, S. E. 1994. Bifunctional effects of tumor necrosis factor alpha (TNF alpha) on the growth of mature and primitive human hematopoietic progenitor cells: Involvement of p55 and p75 TNF receptors, *Blood* 83:3152–3159.

Sachs, L. 1996. The control of hematopoiesis and leukemia: From basic biology to the clinic, *Proc Natl Acad Sci U S A* 93:4742–4749.

Saha, S., Lin, J., De Pablo, J. J. and Palecek, S. P. 2006. Inhibition of human embryonic stem cell differentiation by mechanical strain, *J Cell Physiol* 206:126–137.

Sakagami, H., Kuribayashi, N., Iida, M., Hagiwara, T., Takahashi, H., Yoshida, H., Shiota, F., Ohata, H., Momose, K. and Takeda, M. 1996. The requirement for and mobilization of calcium during induction by sodium ascorbate and by hydrogen peroxide of cell death, *Life Sci* 58:1131–1138.

Sandstrom, C. E., Bender, J. G., Miller, W. M. and Papoutsakis, E. T. 1996. Development of novel perfusion chamber to retain nonadherent cells and its use for comparison of human "mobilized" peripheral blood mononuclear cell cultures with and without irradiated bone marrow stroma, *Biotechnol Bioeng* 50:493–504.

Sandstrom, C. E., Bender, J. G., Papoutsakis, E. T. and Miller, W. M. 1995. Effects of CD34+ cell selection and perfusion on *ex vivo* expansion of peripheral blood mononuclear cells, *Blood* 86:958–970.

Saraya, K. and Reid, C. D. L. 1996. Stem cell factor and the regulation of dendritic cell production from CD34(+) progenitors in bone marrow and cord blood, *Br J Haematol* 93:258–264.

Sattler, M., Winkler, T., Verma, S., Byrne, C. H., Shrikhande, G., Salgia, R. and Griffin, J. D. 1999. Hematopoietic growth factors signal through the formation of reactive oxygen species, *Blood* 93:2928–2935.

Sauer, H., Wartenberg, M. and Hescheler, J. 2001. Reactive oxygen species as intracellular messengers during cell growth and differentiation, *Cell Physiol Biochem* 11:173–186.

Sauvageau, G., Iscove, N. N. and Humphries, R. K. 2004. *In vitro* and *in vivo* expansion of hematopoietic stem cells, *Oncogene* 23:7223–7232.

Schafer, F. Q. and Buettner, G. R. 2001. Redox environment of the cell as viewed through the redox state of the glutathione disulfide/glutathione couple, *Free Radic Biol Med* 30:1191–1212.

Scheding, S., Bergmannn, M., Rathke, G., Vogel, W., Brugger, W. and Kanz, L. 2004. Additional transplantation of *ex vivo* generated megakaryocytic cells after high-dose chemotherapy, *Haematologica* 89:630–631.

Schofield, R. 1978. The relationship between the spleen colony-forming cell and the haemopoietic stem cell, *Blood Cells* 4:7–25.

Schultz, J. S., Naumov, I. M., Vecchini, F., Virji, M. A. and Boggs, S. S. 1992. Enhancement of cell production in long-term bone marrow culture, *Int J Cell Cloning* 10:161–165.

Schulze, H. and Shivdasani, R. A. 2005. Mechanisms of thrombopoiesis, *J Thromb Haemost* 3:1717–1724.

Schwarz, J. B. K., Langwieser, N., Langwieser, N. N., Bek, M. J., Seidl, S., Eckstein, H. H., Lu, B., Schomig, A., Pavenstadt, H. and Zohlnhofer, D. 2009. Novel role of the CXC chemokine receptor 3 in inflammatory response to arterial injury involvement of mTORC1, *Circ Res* 104:189-U194.

Shingo, T., Sorokan, S. T., Shimazaki, T. and Weiss, S. 2001. Erythropoietin regulates the *in vitro* and *in vivo* production of neuronal progenitors by mammalian forebrain neural stem cells, *J Neurosci* 21:9733–9743.

Shirvaikar, N., Reca, R., Jalili, A., Marquez-Curtis, L., Lee, S. F., Ratajczak, M. Z. and Janowska-Wieczorek, A. 2008. CFU-megakaryocytic progenitors expanded *ex vivo* from cord blood maintain their *in vitro* homing potential and express matrix metalloproteinases, *Cytotherapy* 10:182–192.

Siena, S., DiNicola, M., Bregni, M., Mortarini, R., Anichini, A., Lombardi, L., Ravagnani, F., Parmiani, G. and Gianni, A. M. 1995. Massive *ex vivo* generation of functional dendritic cells from mobilized CD34(+) blood progenitors for anticancer therapy, *Exp Hematol* 23:1463–1471.

Simsek, T., Kocabas, F., Zheng, J., Deberardinis, R. J., Mahmoud, A. I., Olson, E. N., Schneider, J. W., Zhang, C. C. and Sadek, H. A. 2010. The distinct metabolic profile of hematopoietic stem cells reflects their location in a hypoxic niche, *Cell Stem Cell* 7:380–390.

Srour, E. F., Abonour, R., Cornetta, K. and Traycoff, C. M. 1999, *Ex vivo* expansion of hematopoietic stem and progenitor cells: Are we there yet? *J Hematother* 8:93–102.

Studer, L., Csete, M., Lee, S. H., Kabbani, N., Walikonis, J., Wold, B. and McKay, R. 2000. Enhanced proliferation, survival, and dopaminergic differentiation of CNS precursors in lowered oxygen, *J Neurosci* 20:7377–7383.

Sullenbarger, B., Bahng, J. H., Gruner, R., Kotov, N. and Lasky, L. C. 2009. Prolonged continuous in vitro human platelet production using three-dimensional scaffolds, *Exp Hematol* 37:101–110.

Takahira, H., Lyman, S. D. and Broxmeyer, H. E. 1996. Flt3 ligand prolongs survival of CD34+++ + human umbilical cord blood myeloid progenitors in serum-depleted culture medium, *Ann Hematol* 72:131–135.

Takubo, K., Goda, N., Yamada, W., Iriuchishima, H., Ikeda, E., Kubota, Y., Shima, H., Johnson, R. S., Hirao, A., Suematsu, M. and Suda, T. 2010. Regulation of the HIF-1alpha level is essential for hematopoietic stem cells, *Cell Stem Cell* 7:391–402.

Tavassoli, M. 1974. Arterial structure of bone-marrow in rabbit with special reference to thin-walled arteries, *Acta Anat* 90:608–616.

Tenen, D. G., Hromas, R., Licht, J. D. and Zhang, D. E. 1997. Transcription factors, normal myeloid development, and leukemia, *Blood* 90:489–519.

ter Huurne, M., Figdor, C. G. and Torensma, R. 2010. Hematopoietic stem cells are coordinated by the molecular cues of the endosteal niche, *Stem Cells Dev* 19:1131–1141.

Thomas, S. R., Millard, R. W., Pratt, R. G., Shiferaw, Y. and Samaratunga, R. C. 1994. Quantitative pO_2 imaging *in vivo* with perfluorocarbon F-19 NMR — tracking oxygen from the airway through the blood to organ tissues, *Artif Cells Blood Substit Immobil Biotechnol* 22:1029–1042.

Timmins, N. E. and Nielsen, L. K. 2009. Blood cell manufacture: Current methods and future challenges, *Trends Biotechnol* 27:415–422.

Timmins, N. E., Palfreyman, E., Marturana, F., Dietmair, S., Luikenga, S., Lopez, G., Fung, Y. L., Minchinton, R. and Nielsen, L. K. 2009. Clinical scale *ex vivo* manufacture of neutrophils from hematopoietic progenitor cells, *Biotechnol Bioeng* 104: 832–840.

Tothova, Z., Kollipara, R., Huntly, B. J., Lee, B. H., Castrillon, D. H., Cullen, D. E., McDowell, E. P., Lazo-Kallanian, S., Williams, I. R., Sears, C., Armstrong, S. A., Passegue, E., DePinho, R. A. and Gilliland, D. G. 2007. FoxOs are critical mediators of hematopoietic stem cell resistance to physiologic oxidative stress, *Cell* 128: 325–339.

Traycoff, C. M., Kosak, S. T., Grigsby, S. and Srour, E. F. 1995. Evaluation of *ex vivo* expansion potential of cord blood and bone marrow hematopoietic progenitor cells using cell tracking and limiting dilution analysis, *Blood* 85:2059–2068.

Tura, O., Barclay, G. R., Roddie, H., Davies, J. and Turner, M. L. 2007. Optimal *ex vivo* expansion of neutrophils from PBSCCD34(+) cells by a combination of SCF, Flt3-L and G-CSF and its inhibition by further addition of TPO, *J Transl Med* 5.

Unwin, R. D., Smith, D. L., Blinco, D., Wilson, C. L., Miller, C. J., Evans, C. A., Jaworska, E., Baldwin, S. A., Barnes, K., Pierce, A., Spooncer, E. and Whetton, A. D. 2006. Quantitative proteomics reveals posttranslational control as a regulatory factor in primary hematopoietic stem cells, *Blood* 107:4687–4694.

Vaupel, P. 1996. Oxygen transport in tumors: Characteristics and clinical implications, *Adv Exp Med Biol* 388:341–351.

Verfaillie, C. M. 2002. Hematopoietic stem cells for transplantation, *Nat Immunol* 3:314–317.

Vitrat, N., Cohen-Solal, K., Pique, C., Le Couedic, J. P., Norol, F., Larsen, A. K., Katz, A., Vainchenker, W. and Debili, N. 1998. Endomitosis of human megakaryocytes are due to abortive mitosis, *Blood* 91:3711–3723.

Wang, G. L., Jiang, B. H. and Semenza, G. L. 1995. Effect of altered redox states on expression and DNA-binding activity of hypoxia-inducible factor 1, *Biochem Biophys Res Commun* 212:550–556.

Wang, G. L. and Semenza, G. L. 1993a. Characterization of hypoxia-inducible factor 1 and regulation of DNA binding activity by hypoxia, *J Biol Chem* 268:21513–21518.

Wang, G. L. and Semenza, G. L. 1993b. General involvement of hypoxia-inducible factor 1 in transcriptional response to hypoxia, *Proc Natl Acad Sci U S A* 90:4304–4308.

Wang, L., Menendez, P., Cerdan, C. and Bhatia, M. 2005. Hematopoietic development from human embryonic stem cell lines, *Exp Hematol* 33:987–996.

Williams, S. F., Lee, W. J., Bender, J. G., Zimmerman, T., Swinney, P., Blake, M., Carreon, J., Schilling, M., Smith, S., Williams, D. E., Oldham, F. and Van Epps, D. 1996. Selection and expansion of peripheral blood CD34+ cells in autologous stem cell transplantation for breast cancer, *Blood* 87:1687–1691.

Wilson, A. and Trumpp, A. 2006. Bone-marrow haematopoietic-stem-cell niches, *Nat Rev Immunol* 6:93–106.

Winkler, I. G., Barbier, V., Wadley, R.,Zannettino, A. C. W., Williams, S. and Levesque, J. P. 2010. Positioning of bone marrow hematopoietic and stromal cells relative to blood flow *in vivo*: Serially reconstituting hematopoietic stem cells reside in distinct nonperfused niches, *Blood* 116:375–385.

Xie, Y. C., Yin, T., Wiegraebe, W., He, *et al.* 2009. Detection of functional haematopoietic stem cell niche using real-time imaging, *Nature* 457:97–U102.

Yamamoto, K., Sokabe, T., Watabe, T., Miyazono, K., Yamashita, J. K., Obi, S., Ohura, N., Matsushita, A., Kamiya, A. and Ando, J. 2005. Fluid shear stress induces differentiation of Flk-1-positive embryonic stem cells into vascular endothelial cells *in vitro*, *Am J Physiol-Heart Circul Physiol* 288:H1915–H1924.

Yan, S. F., Tritto, I., Pinsky, D., Liao, H., Huang, J., Fuller, G., Brett, J., May, L. and Stern, D. 1995. Induction of interleukin 6 (IL-6) by hypoxia in vascular cells. Central role of the binding site for nuclear factor-IL-6, *J Biol Chem* 270:11463–11471.

Yang, H., Miller, W. M. and Papoutsakis, E. T. 2002. Higher pH promotes megakaryocytic maturation and apoptosis, *Stem Cells* 20:320–328.

Zauli, G., Vitale, M., Falcieri, E., Gibellini, D., Bassini, A., Celeghini, C., Columbaro, M. and Capitani, S. 1997. *In vitro* senescence and apoptotic cell death of human megakaryocytes, *Blood* 90:2234–2243.

Zeng, L., Zhang, Y., Chien, S., Liu, X. and Shyy, J. Y. 2003. The role of p53 deacetylation in p21Waf1 regulation by laminar flow, *J Biol Chem* 278:24594–24599.

Zhang, J., Niu, C., Ye, L., Huang, H., He, X., Tong, W. G., Ross, J., Haug, J., Johnson, T., Feng, J. Q., Harris, S., Wiedemann, L. M., Mishina, Y. and Li, L. 2003. Identification of the haematopoietic stem cell niche and control of the niche size, *Nature* 425:836–841.

Zipori, D. and Sasson, T. 1981. Myelopoiesis in the presence of stromal cells from mouse bone marrow: II. Mechanism of glucose dependent colony formation, *Exp Hematol* 9:663–674.

Stem Cell Bioprocessing for Regenerative Medicine

7

Donghui Jing, Abhirath Parikh, and
Emmanuel S. Tzanakakis

7.1. Introduction

The successful isolation of human embryonic stem cells (hESCs) (Thomson *et al.*, 1998) and more recently, the generation of human-induced pluripotent stem cells (hiPSCs) by reprogramming adult cells (Takahashi *et al.*, 2007; Yu *et al.*, 2007) sparked hopes for stem cell-based cures of devastating diseases, like diabetes, Parkinson's and cardiovascular maladies. Human ESCs and iPSCs — collectively termed human pluripotent stem cells (hPSCs) — are characterized by their ability for multi-lineage commitment and their extensive proliferative capacity. Most studies with stem cell populations are routinely performed in static cultures (e.g. dishes, multiwell plates), which are typically small in size and can be easily handled in a laboratory setting. Utilization of such cultures has contributed to the significant advancements in the field of stem cells.

However, most clinical stem cell protocols envisioned require a few billions of hPSCs and their progeny (Kehoe *et al.*, 2010) per patient. For example, about 1–2×10^9 cardiomyocytes are needed to compensate for damaged cardiac tissue after myocardial infarction (Jing *et al.*, 2008). Diabetes patients receiving 20×10^6 insulin-producing β cells/kg body weight may acquire insulin independence (Lock and Tzanakakis, 2007; Ryan *et al.*, 2002). Moreover, a bioartificial liver device with 10^{10} hepatocytes can support fulminant hepatic failure (Tzanakakis *et al.*, 2000). Obtaining these amounts of stem cells will require the design of bioprocesses for the robust, reproducible and tightly controlled expansion of undifferentiated hPSCs and their efficient differentiation to specific progeny.

Central to such bioprocesses are bioreactors providing a suitable environment for the *en masse* self-renewal or differentiation of stem cells. Bioreactor-based processes are utilized in industries for cellular synthesis products such as proteins, antibodies, and vaccines. Unlike these commodities however, hPSCs (or their differentiated derivatives) are themselves the products. This difference imposes separate requirements for downstream processing and unique considerations for bioreactor design. Compared to dish cultures widely used in laboratory experiments, bioreactors are advantageous because of their scalability. Online monitoring and control of bioreactor culture variables is also feasible through automation systems. Various reactor types have been investigated for stem cell cultivation including stirred-suspension vessels (Kehoe *et al.*, 2008; Schroeder *et al.*, 2005; Yin *et al.*, 2007), perfusion bioreactors (Bauwens *et al.*, 2005; Koller *et al.*, 1993; Zhao and Ma, 2005), and wave bags (Eibl *et al.*, 2010). The pros and cons of each culture modality should be taken into account to arrive at a particular reactor design that may be the most suitable for a particular application. Here, we discuss primarily the use of stirred-suspension bioreactors for hPSC cultivation. These vessels are extensively employed in biotechnology and pharmaceutical companies. Thus, one can argue that systems for hPSC production based on stirred-suspension bioreactors would be easier to translate to a commercial setting compared to entirely novel bioreactor designs. Yet, many challenges remain for culturing hPSCs in bioreactors. This is in large part due to the gaps in our knowledge of the effects of the bioreactor environment on stem cell physiology. Of note, issues of downstream processing of stem cell products are not discussed but these are of great importance for designing and developing stem cell bioprocesses.

In this chapter we will focus on (i) recent developments in scalable bioreactors for pluripotent stem cell expansion and differentiation, (ii) our current understanding of the effects of major bioreactor culture variables on stem cell physiology, and (iii) challenges lying ahead. In the last part, we will place emphasis on quantitative approaches linking the fate(s) of stem cell populations to their environment. Such models will be essential not only for predicting the outcome of stem cell cultivation under specific conditions, but also for strategies to control the *en masse* generation of therapeutically useful cells from stem cells.

7.2. Expansion and Differentiation of Embryonic Stem Cell (ESC) in Bioreactors

7.2.1. *Mouse ESCs expansion and differentiation*

Initial studies on the culture of ESCs in stirred-suspension vessels were performed with mouse ESCs (mESCs), which can be maintained without feeder cells but in the presence of leukemia inhibitory factor (LIF) (Ying *et al.*, 2003). Compared to hESCs, mESCs exhibit a shorter doubling time and lower sensitivity to culture conditions, affording experimental convenience as a model system. Mouse ESCs can grow as cell aggregates (Cormier *et al.*, 2006; Fok and Zandstra, 2005; Kehoe *et al.*, 2008; Schroeder *et al.*, 2005; Zandstra *et al.*, 2003; zur Nieden *et al.*, 2007), encapsulated in microbeads (Dang *et al.*, 2004; Magyar *et al.*, 2001), or on microcarriers (Abranches *et al.*, 2007; Fernandes *et al.*, 2007; Fok and Zandstra, 2005).

In contrast to cells seeded as small clumps, inoculating single dispersed cells into stirred-suspension culture vessels may result in aggregates with a more uniform size distribution. The average aggregate size can be controlled largely by the initial cell seeding density (Cormier *et al.*, 2006) and the rate of stirring or the ensued shear (Fok and Zandstra, 2005; Kehoe *et al.*, 2008). Mouse ESCs remain pluripotent when expanded over multiple passages involving dissociation of aggregates and re-inoculation of the dispersed cells into spinner flasks every 4–5 days (zur Nieden *et al.*, 2007). While the aforementioned studies were performed in serum-containing media, Kehoe *et al.* (2008) also showed that expansion of mESCs as aggregates in stirred-suspension vessels is feasible in defined serum-free media.

Besides the expansion of mESCs, this culture mode has also been employed for their differentiation as embryoid bodies (EBs). Mouse ESCs, genetically modified with the insertion of a construct consisting of the alpha myosin heavy chain (αMHC) promoter and the aminoglycoside phosphotransferase (neomycin resistance) gene, were cultured in serum-containing medium with retinoic acid (RA) and no LIF (Zandstra *et al.*, 2003). Cardiomyocyte-like progeny was selected by treating the cells with G418 in a 250-ml spinner flask. In a similar manner, $\sim 1.28 \times 10^9$ MHC positive cells were generated from mESC aggregates cultured in a fully controlled 2-L bioreactor (Schroeder *et al.*, 2005) for 9 days.

The culture of encapsulated mESCs was also reported (Magyar *et al.*, 2001). Mouse ESCs, encapsulated in 1.1–1.6% (v/v) alginate beads, differentiated to smooth and cardiac muscle cells. Encapsulated cells formed EBs and were cultured in dishes. Others have also shown the differentiation of encapsulated mESCs to hepatocytes (Maguire *et al.*, 2006) and insulin-producing cells (Wang *et al.*, 2009) in static cultures. Cell encapsulation can be combined with cultivation in stirred-suspension vessels. For instance, hematopoietic progenitors can be generated from mESCs within agarose capsules in stirred suspension under hypoxia (Dang *et al.*, 2004; Liu and Roy, 2005). Mouse ESCs encapsulated in alginate beads and cultured in high-aspect ratio vessels (HARV) can be differentiated to osteogenic lineages, giving rise to mineralized constructs (Hwang *et al.*, 2009). Our group also demonstrated that mESCs, entrapped in poly-L-lysine (pLL)-coated alginate beads, form aggregates after liquefaction of the bead core (Jing *et al.*, 2010) (see also Section 7.2.2.2). The aggregate size, which is an important parameter affecting the differentiation trajectory of stem cells (Niebruegge *et al.*, 2009), can be controlled by adjusting the mESC loading density (cells/ml of alginate solution). Moreover, pLL-coated alginate capsules are permeable to molecules of 106 kDa (Jing *et al.*, 2010) or even 300 kDa (Awrey *et al.*, 1996). Thus, factors such as activin, Wnt3a, and BMPs can be added to the medium for guiding the differentiation of encapsulated cells. Indeed, mESCs can be coaxed to heart cells in a serum-free medium with BMP4. Utilizing this medium in a stirred-suspension vessel of encapsulated mESCs leads to higher fractions of $Nkx2.5^{+}$ cells than in dish cultures.

The microcarrier stirred-suspension culture has been explored for the expansion and differentiation of mESCs. Microcarrier cultures are characterized by high surface area-to-volume ratio, leading to higher cell densities and economic utilization of media. The variety of available compact microcarriers (e.g. glass, polystyrene) and their potential for surface functionalization makes this culture modality suitable for accommodating diverse types of anchorage-dependent cells. Mouse ESCs on solid beads expand about 190 times over 15 days (Fok and Zandstra, 2005) and express Oct4, Nanog and SSEA-1. The cells also form EBs after their dissociation from the beads and upon differentiation display markers such as Flk-1, CD34, myosin heavy chain (mesoderm) and HNF3β (endoderm). The growth and metabolism of mESCs was also investigated under different inoculum and bead concentrations, and microcarrier types (Cytodex 3, Cultispher S) with or without serum (Abranches *et al.*, 2007; Fernandes *et al.*, 2007). In serum-supplemented media, the increase in mESC concentration was comparable on both bead types (Cytodex 3: 38 $\pm$ twofold vs. Cultispher S: 50 $\pm$ 15-fold). However, the cells grew more extensively (70-fold) on macroporous beads (Cultispher S) under serum-free conditions. This may be attributed to the protection provided by the porous structure to cultured cells against shear. These mESCs were further shown to differentiate into neuronal cells. Differentiation of mESCs in macroporous beads into cardiomyocyte-like cells has also been demonstrated (Akasha *et al.*, 2008).

7.2.2. *Human pluripotent stem cell expansion and differentiation*

7.2.2.1. *Aggregate culture*

Differences in the *in vitro* maintenance and physiology of hPSCs compared to their murine counterparts hamper the direct translation of the methods developed for mESC cultivation in bioreactors to large-scale hPSC culture. Dispersed hESCs, for instance, exhibit poor survival after colony dissociation (Chen *et al.*, 2010; Ohgushi *et al.*, 2010) whereas this is not the case for mESCs. Nonetheless, recent significant advances in hPSC maintenance have helped to overcome some of the major obstacles towards the culture of hPSCs in scalable bioreactors. Unlike initial derivations and cultivation of hESC lines on mouse embryonic fibroblasts (mEFs), nowadays hPSCs can be maintained in defined media without mouse feeder cells (Mallon *et al.*, 2006). Yet, most formulations reported to date, include animal-derived factors, which pose a serious challenge if the cells are intended for clinical use.

Moreover, hPSCs, seeded as clumps in these vessels, form hEBs within which non-specific differentiation transpires. The culture of hESCs as hEBs was investigated in slow-turning lateral vessels (STLV) and HARV (Gerecht-Nir *et al.*, 2004a), which are characterized by low shear. In STLV cultures, the concentration of hESCs increased 70-fold to 3.6×10^7 cells/ml over 28 days. When compared to static cultures, the DNA concentration and total protein content were threefold higher in the STLV. Cell viability remained high (despite an initial decrease) and the expression of markers from all three germ layers was detected. In contrast to the STLV, massive agglomeration of hESCs was observed in HARV. Thus, the low shear in these "microgravity" vessels may minimize its potential (unknown) effects on hPSC gene expression and the balance between self-renewal and differentiation but does not necessarily help to avoid the formation of oversized clusters due to excessive aggregation.

The cultivation of hESCs as aggregates in STLVs, was also compared to that in spinner flasks with either double glass ball bulb-shaped impeller or paddle impeller, Erlenmeyer flasks, and Petri dishes (Yirme *et al.*, 2008). The effects of variables such as the shear at seeding, concentration of inoculated cells, and initial size of hESC clumps were noted. Cells in spinner flasks featuring a glass bulb impeller reached the highest concentration of 5×10^6 cells/ml (6.4-fold expansion) in 10 days when cultured at 75 rpm. Only a 1.2-fold expansion of hESCs was noted for STLV cultures rotated at 16 rpm. Human ESCs cultured in spinner flasks with paddle-type impeller at 105 rpm grew 2.2-fold to 1.7×10^6 viable cells/ml over the same period. Cells of all germ layers were present in the resulting EBs including beating cardiomyocyte-like cells, α-fetoprotein$^+$ cells, CD31$^+$ endothelial cells, and β3-tubulin$^+$ neuroectoderm cells. Despite the lack of details regarding the selection of the agitation/rotation rate (and corresponding shear levels) for each culture modality, these findings demonstrate the impact of bioreactor-operating conditions on the propagation of hPSCs.

Prior to this study, Cameron *et al.* (Cameron *et al.*, 2006) showed that hESCs expand 15-fold (vs. fourfold in static cultures) when cultured in spinner flasks with serum-containing media as EBs for 21 days. These cells turned mainly to hematopoietic progenitors, but cells expressing ectoderm- and endoderm-specific genes were also present.

As already mentioned, seeding hPSCs in stirred-suspension cultures as single cells instead of as clumps is preferable but dispersed hPSCs exhibit low viability. Treating hESCs with a Rho-associated kinase (ROCK) inhibitor (e.g. Y-27632) before and during colony dissociation, enhances the survival of the cells and diminishes their spontaneous differentiation (Watanabe *et al.*, 2007) even in serum-free suspension cultures. Cells treated with Y-27632 form floating aggregates in low-adhesion dishes without loss of their viability and differentiate toward ectoderm. Exposure to Y-27632 also promotes hESC self-renewal (Harb *et al.*, 2008) even after cryopreservation (Martin-Ibanez *et al.*, 2008), suggesting that the observed effect is independent of the hESC line used. In two recent studies, the reduction in viability of dissociated hESCs was attributed to actin–myosin contractility (Chen *et al.*, 2010) or hyperactivation of actomyosin (Ohgushi *et al.*, 2010). In both studies, treatment of colony-dissociated hESCs with blebbistatin rescued the cells from apoptosis. These findings also explain the observed effect of Y-27632 given that the actin–myosin contraction is a downstream target of ROCK regulation.

ROCK inhibition in hESCs may be combined with hESC colony dissociation methods, which limit the risk for unprompted differentiation. Human ESC colonies treated with accutase (a commercially available mix of proteases and DNAses) can be dissociated into single cells and passaged multiple times without significant differentiation (Bajpai *et al.*, 2008). These findings prompted us (Kehoe *et al.*, 2010) and others (Krawetz *et al.*, 2009) to adapt such protocols to the culture of hESCs in stirred-suspension vessels. Human ESCs from colonies dissociated with accutase in the presence of 10μM Y-27632 and seeded in stirred-suspension bioreactors form aggregates which grow for at least 1 week (Figure 7.1a). Soluble Matrigel is added to facilitate the initial clustering of dispersed hESCs and the ROCK inhibitor is kept with mEF-conditioned medium throughout the culture time. This leads to the maintenance of the expression of OCT3/4, NANOG and SSEA-4. The cell concentration increased sixfold over 1 week and viability remained above 80%. Krawetz *et al.* (2009) also reported the propagation of hESCs

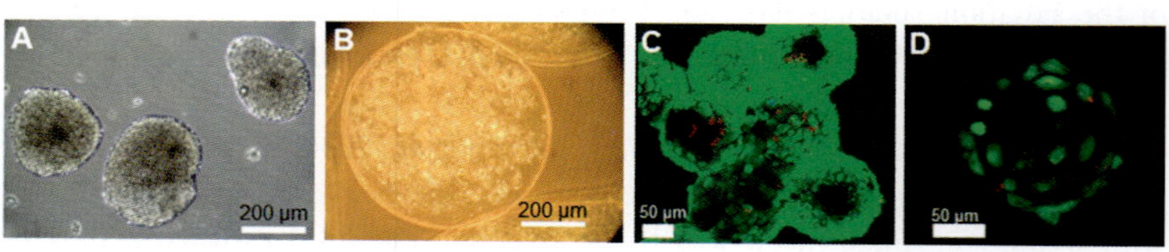

Figure 7.1. Human ESCs cultured in a stirred-suspension bioreactor as (a) aggregates, (b) encapsulated in alginate beads, or (c–d) on microcarriers. Cells on microcarriers are stained with fluorescein diacetate (green-live cells) and propidium iodide (red-dead cells).

as aggregates in stirred-suspension bioreactors following a similar protocol. Cells expanded 25-fold over a 6-day passage and for multiple passages. The authors noted that prolonged exposure to the ROCK inhibitor causes an extended lag in hESC growth. Similarly, hESCs exposed to Matrigel grew slowly over the first passage but their growth rate was restored in subsequent passages. These observations may explain the discrepancy in the fold expansion of hESCs in these studies. Moreover, incubation of hESCs with rapamycin suppresses the emergence of hESC-derived fibroblasts that destabilize hESC aggregates and reduce cell viability. In the bioreactor, hESCs grown for 21 days retained the expression of pluripotency markers, were karyotypically normal and gave rise to teratomas when implanted in SCID mice.

Issues pertinent to how rapamycin, Y-27632 and blebbistatin contribute to the maintenance of hPSC pluripotency warrant further investigation (Samuel and Olson, 2010). Moreover, whether hPSCs within aggregates cultured in stirred-suspension vessels can be directed to particular cell fates while avoiding the nuisances of non-specific and largely uncontrolled EB differentiation is currently an open question. Nonetheless, implementation of this culture modality in future stem cell bioprocesses appears very promising.

7.2.2.2. *Encapsulated cell culture*

Besides the propagation of hPSCs as aggregates, stirred-suspension bioreactors can be employed for culturing encapsulated stem cells (as already mentioned for mESCs). Hydrogels based on agarose, dextran and gelatin have been used to encapsulate cells. Alginate however appears to be the biomaterial of choice due to its biocompatibility, biosafety and adjustable permeability (Orive *et al.*, 2003). Conditions for the preparation of alginate are also benign for cells (e.g., non-cytotoxic solvents and physiological temperature). Interestingly enough, various high-throughput designs are currently available (e.g. electrostatic generator (Bugarski *et al.*, 1993), air-driven multi-nozzle droplet generator (De Vos *et al.*, 1997)) for the encapsulation of a wide range of mammalian cells (Borlongan *et al.*, 2008; Cornolti *et al.*, 2009; Moustafa *et al.*, 2006) even under good manufacturing practice (GMP) conditions (Schwinger *et al.*, 2002).

An important aspect of this system is its flexibility for culturing cells either as a dispersed phase in solid capsules or as aggregates after liquefying the capsule core. When encapsulated in solid capsules, ESCs have the propensity to form multiple, small EBs (Figure 7.1b). This was shown to increase the efficiency of encapsulated mESC commitment to hepatocytes in static cultures (Maguire *et al.*, 2007). However, differentiation along other lineages, e.g. cardiac muscle cells, may

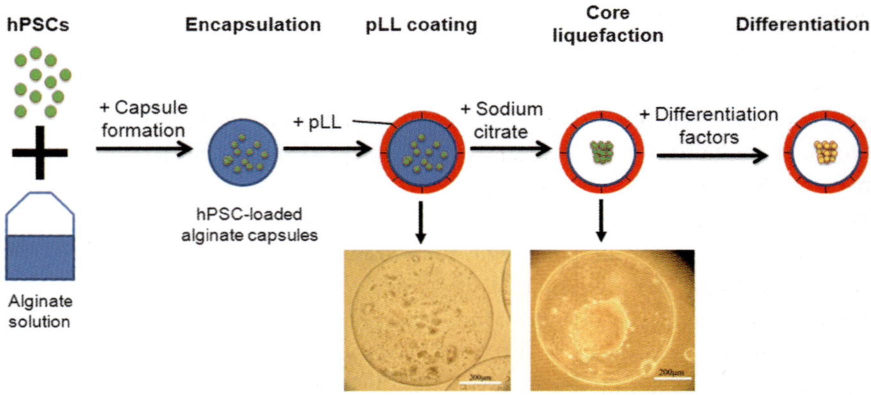

Figure 7.2. A schematic of the encapsulation of hPSCs in pLL-coated alginate capsules for differentiation.

require the formation of EBs with precise control of their size (Niebruegge *et al.*, 2009). To that end, we have shown that hESCs can be encapsulated in pLL-coated alginate beads (Jing *et al.*, 2010). Upon incubation with a divalent ion-chelating agent [e.g. sodium citrate or ethylenediaminetetraacetic acid (EDTA)], the bead core is liquefied without compromising the overall mechanical integrity of the capsules (Figure 7.2). Coating the beads with pLL also permits the use of more dilute alginate matrix with reduced resistance to the transport of nutrients/metabolites in and out of the capsules. In the context of bioreactor culture, encapsulation can shield stem cells from shear, e.g. due to stirring (see Section 7.3). More importantly, the size of the aggregates formed within the liquefied-core alginate capsules can be controlled by adjusting the concentration of hESCs seeded in the alginate solution prior to encapsulation. We have shown that hESCs can be differentiated to cardiomyocyte-like cells after their encapsulation in pLL-coated alginate beads, core liquefaction and culture in spinner flasks.

Implementing this culture modality in bioreactors beyond the laboratory scale will require the development of systems ensuring the high-throughput encapsulation of cells. In fact, setups have been proposed featuring multiple nozzles (De Vos *et al.*, 1997) or rotating jet cutters (Schwinger *et al.*, 2002) for the rapid encapsulation of large quantities of mammalian cells. For instance, 3.2×10^5 mouse fibroblasts/ml alginate can be encapsulated under GMP conditions at a rate of 330 ml of alginate/hr or 5200 capsules/sec although rates of 10,000 capsules/sec are feasible (Schwinger *et al.*, 2002). A 10-L bioreactor can be loaded with 2×10^4 cells/ml in less than 1 hour via two such devices working simultaneously. Of course, the encapsulation process should be optimized to avoid the generation of high shear which can impact the viability and differentiation potential of stem cells.

Stem cell encapsulation can be combined with the entrapment of molecules supporting the self-renewal or commitment of cells along particular lineages. To that end, technologies are constantly improving for the generation of scaffolds loaded with peptides (Ferreira *et al.*, 2007; Garreta *et al.*, 2006), growth factors (Ferreira *et al.*, 2007), DNA (Bolliet *et al.*, 2008; Hosseinkhani *et al.*, 2006), and other bioactive molecules, e.g. dexamethasone (Nuttelman *et al.*, 2006). Such scaffolds can be employed in conjunction with bioreactor cultures for directing the fate of large numbers of stem cells.

Encapsulated cells may be used directly for transplantation reducing post-processing for the removal of the capsule material. The capsule material can protect implanted cells from attacks by the host's immune system. In some instances however, elimination of the encapsulation matrix may be necessary to minimize the risk for inflammation [e.g. due to the pLL (Strand *et al.*, 2001)] or to facilitate coupling of implanted cells with the host tissue. These considerations should be taken into account when designing an encapsulation strategy involving bioreactors for the generation of specific cell types.

7.2.2.3. *Microcarrier culture*

Besides their culture in stirred-suspension bioreactors as aggregates or after encapsulation, cells can be propagated on microbeads. Microcarrier systems have been used for growing anchorage-dependent animal cells [e.g. HEK293 and Vero cells (Mendonca *et al.*, 1993; Wu *et al.*, 2002)] for the production of viral vaccines and therapeutic proteins. Recently, the cultivation of hESCs on different types of microcarriers either in low-adherence plates (Nie *et al.*, 2009; Phillips *et al.*, 2008) or in bioreactors (Fernandes *et al.*, 2009; Lock and Tzanakakis, 2009; Oh *et al.*, 2009) was demonstrated. Nie *et al.* (2009) reported the microcarrier culture of hESCs in dishes for over 11 passages. Human ESCs grow on microcarriers with similar doubling times as on tissue culture plates. The cells preserve their pluripotency as evidenced by the expression of pertinent markers and *in vitro* differentiation toward the three germ layers. Compared to freely suspended hESCs, cells cultured on microcarriers and cryopreserved exhibit a higher recovery after thawing. Interestingly enough, Cytodex 3 beads appear to be superior for the attachment and preservation of hESC viability. This is in contrast to another report (Phillips *et al.*, 2008) where hESCs on these beads exhibited the poorest growth with little or no recovery of viable cells after 2–3 days. Such discrepancies may be due to the different hESC lines used, and the different methods for bead preparation (e.g. seeding with mEFs vs. coating with Matrigel) and cell loading.

We provided the first account (Lock and Tzanakakis, 2009) of successful expansion and differentiation of hESCs on microcarriers in stirred-suspension bioreactors (Figures 7.1c and 7.1d). In a subsequent report, we also demonstrated for the first time, the culture of hiPSCs in microcarrier stirred-suspension culture vessels (Kehoe *et al.*, 2010). Human H9 and H1 ESCs seeded at 5–20 × 10^4 cells/ml populate Matrigel-coated polystyrene beads with a 23–30% efficiency (Lock and Tzanakakis, 2009) compared to only ~7% observed in dish cultures (Forsyth *et al.*, 2006). The growth rate of hESCs on microcarriers greatly depends on the agitation rate (and induced shear) in the spinner flasks. The doubling times of hESCs on microcarriers cultured at 45 and 60 rpm are similar to dish cultures (~36 hours) but growth is severely retarded with more vigorous stirring. Pluripotent hESCs on microcarriers maintain a constant expression of OCT3/4A, NANOG, REX1, SSEA-4 and TRA-1–81 during their expansion. Furthermore, the expression of genes, which is a characteristic of germ layer differentiation, is either absent or minimal. The same methodology (Kehoe *et al.*, 2010) was applied successfully to the propagation of B12-3 hiPSCs (Huangfu *et al.*, 2008) on microcarriers in spinner flasks. A ~20-fold increase in the live hiPSC concentration is achieved over 8 days of self-renewal. More recent reports by others (Fernandes *et al.*, 2009; Oh *et al.*, 2009) investigating the culture of hESCs in microcarrier stirred-suspension cultures agree with our findings, lending support to the utilization of this culture modality in future scalable bioprocesses for stem cells.

Besides the self-renewal of hESCs in a microcarrier culture, guiding their fate along specific lineages is also possible (Lock and Tzanakakis, 2009). After propagating undifferentiated hESCs for a week on microcarriers, factors can be introduced promoting differentiation to definitive endoderm (DE) cells, which are precursors of pancreatic islet cells and hepatocytes. Endoderm differentiation of H1 and H9 hESCs is prompted in medium containing activin and Wnt3a in 0–2% serum or serum replacer. Cells co-expressing FOXA2 and SOX17 (characterizing DE cells) emerge in the bioreactor with efficiencies exceeding 80%, compared to ~63% in dish cultures.

Thus, under carefully selected conditions, the bioreactor environment facilitates the recapitulation of aspects of *in vivo* cell differentiation and promotes the commitment of stem cells with high efficiency. Ongoing work focuses on directing hPSCs in microcarrier bioreactor cultures further along the pancreatic endocrine cell lineage, as well as toward other therapeutically useful cell types including cardiomyocytes. Our results illustrate that microcarrier bioreactors are attractive for the *en masse* production of therapeutically useful cells from hESCs and hiPSCs.

7.3. Agitation/Shear and Stem Cell Culture

Cells cultured in bioreactors with agitation are exposed to hydrodynamic forces, which even at low intensity may transduce effects via mechanisms that are currently largely unknown. Thus, the agitation and ensuing shear are important variables in the operation of stem cell bioreactors.

For cell aggregate culture systems, stirring-induced shear may help to control the size of clusters by preventing the formation of oversized clusters but high shear levels can be damaging for cells. Similarly, clustering ("bridging") of cell-populated microcarriers can be minimized through appropriately selected shear levels. Cells encapsulated in various matrices (e.g. alginate beads) are protected from shear when cultured in stirred-suspension bioreactors although high enough shear can compromise the mechanical integrity of the scaffold.

The size of cell aggregates, above which damage is likely to occur, can be estimated considering a model flow field of isotropic turbulence in the reactor. Fluid flow around the impeller generates eddies of decreasing size down to the range of viscous dissipation. The size η of the smallest eddies [also termed Kolmogorov eddies (Papoutsakis, 1991)] is calculated as

$$\eta = \left(\frac{\nu^3}{\epsilon}\right)^{\frac{1}{4}} \tag{1}$$

where ν is the kinematic viscosity of the medium. The viscous energy dissipation per unit fluid mass ϵ is determined as

$$\epsilon = N_P \frac{D_i^5 N^3}{V_d} \tag{2}$$

N_P and V_d are the dimensionless power number and the dissipation volume, respectively. The Kolmogorov eddy size decreases as the agitation speed increases. Aggregates with similar or larger size compared to these eddies are damaged the most and the same applies to cells cultured on microcarriers (Cherry and Papoutsakis, 1988; Cherry and Papoutsakis, 1989; Croughan *et al.*, 1987). Cells on microcarriers may also be damaged by bead-to-bead collisions although their effects are more challenging to quantify. Compared to cell-cultured aggregates or on microcarriers, single cells in suspension can be damaged by cell-size eddies. However, in the absence of bubbles (e.g. due to sparging), the generation of such small eddies requires impractically intense stirring.

An estimate of the maximum shear stress (Cherry and Kwon, 1990) $\tau_{\max}$ which the cells experience is given by

$$\tau_{\max} = 5.33\rho(\epsilon v)^{\frac{1}{2}} \tag{3}$$

Human embryonic kidney cells cultured on surfaces (Stathopoulos and Hellums, 1985) are removed at ~0.65 Pa whereas significant damage ensues at 26 dyn/cm^2. Others (Cherry and Kwon, 1990) have suggested that shear stresses of 15–30 dyn/cm^2 reduce cell viability, although such a range is heavily cell type-specific.

Mouse ESCs have been expanded (Cormier *et al.*, 2006) as aggregates in spinner flasks at 80–100 rpm corresponding to a maximum shear stress range between 4.5–6.1 dyn/cm^2. At 120 rpm or 7.8 dyn/cm^2, mESCs exhibit minimal growth and extensive damage. For comparison, STLVs used for ESC culture exhibit very low shear stress levels (Begley and Kleis, 2000) (~10^{-2} dyn/cm^2), but the cells form fewer and mostly oversized aggregates, leading to spontaneous differentiation more readily. Thus, conditions should be carefully selected to maintain the shear in the bioreactor within a range that is well tolerated by the cells while it allows for a better control of aggregation. Moreover, agents traditionally used for reducing shear in bioreactor cultures [reviewed in Kehoe *et al.* (2010) and references therein] have not been tested in stem cell cultures to date.

The effects of shear on the differentiation potential of stem cells are less well understood. Mouse ESCs exposed to 10 dyn/cm^2 raise their histone acetylase activity, resulting in remodeling of their chromatin structure with concomitant increases in cardiac- and vascular-specific markers (Illi *et al.*, 2005). Even mESC-derived Flk-1$^+$ cells subjected to shear stress of 1.5–10 dyn/cm^2 exhibit increased proliferation and expression of endothelial proteins (Flt-1, VE-cadherin and PECAM-1) (Yamamoto *et al.*, 2005). Human ESCs cultured under cyclic strain retain their pluripotency (Saha *et al.*, 2006), most likely due to TGF-β/Activin/Nodal signaling induction (Saha *et al.*, 2008). In addition, preferential differentiation towards cardiac cells has been noted for mESCs cultured in rotary orbital suspension and spinner flasks (Sargent *et al.*, 2009; zur Nieden *et al.*, 2007).

Discoveries on the mechanotransduction of stem cells will be expedited by microfluidic technologies (Kamei *et al.*, 2009; Kim *et al.*, 2006). Microfluidic platforms allow for precise control of the microenvironment, including the fluid flow and ensuing shear patterns while changes in stem cell phenotype are monitored. For instance, mESCs cultured in a microfluidic device under flow-induced shear stress (0.016–16 dyn/cm^2) upregulate the expression of the epiblast marker Fgf5 (Toh and Voldman, 2010). The variability of stem cell lines (e.g. genetic background, propensity for differentiation) translates to differences in their response to external stimuli including (most likely) shear. Thus, such devices could be used to determine a range of shear appropriate for expansion or differentiation to a particular lineage of a given stem cell type. Then, conditions for bioreactor culture can be selected accordingly.

7.4. Oxygen Tension and Stem Cell Culture

Stem cells in the reproductive track experience 60 mm Hg of O_2 tension (pO_2; in hamsters and rabbits), and 24 mm Hg during blastocyst formation and implantation (Fischer and Bavister, 1993). The uterus of rhesus monkeys is also characterized by a low pO_2 of 11 mm Hg. Why such low pO_2 exists in the natural environment of ESCs is unclear. Yet, this environment is hypothesized to protect the stem cells from O_2 toxicity while inducing the modulation of genes, orchestrating the earliest steps of development.

Hypoxia promotes differentiation of various stem/progenitor cells (see also Chapter 6). In mESCs, low pO_2 negatively regulates the LIF-STAT3 signaling pathway via the hypoxia inducible factor (HIF)-1α gene (Jeong *et al.*, 2007). This is in agreement with the increase in the efficiency of mESC (Bauwens *et al.*, 2005) and hESC (Niebruegge *et al.*, 2009) differentiation to cardiomyocytes under hypoxic conditions (4% O_2) in bioreactors. In addition, fewer spontaneous chromosomal aberrations are observed (Forsyth *et al.*, 2006) in hESCs cultured at 2% instead of 21% O_2. Human ESCs maintained under 4% or 20% O_2 differentially regulate (mainly HIF-controlled) genes encoding enzymes for carbohydrate metabolism and cellular redox state, but pluripotency genes such as OCT4 and NANOG remain mostly unaltered (Westfall *et al.*, 2008). Stem cells at 20% O_2 also show upregulation of genes associated with lineage specification, suggesting that hypoxic conditions favor the maintenance of pluripotency. In contrast, another report (Chen *et al.*, 2009) indicated that 5% pO_2 is not advantageous for preventing hESCs from differentiating compared to their culture at normoxia. Changes brought about by O_2 levels on the stem cell phenotype are still under intense investigation.

It is thus obvious that O_2 regulation is essential for the culture of ESCs in bioreactors. Cells within clusters experience limited O_2 transfer with increasing aggregate size. Oxygen diffusion models developed for multicellular spheroids (Adam and Maggelakis, 1990; Casciari *et al.*, 1992; Grossmann, 1984) can be applied to stem cell clusters (Gassmann *et al.*, 1996). In addition, the diverse ultrastructural attributes of these clusters should also be considered. Stem cell clusters may undergo compaction as observed for other cell types (Tzanakakis *et al.*, 2001), contain cysts or form external layers of extracellular matrix with variable thickness (Sachlos and Auguste, 2008). The transfer of O_2 from the bulk medium should be at least equal to the specific O_2 consumption rate (OCR) of ESCs. For H9 cell aggregates, the measured OCR is 1.7×10^{-18} mol O_2/cells (Kehoe *et al.*, 2010). The OCR strongly depends on culture conditions and therefore its value should be determined experimentally for a particular hPSC line and culture system.

The transfer of O_2 in the bulk medium also depends on the mode of aeration. Larger vessels employ spargers for forced oxygenation since headspace aeration is not sufficient to maintain proper levels of dissolved O_2. However, the bubbles due to sparging contribute to the agitation-induced shear stress inside the bioreactor. The cell damage caused by cell-to-bubble interactions in the bulk is not as significant (Michaels *et al.*, 1996) as that induced by bubble coalescence and breakup close to the air-medium interphase (Papoutsakis, 1991). This problem can be tackled with adjustments to the sparging gas flow rate and the use of surface tension-lowering additives (e.g. pluronics). A detailed analysis of the effects of such additives on stem cell proliferation and pluripotency is not available yet. Mouse ESCs cultured as EBs in a 2-L stirred tank with sparging and in the presence of 0.0125% Antifoam C retain their viability and adopt a cardiomyocyte-like phenotype (Schroeder *et al.*, 2005). Whether expansion of undifferentiated stem cells in a bioreactor with active (instead of headspace) aeration in the presence of surface active agents is feasible remains an open question.

7.5. Challenges

Previous advances in the area of animal cell culture have been pivotal in the development of methods for the cultivation of stem cells and their derivatives in bioreactors. However, as the use of stem cells is moving closer to clinical applications, unique challenges come in focus. In typical cell culture processes major emphasis is placed on maximizing the yield of high-quality secretable products such as recombinant proteins, vaccines, and antibodies whereas the cells are commonly discarded. In contrast, the cellular component itself is the actual product in stem cell bioprocesses. Therefore, emphasis in designing downstream processes should be placed on the isolation, purification, and polishing of stem cell progeny rather than metabolites. New technologies characterized by high fidelity and throughput will be necessary for sorting the desired cell type in clinically relevant quantities while eliminating undifferentiated (and potential tumorigenic) cells. Real-time monitoring of stem cells, cultured in the bioreactor for markers specific to a particular lineage, and chromosomal abnormalities, which often emerge with prolonged culture (Draper *et al.*, 2004; Maitra *et al.*, 2005), do not exist yet. One possibility is to use microfluidic chips specifically designed for online monitoring of the bioreactor culture.

The engineering of large-scale bioprocesses based on stirred-suspension bioreactors can be facilitated by well-established design rules and heuristics for the scale-up of this culture modality. Yet, such bioprocesses employed for the production of clinical-grade stem cell products should be GMP compliant. The bioreactor with its process monitoring and control systems, the maintenance of sterile conditions, and the management for efficient operation are all part of GMP protocols. Such protocols also apply to the cellular component involved in the process. However, most stem cell lines derived in recent years have been exposed to animal cells or proteins. Native hESCs exhibit low immunogenicity (Drukker *et al.*, 2006; Li *et al.*, 2004) but their proximity/contact with animal cells may result in adopting immunogenic molecules (Martin *et al.*, 2005). Moreover, the effects of the culture environment on the immunoprofile of stem cells are unclear. It should be noted that derivation of hiPSCs without the use of animal-derived components (Hayashi *et al.*, 2010; Rodriguez-Piza *et al.*, 2010) is expected to alleviate some of the aforementioned issues (see Section 7.5.1) and facilitate the generation of GMP-conforming stem cell lines.

The cost per unit product decreases as the bioreactor size for stem cell products increases. However, small scale may be sufficient for generate patient-specific cells. If so, many of the scale-up issues for stirred-suspension bioreactors become less relevant. Bioreactors with working volumes of a few hundred mL to a few liters may be sufficient to produce the cell quantities needed per patient. Still, the development of defined media and substrates for the culture of human pluripotent stem cells and the modeling of stem cell populations are deemed significant, irrespective of the scale of the process and are discussed below.

7.5.1. *Chemically defined media and substrates for stem cell culture*

Human ESCs were originally (Thomson *et al.*, 1998) derived and propagated on layers of mEFs, rendering them unsuitable for clinical use. Replacement of mEFs by human fibroblast lines was suggested as a way to circumvent problems stemming from the presence of xenogeneic cells in hESC cultures. Human fibroblasts (Genbacev *et al.*, 2005; Richards *et al.*, 2002) including commercially available cell lines have been used as feeder cells for the derivation and culture of hESCs for over 20–30 passages (Cheng *et al.*, 2003) in media supplemented with a serum replacer (Price *et al.*, 1998). It should be noted however that most of these mEF surrogates are maintained in serum. Moreover, the use of any feeder cells (human or other) still requires their separation from the stem cells or their derivatives — quite an onerous task as production moves to large scale.

The introduction of Matrigel for coating surfaces for hESC and hiPSC growth has been a major step toward the elimination of feeder cells. Pluripotent cells can be cultured in mEF-conditioned medium supplemented with basic fibroblast growth factor (bFGF) (Rosler *et al.*, 2004) and serum replacer. Even bFGF alone in non-conditioned medium supports the expansion of undifferentiated H9 and H7 hESCs for 15 passages (Xu *et al.*, 2005a). Aberrant differentiation of cultured hESCs is further limited with the inclusion of Noggin, which suppresses bone morphogenetic protein (BMP) signaling (Xu *et al.*, 2005b).

Nonetheless, human stem cells grown on Matrigel with pro-self-renewal factors tend to differentiate over multiple passages (Ullmann *et al.*, 2007). Also, Matrigel is extracted from mouse Engelbreth-Holm-Swarm tumors (Kleinman *et al.*, 1982). We have also shown that soluble Matrigel promotes the formation of aggregates in suspension after dissociation of hESC colonies into single cells (Kehoe *et al.*, 2010). Similarly, covering compact microcarriers with Matrigel aids the intercellular adhesion of hESCs and hiPSCs cultured in stirred-suspension vessels. Various approaches have been suggested for the elimination of xenogeneic matrices. For instance, the surface of microcarriers can be customized via functionalization (Kato *et al.*, 2003) with various synthetic peptides (Varani *et al.*, 1993) or extracellular matrix (ECM) molecules to accommodate the adhesion, self-renewal, and differentiation needs of hPSCs. Much information needed for the customization of the microcarrier surface can be obtained from studies on ECM biology and engineering, especially in the context of stem cell self-renewal and differentiation. Recent reports show that stem cells can be cultured on surfaces coated with synthetic peptides and polymers (Melkoumian *et al.*, 2010; Rodin *et al.*, 2010; Villa-Diaz *et al.*, 2010). The results from these studies prove that artificial substrates can be engineered for the maintenance and differentiation of stem cells but questions remain. For example, Melkoumian *et al.* (2010) concluded that hESCs did not adhere well on surfaces covered with laminin-derived peptides whereas Rodin *et al.* (2010) demonstrated that hESCs can be cultured on laminin-511 for more than 20 passages without changes in their proliferation and expression of pluripotency markers. Does a universally suitable matrix exist for stem cell maintenance? Or, should matrices be customized for each hPSC line? Further investigations are warranted to answer these questions.

Besides substrates free of xenogeneic components, the media for growing hPSCs should be defined. To that end, Ludwig *et al.* (2006) showed that hESCs can be sustained for up to 6 months in the absence of feeder cells on human matrix components with medium containing only recombinant or human proteins. Others (Yao *et al.*, 2006) demonstrated the maintenance of hESCs and their directed differentiation in a chemically defined medium with the B27 and N2 supplements.

Media for the cultivation of hPSCs should not only be free of animal products, but also inexpensive, for future bioprocesses to be economically appealing. Growth factors and cytokines used for differentiation are costly, making their use in large scales prohibitive. The use of these factors can be minimized or eliminated, for example, by utilizing natural and synthetic small molecules supporting stem cell self-renewal or differentiation. The synthesis and purification of such moieties should be economical. Small molecules could be used to target signal transduction pathways such as the Wnt, Hedgehog, retinoid and the NF-κB, which regulate stem cell fate decisions (Fraichard *et al.*, 1995; Graichen *et al.*, 2008; Li *et al.*, 2008; Passier *et al.*, 2005; Pevsner-Fischer *et al.*, 2007; Wada *et al.*, 2009; Wobus *et al.*, 1997; Xiong *et al.*, 2005; Zhu *et al.*, 2009). For example, mESCs and hESCs remain undifferentiated in culture when treated with 6-bromoindirubin-3′-oxime (BIO), which activates the canonical Wnt/β-catenin pathway by selectively inhibiting GSK-3β (Sato *et al.*, 2004). Mouse ESCs maintain their pluripotent state when treated with IQ-1, which prevents β-catenin from switching between transcriptional coactivators CBP and p300 (Miyabayashi *et al.*, 2007). Small molecules targeting the Wnt or Hedgehog pathway also promote differentiation of stem cells to cardiac (Tseng *et al.*, 2006), hematopoietic (Naito *et al.*, 2006), neuronal (Ding *et al.*, 2003), and bone (Wu *et al.*, 2004) cell phenotypes. Recent advances in high-throughput screening technologies and combinatorial chemistry have made easier the analysis of small molecule libraries (Peterson *et al.*, 2000) for the identification of molecular interactions leading to particular stem cell responses [see reviews (Ding and Schultz, 2004; McNeish, 2007)]. Structural and functional information for such small molecules can be found at the public database PubChem (http://pubchem.ncbi.nlm.nih.gov/), established as part of the Molecular Libraries Roadmap Initiative by the National Institutes of Health (NIH).

7.5.2. *Modeling of stem cell populations*

Quantitative models are integral to the design, operation and maintenance of bioprocesses. Such models link state parameters (e.g. biomass content, antibody or protein concentration, virus titer etc.) to adjustable culture variables (e.g. dilution rate, pH, pO_2, substrate concentrations etc.). The equations involved combine material and energy balances with models for cell growth, oxygen and nutrient consumption, and metabolite production (bioreactions). Although process modeling is not a new area, stem cells bioprocesses present distinct complexities. Frameworks for stem cell bioprocesses should account for the fate changes of individual stem cells, rather than average changes in the population as a whole. Various deterministic, stochastic, and hybrid models have been proposed for stem cell populations and have been reviewed elsewhere (O'Neill and Schaffer, 2004; Viswanathan and Zandstra, 2003; Yener *et al.*, 2008; see also Chapter 2). Discussion in this chapter concentrates on population balance equation (PBE)-based models, which have been applied to bioreactor cultures of animal cells and can be extended by building in features relevant to stem cells.

The physiological state of cultured stem cells is sensitive to a variety of exogenous signals such as soluble factors (e.g. growth factors, signaling ligands), cell–cell and cell–substratum (including extracellular matrix) interactions, extracellular matrix and mechanotransduction. Additional parameters (e.g. agitation rate, cell seeding density) come into play when the stem cells are cultivated in bioreactors. Segregated PBE models can describe the diversity (e.g. stem cell mass, age and differentiation status) of stem cell ensembles through frequency functions, representing the probability that a randomly selected cell is in a particular state. Models of PBEs have been studied in processes including aerosol dynamics, particle aggregation, crystallization, protein precipitation, and microbial and animal cell population dynamics (in culture or *in vivo*) (Fredrickson, 1967; Fredrickson and Tsuchiya, 1963; Koch and Schaechter, 1962; Nielsen *et al.*, 1998a; Nielsen *et al.*, 1998b; Sherer *et al.*, 2008). PBE-based approaches have been utilized to predict and control the property distributions of cells cultured in bioreactors (Daoutidis and Henson, 2002; Mantzaris and Daoutidis, 2004). A general cell PBE can be written for the number distribution function $F(\tilde{x}, t)$:

$$\frac{\partial F(\tilde{x},t)}{\partial t} + \nabla_{\tilde{x}} \cdot [r(\tilde{x},\tilde{C})F(\tilde{x},t)] = 2\int_{\tilde{x}}^{\tilde{x}_{\max}} \Gamma(\tilde{z},\tilde{C})p(\tilde{z},\tilde{x},\tilde{C})F(\tilde{z},t)d\tilde{z} - [D(\tilde{x},\tilde{C}) + \Gamma(\tilde{x},\tilde{C}) + \Theta(\tilde{x},\tilde{C})]F(\tilde{x},t) \quad (4)$$

with $F(\tilde{x}, t)d\tilde{x}$ representing the number of cells (per unit of culture volume) having a physiological state representation between $\tilde{x}$ and $\tilde{x} + d\tilde{x}$ at time t. Each cell is characterized by a set of variables (e.g. mass, protein, etc.), contained in the vector $\tilde{x}$ and $\tilde{x}_{\max}$ corresponds to the maximum values of these variables. The vector $\tilde{C}$ encompasses the concentration of nutrients, cytokines, and other factors, which affect cell proliferation, differentiation and death. The birth state distribution $p(\tilde{z}, \tilde{x}, \tilde{C})$ describes how the cellular material is partitioned among daughter cells. The quantity $r(\tilde{x}, \tilde{C})$ is the single-cell growth rate (J. M. Eakman, 1966; Mantzaris and Daoutidis, 2004) and the functions $\Gamma(\tilde{x}, \tilde{C})$ and $\Theta(\tilde{x}, \tilde{C})$ are the cell division and death intensities, respectively. Cell differentiation is represented by $D(\tilde{x}, \tilde{C})$ and additional terms can be incorporated depending on the specific culture system (e.g., dilution rate for bioreactors with continuous operation). The PBEs can be coupled to equations describing the variation of the elements of $\tilde{C}$ in the culture (e.g. substrate consumption).

We have applied a mass-structured PBE model to describe the aggregate number distribution of stem cells cultured in a stirred-suspension bioreactor (Kehoe *et al.*, 2010). Similar approaches have been implemented for the aggregation of platelets and other cell types (Anderson and Petersen, 1967; Belval and Hellums, 1986; Laurenzi and Diamond, 1999). The overall balance, written after Eq. (4) is:

$$\frac{\partial F(m,t)}{\partial t} = \int_{m_0}^{\frac{m}{2}} F(m_c,t)F(m',t)K(m_c|m')dm' - \int_{m_0}^{\infty} F(m,t)F(m',t)K(m|m')dm' - \frac{\partial[r(m,S)F(m,t)]}{\partial m} \quad (5)$$

with $F(m,t)dm$ corresponding to the mean number of aggregates with mass between m and $m + dm$ in a unit volume of the culture at time t. The first term on the right side describes the formation of ESC aggregates with mass m (m-aggregates) due to aggregation of clusters with masses $m' < m$ and $m_c = m - m'$ (m_o refers to the mass of a single cell). The last two terms represent the "loss" of ESC m-clusters due to: (i) their aggregation with m'-clusters to form larger aggregates, and (ii) cell proliferation ($r(m,S)$), which depends on the concentration of substrate S. The PBE can be expanded to include other phenomena influencing the process such as aggregate breakage (Huang and Hellums, 1993; Ramkrishna, 2000)). The kernel $K(m|m')$, which captures the probability of adhesion between aggregates with masses m and m', was originally reported for spherical body collisions (von Smoluchowski, 1917) but other forms have also been considered for aggregation processes, especially under different flow fields (Aldous, 1999; Smit *et al.*, 1994). The aggregation kernel is generally in the form

$$K(m|m') = \beta(\tilde{\rho})\lambda(m,m') \quad (6)$$

where $\beta(\tilde{\rho})$ encompasses the efficiency of collisions dependent on multiple parameters described by the vector $\tilde{\rho}$ (e.g. the masses m and m', time t, shear rate). The function $\lambda(m,m')$ is the kernel of orthokinetic aggregation of the m- and m'-clusters.

Constraints on the aggregate size of stem cells cultured in a stirred-suspension vessel can be applied based on, for example, the transport of O_2 or factors important for stem cell self-renewal (e.g. bFGF) or directed differentiation. Thus, the aforementioned PBE model can be used for predicting an optimal distribution of stem cell aggregate sizes dependent on bioreactor-operating variables and time.

Stem cell populations undergoing differentiation can be modeled by PBEs written for each cell type. A mass-structured PBE-based model for the differentiation of mesenchymal stem cells into bone cells (Pisu *et al.*, 2007) included equations for three density functions for undifferentiated mesenchymal cells, chondrocytes, and osteoblasts, as well as for the temporal profiles of extracellular matrix components (glycosaminoglycans, collagen) and growth factors (BMPs and other TGF-β ligands). Utilization of cell mass as a PBE-independent variable is advantageous because the first moment of the population density function is the biomass. Also, the distribution of masses in a cell population can be determined by flow cytometry. To date however, there are scarce (if any) published data on stem cell size (or mass) distribution (Pisu *et al.*, 2007) which can be used for PBE modeling. Acquiring reliable mass-based distributions of stem/progenitor cell populations can also be challenging. For example, hESCs tend to cluster making difficult to distinguish mitotic/dividing cells based on their size by flow cytometry.

Introducing additional state variables to depict distinct cell populations can increase the predictive power and accuracy of a PBE model (Figure 7.3). A population of hPSCs directed along a particular lineage, e.g. mesoderm, can be described by the following equations:

- hPSC subpopulation:

$$\begin{aligned}\frac{\partial F_1(\tilde{x},t)}{\partial t} &+ \nabla_{\tilde{x}} \cdot \sum_{i=1}^{k} [r_i(x_i,\tilde{C})F_1(\tilde{x},t)] \\ &= 2\int_{\tilde{x}}^{\tilde{x}_{\max}} \Gamma_1(\tilde{z},\tilde{C})F_1(\tilde{z},t)p_1(\tilde{z},\tilde{x},\tilde{C})d\tilde{z} \\ &\quad - [\Gamma_1(\tilde{x},\tilde{C}) + D_1(\tilde{x},\tilde{C}) + \Xi_1(\tilde{x},\tilde{C})]F_1(\tilde{x},t) \qquad (7)\end{aligned}$$

- Mesoderm subpopulation:

$$\begin{aligned}\frac{\partial F_2(\tilde{x},t)}{\partial t} &+ \nabla_{\tilde{x}} \cdot \Sigma_{i=1}^{k} [r_i(x_i,\tilde{C})F_2(\tilde{x},t)] \\ &= 2\int_{\tilde{x}}^{\tilde{x}_{\max}} \Gamma_2(\tilde{z},\tilde{C})F_2(\tilde{z},t)p_2(\tilde{z},\tilde{x},\tilde{C})d\tilde{z} \\ &\quad - [\Gamma_2(\tilde{x},\tilde{C}) + \Xi_2(\tilde{x},\tilde{C})] \cdot F_2(\tilde{x},t) \\ &\quad + \alpha D_1(\tilde{x},\tilde{C}) \cdot F_1(\tilde{x},t) - D_2(\tilde{x},\tilde{C}) \cdot F_2(\tilde{x},t) \qquad (8)\end{aligned}$$

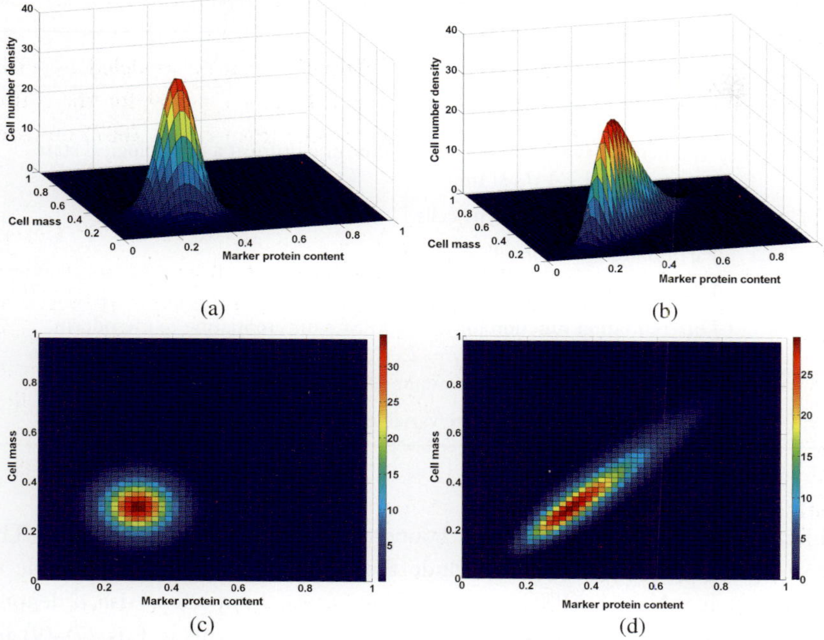

Figure 7.3. An example of a PBE model [Eq. (4)] describing a stem cell population and taking into account the growth ($\frac{m\cdot\ln(2)}{t_d}$), birth, division (left-truncated normalized Gaussian expression $\Gamma(m) = \frac{e^{-\left(\frac{m-m_c}{2\sigma}\right)^2}}{\sigma\sqrt{2\pi}\left(1-erf\left(\frac{m-m_c}{2\sigma}\right)\right)}$, and partitioning (a symmetrical beta distribution) of stem cells as well as the accumulation rate of the marker protein (assumed here to be similar to that for stem cell growth for illustrative purposes). (a) Initial distribution for cells with respect to cell mass and marker protein content (e.g. Nanog). (b) Steady state/time invariant distribution after multiple stem cell population doublings. Lower row graphs depict the density plots corresponding to the upper row graphs in (a) and (b). The values for cell mass and protein content have been normalized to the [0, 1] range. The color bars correspond to the values of cell number density.

- All other cells:

$$\begin{aligned}
&\frac{\partial F_3(\tilde{x},t)}{\partial t} + \nabla_{\tilde{x}} \cdot \Sigma_{i=1}^{k}[r_i(x_i,\tilde{C})F_3(\tilde{x},t)] \\
&\quad = 2\int_{\tilde{x}}^{\tilde{x}_{\max}} \Gamma_3(\tilde{z},\tilde{C})F_3(\tilde{z},t)p_3(\tilde{z},\tilde{x},\tilde{C})d\tilde{z} \\
&\qquad -[\Gamma_3(\tilde{x},\tilde{C}) + \Xi_3(\tilde{x},\tilde{C})]\cdot F_3(\tilde{x},t) \\
&\qquad +(1-\alpha)D_1(\tilde{x},\tilde{C})\cdot F_1(\tilde{x},t) + D_2(\tilde{x},\tilde{C})\cdot F_2(\tilde{x},t)
\end{aligned} \tag{9}$$

Table 7.1. Description of terms in Eqs. (7)–(9).

Term	Description
i	i-type cells: 1: hPSCs, 2: Mesoderm cells, 3: All other cells.
$F_i(\tilde{x}, t)d\tilde{x}$	Number of i-type cells per unit volume of culture at a physiological state between $\tilde{x}$ and $\tilde{x} + d\tilde{x}$ at time t.
$\Gamma_i(\tilde{x})$	Division intensity of i-type cells.
$p_i(\tilde{x}, \tilde{x}', \tilde{C})$	Partition function of cells of i-type.
$\Xi_i(\tilde{x})$	Death intensity of i-type cells.
$D_j(\tilde{x}, \tilde{C})$	Differentiation function for $j = 1$: hPSCs differentiating to mesoderm (a fraction) and other cells, $j = 2$: Mesoderm cells differentiating to other cells. No differentiation was assumed for other cells.
$r_j(x_j, \tilde{C})$	Rate of accumulation of state variable $x_j, j = 1, 2 \ldots k$.

subjected to appropriate initial and boundary or containment conditions. The physiological state vector $\tilde{x}$ may include the cell mass, and markers specific to each fate such as Nanog (hPSCs) and Brachyury (mesoderm cells). Also, α denotes the fraction of hPSCs differentiating to mesoderm. The terms in Eqs. (7)–(9) are described in Table 7.1. As in other reports (Pisu *et al.*, 2007), additional mass balance equations can be written for changes in substrates or growth factors (components of $\tilde{C}$).

Although this approach can be strictly deterministic, it also allows the incorporation of stochastic terms. This is a significant advantage over other types of models for stem cell populations. Besides extrinsic stimuli (e.g. soluble factors), intrinsic signals from gene transcription, protein processing and intracellular signaling networks affect cell self-renewal or differentiation along a particular lineage. The behavior of these networks is largely influenced by noise. Recent reports (Chang *et al.*, 2008; Kalmar *et al.*, 2009) have surfaced on the key role of intrinsic noise on stem cell fate, providing a basis for the observed heterogeneity in stem cell populations and the apparent stochasticity in stem cell fate decisions. Interesting findings are expected in the near future from several research groups active in this area.

Application of PBE models to stem cell populations (and specifically in bioreactor culture systems) has been limited for various reasons. First, the algorithmic complexity increases substantially as the vector of state variables $\tilde{x}$ is expanded. New methods must be developed for solving PBEs as the implementation of commonly used finite difference methods becomes cumbersome with increasing dimensionality. The inclusion of stochastic terms also dictates the need for more robust methods for PBE models for stem cells. As a result, the computational requirements increase significantly although this concern is alleviated considering the rapid increase in available computing power. Second, there are no known general forms of the various single-cell functions involved in PBEs (e.g., for differentiation $D(\tilde{x}, \tilde{C})$ or partition $p(\tilde{z}, \tilde{x}, \tilde{C})$). Identification of such functions requires measurements at a single-cell level. Rapid estimation of parameters in distributions of single-cell properties (Srienc, 1999) can be achieved by flow cytometry. Thus far, flow cytometric analysis in this context has yielded information about the position of cells in the cell cycle in relation to their size (or mass) (Darzynkiewicz *et al.*, 2004; Kacmar and Srienc, 2005). Other state variables besides the cell mass will be necessary to describe populations of stem cells either in culture or in tissues. As described above, subpopulations of cells may be ascribed state variables representing the expression of markers typical of pluripotency (e.g. OCT3/4A, NANOG) or of specific lineages (e.g. Brachyury, SOX17, PAX6). Assessing the distribution of these state variables will necessitate the development of highly specific antibodies and the engineering of stem cell lines carrying reporter genes coupled to lineage-specific gene promoters. As these tools become widely available, PBE-based models are expected to become more tractable for stem cell applications.

7.6. Summary

The recent progress in the exploration of stirred bioreactors for PSC self-renewal and/or differentiation demonstrates clearly the potential of this culture modality for large-scale stem cell cultivation. These advancements also exemplify the importance and difficulties of selecting appropriate physicochemical parameters such as agitation/shear, oxygen tension, and media composition. Lessons from contemporary unit operations for the generation of biologics will aid greatly the engineering of stem cell bioprocesses. Still, the distinct characteristics of stem cells require a fresh look at the design and operation of such bioprocesses.

Some of the major challenges in the field have also been discussed. Overcoming these challenges will require concerted approaches based on engineering (e.g. bioreactor design, stem cell population modeling) and biological principles, especially given the gaps in our knowledge of how the bioprocess environment affects stem cell physiology.

Bibliography

Abranches, E., Bekman, E., Henrique, D. and Cabral, J. M. 2007. Expansion of mouse embryonic stem cells on microcarriers, *Biotechnol Bioeng* 96:1211–1221.

Adam, J. and Maggelakis, S. 1990. Diffusion regulated growth characteristics of a spherical prevascular carcinoma, *Bull Math Biol* 52:549–582.

Akasha, A. A., Sotiriadou, I., Doss, M. X., Halbach, M., Winkler, J., Baunach, J. J., Katsen-Globa, A., Zimmermann, H., Choo, Y., Hescheler, J. and Sachinidis, A. 2008. Entrapment of embryonic stem cells-derived cardiomyocytes in macroporous biodegradable microspheres: Preparation and characterization, *Cell Physiol Biochem* 22:665–672.

Aldous, D. J. 1999. Deterministic and stochastic models for coalescence (aggregation and coagulation): A review of the mean-field theory for probabilists, *Bernoulli* 5:3–48.

Anderson, E. C. and Petersen, D. F. 1967. Cell growth and division. II. Experimental studies of cell volume distributions in mammalian suspension cultures, *Biophys J* 7:353–364.

Awrey, D. E., Tse, M., Hortelano, G. and Chang, P. L. 1996. Permeability of alginate microcapsules to secretory recombinant gene products, *Biotechnol Bioeng* 52:472–484.

Bajpai, R., Lesperance, J., Kim, M. and Terskikh, A. V. 2008. Efficient propagation of single cells Accutase-dissociated human embryonic stem cells, *Mol Reprod Dev* 75:818–827.

Bauwens, C., Yin, T., Dang, S., Peerani, R. and Zandstra, P. W. 2005. Development of a perfusion fed bioreactor for embryonic stem cell-derived cardiomyocyte generation: Oxygen-mediated enhancement of cardiomyocyte output, *Biotechnol Bioeng* 90:452–461.

Begley, C. M. and Kleis, S. J. 2000. The fluid dynamic and shear environment in the NASA/JSC rotating-wall perfused-vessel bioreactor, *Biotechnol Bioeng* 70:32–40.

Belval, T. K. and Hellums, J. D. 1986. Analysis of shear-induced platelet aggregation with population balance mathematics, *Biophys J* 50:479–487.

Bolliet, C., Bohn, M. C. and Spector, M. 2008. Non-viral delivery of the gene for glial cell line-derived neurotrophic factor to mesenchymal stem cells *in vitro* via a collagen scaffold, *Tissue Eng. Part C, Methods* 14:207–219.

Borlongan, C. V., Thanos, C. G., Skinner, S. J., Geaney, M. and Emerich, D. F. 2008. Transplants of encapsulated rat choroid plexus cells exert neuroprotection in a rodent model of Huntington's disease, *Cell Transplant* 16:987–992.

Bugarski, B., Smith, J., Wu, J. and Goosen, M. F. A. 1993. Methods for animal-cell immobilization using electrostatic droplet generation, *Biotechnol Tech* 7:677–682.

Cameron, C. M., Hu, W. S. and Kaufman, D. S. 2006. Improved development of human embryonic stem cell-derived embryoid bodies by stirred vessel cultivation, *Biotechnol Bioeng* 94:938–948.

Casciari, J. J., Sotirchos, S. V. and Sutherland, R. M. 1992. Mathematical modelling of microenvironment and growth in EMT6/Ro multicellular tumour spheroids, *Cell Prolif* 25:1–22.

Chang, H. H., Hemberg, M., Barahona, M., Ingber, D. E. and Huang, S. 2008. Transcriptome-wide noise controls lineage choice in mammalian progenitor cells, *Nature* 453:544–547.

Chen, G., Hou, Z., Gulbranson, D. R. and Thomson, J. A. 2010. Actin-myosin contractility is responsible for the reduced viability of dissociated human embryonic stem cells, *Cell Stem Cell* 7:240–248.

Chen, H. F., Kuo, H. C., Chen, W., Wu, F. C., Yang, Y. S. and Ho, H. N. 2009. A reduced oxygen tension (5%) is not beneficial for maintaining human embryonic stem cells in the undifferentiated state with short splitting intervals, *Hum Reprod* 24:71–80.

Cheng, L., Hammond, H., Ye, Z., Zhan, X. and Dravid, G. 2003. Human adult marrow cells support prolonged expansion of human embryonic stem cells in culture, *Stem Cells* 21:131–142.

Cherry, R. S. and Kwon, K. Y. 1990. Transient shear stresses on a suspension cell in turbulence, *Biotechnol Bioeng* 36:563–571.

Cherry, R. S. and Papoutsakis, E. T. 1988. Physical mechanisms of cell damage in microcarrier cell culture bioreactors, *Biotechnol Bioeng* 32:1001–1014.

Cherry, R. S. and Papoutsakis, E. T. 1989. Growth and death rates of bovine embryonic kidney cells in turbulent microcarrier bioreactors. *Bioprocess and Biosystems Engineering* 4:81–89.

Cormier, J. T., zur Nieden, N. I., Rancourt, D. E. and Kallos, M. S. 2006. Expansion of undifferentiated murine embryonic stem cells as aggregates in suspension culture bioreactors, *Tissue Eng* 12:3233–3245.

Cornolti, R., Figliuzzi, M. and Remuzzi, A. 2009. Effect of micro- and macroencapsulation on oxygen consumption by pancreatic islets, *Cell Transplant* 18:195–201.

Croughan, M. S., Hamel, J. F. and Wang, D. I. 1987. Hydrodynamic effects on animal cells grown in microcarrier cultures, *Biotechnol Bioeng* 29:130–141.

Dang, S. M., Gerecht-Nir, S., Chen, J., Itskovitz-Eldor, J. and Zandstra, P. W. 2004. Controlled, scalable embryonic stem cell differentiation culture, *Stem Cells* 22:275–282.

Daoutidis, P. and Henson, M. A. 2002. Dynamics and control of cell populations in continuous bioreactors, *AIChE Symposium Series* 326:274–289.

Darzynkiewicz, Z., Crissman, H. and Jacobberger, J. W. 2004. Cytometry of the cell cycle: Cycling through history, *Cytometry A* 58:21–32.

De Vos, P., De Haan, B. J. and Van Schilfgaarde, R. 1997. Upscaling the production of microencapsulated pancreatic islets, *Biomaterials* 18:1085–1090.

Ding, S. and Schultz, P. G. 2004. A role for chemistry in stem cell biology, *Nat Biotechnol* 22:833–840.

Ding, S., Wu, T. Y., Brinker, A., Peters, E. C., Hur, W., Gray, N. S. and Schultz, P. G. 2003. Synthetic small molecules that control stem cell fate, *Proc Natl Acad Sci U S A* 100:7632–7637.

Draper, J. S., Smith, K., Gokhale, P., Moore, H. D., Maltby, E., Johnson, J., Meisner, L., Zwaka, T. P., Thomson, J. A. and Andrews, P. W. 2004. Recurrent gain of chromosomes 17q and 12 in cultured human embryonic stem cells, *Nat Biotechnol* 22:53–54.

Drukker, M., Katchman, H., Katz, G., Even-Tov Friedman, S., Shezen, E., Hornstein, E., Mandelboim, O., Reisner, Y. and Benvenisty, N. 2006. Human embryonic stem cells and their differentiated derivatives are less susceptible to immune rejection than adult cells, *Stem Cells* 24:221–229.

Eakman, J. M., Fredrickson, A. G. and Tsuchiya, H. M. 1966. Statistics and dynamics of microbial cell populations, *Chem. Eng. Prog. Symp. Series* 62:37–49.

Eibl, R., Werner, S. and Eibl, D. 2010. Bag bioreactor based on wave-induced motion: Characteristics and applications, *Adv Biochem Eng Biotechnol* 115:55–87.

Fernandes, A. M., Fernandes, T. G., Diogo, M. M., da Silva, C. L., Henrique, D. and Cabral, J. M. 2007. Mouse embryonic stem cell expansion in a microcarrier-based stirred culture system, *J Biotechnol* 132:227–236.

Fernandes, A. M., Marinho, P. A., Sartore, R. C., Paulsen, B. S., Mariante, R. M., Castilho, L. R. and Rehen, S. K. 2009. Successful scale-up of human embryonic stem cell production in a stirred microcarrier culture system, *Braz J Med Biol Res* 42:515–522.

Ferreira, L. S., Gerecht, S., Fuller, J., Shieh, H. F., Vunjak-Novakovic, G. and Langer, R. 2007. Bioactive hydrogel scaffolds for controllable vascular differentiation of human embryonic stem cells, *Biomaterials* 28:2706–2717.

Fischer, B. and Bavister, B. D. 1993. Oxygen tension in the oviduct and uterus of rhesus monkeys, hamsters and rabbits, *J Reprod Fertil* 99:673–679.

Fok, E. Y. and Zandstra, P. W. 2005. Shear-controlled single-step mouse embryonic stem cell expansion and embryoid body-based differentiation, *Stem Cells* 23:1333–1342.

Forsyth, N. R., Musio, A., Vezzoni, P., Simpson, A. H., Noble, B. S. and McWhir, J. 2006. Physiologic oxygen enhances human embryonic stem cell clonal recovery and reduces chromosomal abnormalities, *Cloning Stem Cells* 8:16–23.

Fraichard, A., Chassande, O., Bilbaut, G., Dehay, C., Savatier, P. and Samarut, J. 1995. *In vitro* differentiation of embryonic stem cells into glial cells and functional neurons, *J Cell Sci* 108(Pt 10):3181–3188.

Fredrickson, A. G., Ramkrishna, D. and Tsuchiya, H. M. 1967. Statistics and dynamics of procaryotic cell populations, *Mathematical Bioscience* 1:327–374.

Fredrickson, A. G. and Tsuchiya, H. M. 1963. Continuous propagation of microorganisms, *AIChE Journal* 9:459–468.

Garreta, E., Genove, E., Borros, S. and Semino, C. E. 2006. Osteogenic differentiation of mouse embryonic stem cells and mouse embryonic fibroblasts in a three-dimensional self-assembling peptide scaffold, *Tissue Eng* 12:2215–2227.

Gassmann, M., Fandrey, J., Bichet, S., Wartenberg, M., Marti, H. H., Bauer, C., Wenger, R. H. and Acker, H. 1996. Oxygen supply and oxygen-dependent gene expression in differentiating embryonic stem cells, *Proc Natl Acad Sci USA* 93:2867–2872.

Genbacev, O., Krtolica, A., Zdravkovic, T., Brunette, E., Powell, S., Nath, A., Caceres, E., McMaster, M., McDonagh, S., Li, Y., Mandalam, R., Lebkowski, J. and Fisher, S. J. 2005. Serum-free derivation of human embryonic stem cell lines on human placental fibroblast feeders, *Fertil Steril* 83:1517–1529.

Gerecht-Nir, S., Cohen, S. and Itskovitz-Eldor, J. 2004a. Bioreactor cultivation enhances the efficiency of human embryoid body (hEB) formation and differentiation, *Biotechnol Bioeng* 86:493–502.

Gerecht-Nir, S., Cohen, S., Ziskind, A. and Itskovitz-Eldor, J. 2004b. Three-dimensional porous alginate scaffolds provide a conducive environment for generation of well-vascularized embryoid bodies from human embryonic stem cells, *Biotechnol Bioeng* 88:313–320.

Graichen, R., Xu, X., Braam, S. R., Balakrishnan, T., Norfiza, S., Sieh, S., Soo, S. Y., Tham, S. C., Mummery, C., Colman, A., Zweigerdt, R. and Davidson, B. P. 2008. Enhanced cardiomyogenesis of human embryonic stem cells by a small molecular inhibitor of p38 MAPK, *Differentiation* 76:357–370.

Grossmann, U. 1984. Profiles of oxygen partial pressure and oxygen consumption inside multicellular spheroids, *Recent Results Cancer Res* 95:150–161.

Harb, N., Archer, T. K. and Sato, N. 2008. The Rho-Rock-Myosin signaling axis determines cell-cell integrity of self-renewing pluripotent stem cells, *PLoS One* 3:e3001.

Hayashi, Y., Chan, T., Warashina, M., Fukuda, M., Ariizumi, T., Okabayashi, K., Takayama, N., Otsu, M., Eto, K., Furue, M. K., Michiue, T., Ohnuma, K., Nakauchi, H. and M. Asashima. 2010. Reduction of N-glycolylneuraminic acid in human induced pluripotent stem cells generated or cultured under feeder- and serum-free defined conditions, *PLoS One* 5:e14099.

Hosseinkhani, H., Yamamoto, M., Inatsugu, Y., Hiraoka, Y., Inoue, S., Shimokawa, H. and Tabata, Y. 2006. Enhanced ectopic bone formation using a combination of plasmid DNA impregnation into 3-D scaffold and bioreactor perfusion culture, *Biomaterials* 27:1387–1398.

Huang, P. Y. and Hellums, J. D. 1993. Aggregation and disaggregation kinetics of human blood platelets: Part II. Shear-induced platelet aggregation, *Biophys J* 65:344–353.

Huangfu, D., Osafune, K., Maehr, R., Guo, W., Eijkelenboom, A., Chen, S., Muhlestein, W. and Melton, D. A. 2008. Induction of pluripotent stem cells from primary human fibroblasts with only Oct4 and Sox2, *Nat Biotechnol* 26:1269–1275.

Hwang, Y. S., Cho, J., Tay, F., Heng, J. Y., Ho, R., Kazarian, S. G., Williams, D. R., Boccaccini, A. R., Polak, J. M. and Mantalaris, A. 2009. The use of murine embryonic stem cells, alginate encapsulation, and rotary microgravity bioreactor in bone tissue engineering, *Biomaterials* 30:499–507.

Illi, B., Scopece, A., Nanni, S., Farsetti, A., Morgante, L., Biglioli, P., Capogrossi, M. C. and Gaetano, C. 2005. Epigenetic histone modification and cardiovascular lineage programming in mouse embryonic stem cells exposed to laminar shear stress, *Circ Res* 96:501–508.

Jeong, C. H., Lee, H. J., Cha, J. H., Kim, J. H., Kim, K. R., Yoon, D. K. and Kim, K. W. 2007. Hypoxia-inducible factor-1 alpha inhibits self-renewal of mouse embryonic stem cells *in vitro* via negative regulation of the leukemia inhibitory factor-STAT3 pathway, *J Biol Chem* 282:13672–13679.

Jing, D., Parikh, A., Canty, Jr., J. M. and Tzanakakis, E. S. 2008. Stem cells for heart cell therapies, *Tissue Eng Part B Rev* 14:393–406.

Jing, D., Parikh, A. and Tzanakakis, E. S. 2010. Cardiac cell generation from encapsulated embryonic stem cells in static and scalable culture systems, *Cell Transplant* 19:1397–1412.

Kacmar, J. and Srienc, F. 2005. Dynamics of single cell property distributions in Chinese hamster ovary cell cultures monitored and controlled with automated flow cytometry, *J Biotechnol* 120:410–420.

Kalmar, T., Lim, C., Hayward, P., Munoz-Descalzo, S., Nichols, J., Garcia-Ojalvo, J. and Martinez Arias, A. 2009. Regulated fluctuations in nanog expression mediate cell fate decisions in embryonic stem cells, *PLoS Biol* 7:e1000149.

Kamei, K., Guo, S., Yu, Z. T., Takahashi, H., Gschweng, E., Suh, C., Wang, X., Tang, J., McLaughlin, J., Witte, O. N., Lee, K. B. and Tseng, H. R. 2009. An integrated microfluidic culture device for quantitative analysis of human embryonic stem cells, *Lab Chip* 9:555–563.

Kato, D., Takeuchi, M., Sakurai, T., Furukawa, S., Mizokami, H., Sakata, M., Hirayama, C. and Kunitake, M. 2003. The design of polymer microcarrier surfaces for enhanced cell growth, *Biomaterials* 24:4253–4264.

Kehoe, D. E., Jing, D., Lock, L. T. and Tzanakakis, E. M. 2010. Scalable Stirred-suspension bioreactor culture of human pluripotent stem cells, *Tissue Eng Part A* 16:405–421.

Kehoe, D. E., Lock, L. T., Parikh, A. and Tzanakakis, E. S. 2008. Propagation of embryonic stem cells in stirred suspension without serum, *Biotechnol Prog* 24:1342–1352.

Kim, L., Vahey, M. D., Lee, H. Y. and Voldman, J. 2006. Microfluidic arrays for logarithmically perfused embryonic stem cell culture, *Lab Chip* 6:394–406.

Kleinman, H. K., McGarvey, M. L., Liotta, L. A., Robey, P. G., Tryggvason, K. and Martin, G. R. 1982. Isolation and characterization of type IV procollagen, laminin, and heparan sulfate proteoglycan from the EHS sarcoma, *Biochemistry* 21:6188–6193.

Koch, A. L. and M. Schaechter. 1962. A model for statistics of the cell division process, *J Gen Microbiol* 29:435–454.

Koller, M. R., Bender, J. G., Miller, W. M. and Papoutsakis, E. T. 1993. Expansion of primitive human hematopoietic progenitors in a perfusion bioreactor system with IL-3, IL-6, and stem cell factor, *Biotechnology (NY)* 11:358–363.

Krawetz, R., Taiani, J. T., Liu, S., Meng, G., Li, X., Kallos, M. S. and Rancourt, D. 2009. Large-scale expansion of pluripotent human embryonic stem cells in stirred suspension bioreactors, *Tissue Eng Part C Methods* 16(4): 573–582.

Laurenzi, I. J. and Diamond, S. L. 1999. Monte Carlo simulation of the heterotypic aggregation kinetics of platelets and neutrophils, *Biophys J* 77:1733–1746.

Li, L., Baroja, M. L., Majumdar, A., Chadwick, K., Rouleau, A., Gallacher, L., Ferber, I., Lebkowski, J., Martin, T., Madrenas, J. and Bhatia, M. 2004. Human embryonic stem cells possess immune-privileged properties, *Stem Cells* 22:448–456.

Li, X. J., Hu, B. Y., Jones, S. A., Zhang, Y. S., Lavaute, T., Du, Z. W. and Zhang, S. C. 2008. Directed differentiation of ventral spinal progenitors and motor neurons from human embryonic stem cells by small molecules, *Stem Cells* 26:886–893.

Liu, H. and Roy, K. 2005. Biomimetic three-dimensional cultures significantly increase hematopoietic differentiation efficacy of embryonic stem cells, *Tissue Eng* 11:319–330.

Lock, L. T. and Tzanakakis, E. S. 2007. Stem/Progenitor cell sources of insulin-producing cells for the treatment of diabetes, *Tissue Eng* 13:1399–1412.

Lock, L. T. and Tzanakakis, E. S. 2009. Expansion and differentiation of human embryonic stem cells to endoderm progeny in a microcarrier stirred-suspension culture, *Tissue Eng Part A* 15:2051–2063.

Ludwig, T. E., Levenstein, M. E., Jones, J. M., Berggren, W. T., Mitchen, E. R., Frane, J. L., Crandall, L. J., Daigh, C. A., Conard, K. R., Piekarczyk, M. S., Llanas, R. A. and Thomson, J. A. 2006. Derivation of human embryonic stem cells in defined conditions, *Nat Biotechnol* 24:185–187.

Maguire, T., Davidovich, A. E., Wallenstein, E. J., Novik, E., Sharma, N., Pedersen, H., Androulakis, I. P., Schloss, R. and Yarmush, M. 2007. Control of hepatic differentiation via cellular aggregation in an alginate microenvironment, *Biotechnol Bioeng* 98:631–644.

Maguire, T., Novik, E., Schloss, R. and Yarmush, M. 2006. Alginate-PLL microencapsulation: Effect on the differentiation of embryonic stem cells into hepatocytes, *Biotechnol Bioeng* 93:581–591.

Magyar, J. P., Nemir, M., Ehler, E., Suter, N., Perriard, J. C. and Eppenberger, H. M. 2001. Mass production of embryoid bodies in microbeads, *Ann N Y Acad Sci* 944:135–143.

Maitra, A., Arking, D. E., Shivapurkar, *et al.* 2005. Genomic alterations in cultured human embryonic stem cells, *Nat Genet* 37:1099–1103.

Mallon, B. S., Park, K. Y., Chen, K. G., Hamilton, R. S. and McKay, R. D. 2006. Toward xeno-free culture of human embryonic stem cells, *Int J Biochem Cell Biol* 38:1063–1075.

Mantzaris, N. V. and Daoutidis, P. 2004. Cell population balance modeling and control in continuous bioreactors, *Journal of Process Control* 14:775–784.

Martin-Ibanez, R., Unger, C., Stromberg, A., Baker, D., Canals, J. M. and Hovatta, O. 2008. Novel cryopreservation method for dissociated human embryonic stem cells in the presence of a ROCK inhibitor, *Hum Reprod* 23:2744–2754.

Martin, M. J., Muotri, A., Gage, F. and Varki, A. 2005. Human embryonic stem cells express an immunogenic nonhuman sialic acid, *Nature medicine* 11:228–232.

McNeish, J. D. 2007. Stem cells as screening tools in drug discovery, *Curr Opin Pharmacol* 7:515–520.

Melkoumian, Z., Weber, J. L., Weber, D. M., Fadeev, A. G., Zhou, Y., Dolley-Sonneville, P., Yang, J., Qiu, L., Priest, C. A., Shogbon, C., Martin, A. W., Nelson, J., West, P., Beltzer, J. P., Pal, S. and Brandenberger, R. 2010. Synthetic peptide-acrylate surfaces for long-term self-renewal and cardiomyocyte differentiation of human embryonic stem cells, *Nat Biotechnol* 28:606–610.

Mendonca, R. Z., Ioshimoto, L. M., Mendonca, R. M., De-Franco, M., Valentini, E. J., Becak, W., Raw, I. and Pereira, C. A. 1993. Preparation of human rabies vaccine in VERO cell culture using a microcarrier system, *Braz J Med Biol Res* 26:1305–1317.

Michaels, J. D., Mallik, A. K. and Papoutsakis, E. T. 1996. Sparging and agitation-induced injury of cultured animals cells: Do cell-to-bubble interactions in the bulk liquid injure cells? *Biotechnol Bioeng* 51:399–409.

Miyabayashi, T., Teo, J. L., Yamamoto, M., McMillan, M., Nguyen, C. and Kahn, M. 2007. Wnt/beta-catenin/CBP signaling maintains long-term murine embryonic stem cell pluripotency, *Proc Natl Acad Sci U S A* 104:5668–5673.

Moustafa, T., Girod, S., Tortosa, F., Li, R., Sol, J. C., Rodriguez, F., Bastide, R., Lazorthes, Y. and Sallerin, B. 2006. Viability and functionality of bovine chromaffin cells encapsulated into alginate-PLL microcapsules with a liquefied inner core, *Cell Transplant* 15:121–133.

Naito, A. T., Shiojima, I., Akazawa, H., Hidaka, K., Morisaki, T., Kikuchi, A., and Komuro, I. 2006. Developmental stage-specific biphasic roles of Wnt/beta-catenin signaling in cardiomyogenesis and hematopoiesis, *Proc Natl Acad Sci USA* 103:19812–19817.

Nie, Y., Bergendahl, V., Hei, D. J., Jones, J. M. and Palecek, S. P. 2009. Scalable culture and cryopreservation of human embryonic stem cells on microcarriers, *Biotechnol Prog* 25:20–31.

Niebruegge, S., Bauwens, C. L., Peerani, R., Thavandiran, N., Masse, S., Sevaptisidis, E., Nanthakumar, K., Woodhouse, K., Husain, M., Kumacheva, E. and Zandstra, P. W. 2009. Generation of human embryonic stem cell-derived mesoderm and cardiac cells using size-specified aggregates in an oxygen-controlled bioreactor, *Biotechnol Bioeng* 102:493–507.

Nielsen, L. K., Bender, J. G., Miller, W. M. and Papoutsakis, E. T. 1998a. Population balance model of *in vivo* neutrophil formation following bone marrow rescue therapy, *Cytotechnology* 28:157–162.

Nielsen, L. K., Papoutsakis, E. T. and Miller, W. M. 1998b. Modeling *ex vivo* hematopoiesis using chemical engineering metaphors, *Chem Eng Sci* 53:1913–1925.

Nuttelman, C. R., Tripodi, M. C. and Anseth, K. S. 2006. Dexamethasone-functionalized gels induce osteogenic differentiation of encapsulated hMSCs, *J Biomed Mater Res A* 76:183–195.

O'Neill, A. and Schaffer, D. V. 2004. The biology and engineering of stem-cell control, *Biotechnol Appl Bioc* 40:5–16.

Oh, S. K., Chen, A. K., Mok, Y., Chen, X., Lim, U. M., Chin, A., Choo, A. B. and Reuveny, S. 2009. Long-term microcarrier suspension cultures of human embryonic stem cells, *Stem Cell Res* 2:219–230.

Ohgushi, M., Matsumura, M., Eiraku, M., Murakami, K., Aramaki, T., Nishiyama, A., Muguruma, K., Nakano, T., Suga, H., Ueno, M., Ishizaki, T., Suemori, H., Narumiya, S., Niwa, H. and Sasai, Y. 2010. Molecular pathway and cell state responsible for dissociation-induced apoptosis in human pluripotent stem cells, *Cell Stem Cell* 7:225–239.

Orive, G., Hernandez, R. M., Gascon, A. R., Calafiore, R., Chang, T. M., De Vos, P., Hortelano, G., Hunkeler, D., Lacik, I., Shapiro, A. M. and Pedraz, J. L. 2003. Cell encapsulation: Promise and progress, *Nat Med* 9:104–107.

Papoutsakis, E. T. 1991. Fluid-mechanical damage of animal cells in bioreactors, *Trends Biotechnol* 9:427–437.

Passier, R., Oostwaard, D. W., Snapper, J., Kloots, J., Hassink, R. J., Kuijk, E., Roelen, B., de la Riviere, A. B. and Mummery, C. 2005. Increased cardiomyocyte differentiation from human embryonic stem cells in serum-free cultures, *Stem Cells* 23:772–780.

Peterson, R. T., Link, B. A., Dowling, J. E. and Schreiber, S. L. 2000. Small molecule developmental screens reveal the logic and timing of vertebrate development, *Proc Natl Acad Sci USA* 97:12965–12969.

Pevsner-Fischer, M., Morad, V., Cohen-Sfady, M., Rousso-Noori, L., Zanin-Zhorov, A., Cohen, S., Cohen, I. R. and Zipori, D. 2007. Toll-like receptors and their ligands control mesenchymal stem cell functions, *Blood* 109:1422–1432.

Phillips, B. W., Horne, R., Lay, T. S., Rust, W. L., Teck, T. T. and Crook, J. M. 2008. Attachment and growth of human embryonic stem cells on microcarriers, *J Biotechnol* 138:24–32.

Pisu, M., Concas, A. and Cao, G. 2007. A novel simulation model for stem cells differentiation, *J Biotechnol* 130:171–182.

Price, P. J., Goldsborough, M. D. and Tilkins, M. L. 1998, *Embryonic Stem Cell Serum Replacement. Vol.* WO/1998/030679, USA.

Ramkrishna, D. 2000, *Population Balances: Theory and Applications to Particulate Systems in Engineering*. Academic Press. pp. 355.

Richards, M., Fong, C. Y., Chan, W. K., Wong, P. C. and Bongso, A. 2002. Human feeders support prolonged undifferentiated growth of human inner cell masses and embryonic stem cells, *Nat Biotechnol* 20:933–936.

Rodin, S., Domogatskaya, A., Strom, S., Hansson, E. M., Chien, K. R., Inzunza, J., Hovatta, O. and K. Tryggvason. 2010. Long-term self-renewal of human pluripotent stem cells on human recombinant laminin-511, *Nat Biotechnol* 28:611–615.

Rodriguez-Piza, I., Richaud-Patin, Y., Vassena, R., Gonzalez, F., Barrero, M. J., Veiga, A., Raya, A. and Belmonte, J. C. 2010. Reprogramming of human fibroblasts to induced pluripotent stem cells under xeno-free conditions, *Stem Cells* 28:36–44.

Rosler, E. S., Fisk, G. J., Ares, X., Irving, J., Miura, T., Rao, M. S. and Carpenter, M. K. 2004. Long-term culture of human embryonic stem cells in feeder-free conditions, *Dev Dyn* 229:259–274.

Ryan, E. A., Lakey, J. R., Paty, B. W., Imes, S., Korbutt, G. S., Kneteman, N. M., Bigam, D., Rajotte, R. V. and Shapiro, A. M. 2002. Successful islet transplantation: Continued insulin reserve provides long-term glycemic control, *Diabetes* 51:2148–2157.

Sachlos, E. and Auguste, D. T. 2008. Embryoid body morphology influences diffusive transport of inductive biochemicals: A strategy for stem cell differentiation, *Biomaterials* 29:4471–4480.

Saha, S., Ji, L., de Pablo, J. J. and Palecek, S. P. 2006. Inhibition of human embryonic stem cell differentiation by mechanical strain, *J Cell Physiol* 206:126–137.

Saha, S., Ji, L., de Pablo, J. J. and Palecek, S. P. 2008. TGFbeta/Activin/Nodal pathway in inhibition of human embryonic stem cell differentiation by mechanical strain, *Biophys J* 94:4123–4133.

Samuel, M. S. and Olson, M. F. 2010. Dying alone: A tale of rho, *Cell Stem Cell* 7:135–136.

Sargent, C. Y., Berguig, G. Y. and McDevitt, T. C. 2009. Cardiomyogenic differentiation of embryoid bodies is promoted by rotary orbital suspension culture, *Tissue Eng Part A* 15:331–342.

Sato, N., Meijer, L., Skaltsounis, L., Greengard, P. and Brivanlou, A. H. 2004. Maintenance of pluripotency in human and mouse embryonic stem cells through activation of Wnt signaling by a pharmacological GSK-3-specific inhibitor, *Nat Med* 10:55–63.

Schroeder, M., Niebruegge, S., Werner, A., Willbold, E., Burg, M., Ruediger, M., Field, L. J., Lehmann, J. and Zweigerdt, R. 2005. Differentiation and lineage selection of mouse embryonic stem cells in a stirred bench scale bioreactor with automated process control, *Biotechnol Bioeng* 92:920–933.

Schwinger, C., Koch, S., Jahnz, U., Wittlich, P., Rainov, N. G. and Kressler, J. 2002. High throughput encapsulation of murine fibroblasts in alginate using the JetCutter technology, *J Microencapsul* 19:273–280.

Sherer, E., Tocce, E., Hannemann, R. E., Rundell, A. E. and Ramkrishna, D. 2008. Identification of age-structured models: cell cycle phase transitions, *Biotechnol Bioeng* 99:960–974.

Smit, D. J., Hounslow, M. J. and Paterson, W. R. 1994. Aggregation and gelation- I. Analytical solutions for CST and batch operation, *Chem Eng Sci* 49:1025–1035.

Srienc, F. 1999. Cytometric data as the basis for rigorous models of cell population dynamics, *J Biotechnol* 71:233–238.

Stathopoulos, N. A. and Hellums, J. D. 1985. Shear stress effects on human embryonic kidney cells in Vitro, *Biotechnol Bioeng* 27:1021–1026.

Strand, B. L., Ryan, T. L., In't Veld, P., Kulseng, B., Rokstad, A. M., Skjak-Brek, G. and Espevik, T. 2001. Poly-L-Lysine induces fibrosis on alginate microcapsules via the induction of cytokines, *Cell Transplant* 10:263–275.

Takahashi, K., Tanabe, K., Ohnuki, M., Narita, M., Ichisaka, T., Tomoda, K. and Yamanaka, S. 2007. Induction of pluripotent stem cells from adult human fibroblasts by defined factors, *Cell* 131:861–872.

Thomson, J. A., Itskovitz-Eldor, J., Shapiro, S. S., Waknitz, M. A., Swiergiel, J. J., Marshall, V. S. and Jones, J. M. 1998. Embryonic stem cell lines derived from human blastocysts, *Science* 282:1145–1147.

Toh, Y. C. and Voldman, J. 2010. Fluid shear stress primes mouse embryonic stem cells for differentiation in a self-renewing environment via heparan sulfate proteoglycans transduction, *Faseb J*.

Tseng, A. S., Engel, F. B. and Keating, M. T. 2006. The GSK-3 inhibitor BIO promotes proliferation in mammalian cardiomyocytes, *Chem Biol* 13:957–963.

Tzanakakis, E. S., Hansen, L. K. and Hu, W. S. 2001. The role of actin filaments and microtubules in hepatocyte spheroid self-assembly. *Cell Motil Cytoskeleton* 48:175–189.

Tzanakakis, E. S., Hess, D. J., Sielaff, T. D. and Hu, W. S. 2000. Extracorporeal tissue engineered liver-assist devices, *Annu Rev Biomed Eng* 2:607–632.

Ullmann, U., In't Veld, P., Gilles, C., Sermon, K., De Rycke, M., Van de Velde, H., Van Steirteghem, A. and I. Liebaers. 2007. Epithelial-mesenchymal transition process in human embryonic stem cells cultured in feeder-free conditions, *Mol Hum Reprod* 13:21–32.

Varani, J., Inman, D. R., Fligiel, S. E. and Hillegas, W. J. 1993. Use of recombinant and synthetic peptides as attachment factors for cells on microcarriers, *Cytotechnology* 13:89–98.

Villa-Diaz, L. G., Nandivada, H., Ding, J., Nogueira-de-Souza, N. C., Krebsbach, P. H., O'Shea, K. S., Lahann, J. and Smith, G. D. 2010. Synthetic polymer coatings for long-term growth of human embryonic stem cells, *Nat Biotechnol* 28:581–583.

Viswanathan, S. and Zandstra, P. W. 2003. Towards predictive models of stem cell fate, *Cytotechnology* 41:75–92.

von Smoluchowski, M. 1917. Versuch einer mathematischen Theorie der Koagulationskinetik kolloider Lösungen, *Zeitschrift f. Physik. Chemie* 92:129–168.

Wada, T., Honda, M., Minami, I., Tooi, N., Amagai, Y., Nakatsuji, N. and Aiba, K. 2009. Highly efficient differentiation and enrichment of spinal motor neurons derived from human and monkey embryonic stem cells, *PLoS One* 4:e6722.

Wang, N., Adams, G., Buttery, L., Falcone, F. H. and Stolnik, S. 2009. Alginate encapsulation technology supports embryonic stem cells differentiation into insulin-producing cells, *J Biotechnol.*

Watanabe, K., Ueno, M., Kamiya, D., Nishiyama, A., Matsumura, M., Wataya, T., Takahashi, J. B., Nishikawa, S., Muguruma, K. and Sasai, Y. 2007. A ROCK inhibitor permits survival of dissociated human embryonic stem cells, *Nat Biotechnol* 25:681–686.

Westfall, S. D., Sachdev, S., Das, P., Hearne, L. B., Hannink, M., Roberts, R. M. and Ezashi, T. 2008. Identification of oxygen-sensitive transcriptional programs in human embryonic stem cells, *Stem Cells Dev* 17:869–881.

Wobus, A. M., Kaomei, G., Shan, J., Wellner, M. C., Rohwedel, J., Ji, G., Fleischmann, B., Katus, H. A., Hescheler, J. and Franz, W. M. 1997. Retinoic acid accelerates embryonic stem cell-derived cardiac differentiation and enhances development of ventricular cardiomyocytes, *J Mol Cell Cardiol* 29:1525–1539.

Wu, S. C., Huang, G. Y. and Liu, J. H. 2002. Production of retrovirus and adenovirus vectors for gene therapy: A comparative study using microcarrier and stationary cell culture, *Biotechnol Prog* 18:617–622.

Wu, X., Walker, J., Zhang, J., Ding, S. and Schultz, P. G. 2004. Purmorphamine induces osteogenesis by activation of the hedgehog signaling pathway, *Chem Biol* 11: 1229–1238.

Xiong, C., Xie, C. Q., Zhang, L., Zhang, J., Xu, K., Fu, M., Thompson, W. E., Yang, L. J. and Chen, Y. E. 2005. Derivation of adipocytes from human embryonic stem cells, *Stem Cells Dev* 14:671–675.

Xu, C., Rosler, E., Jiang, J., Lebkowski, J. S., Gold, J. D., O'Sullivan, C., Delavan-Boorsma, K., Mok, M., Bronstein, A. and Carpenter, M. K. 2005a. Basic fibroblast growth factor supports undifferentiated human embryonic stem cell growth without conditioned medium, *Stem Cells* 23:315–323.

Xu, R. H., Peck, R. M., Li, D. S., Feng, X., Ludwig, T. and Thomson, J. A. 2005b. Basic FGF and suppression of BMP signaling sustain undifferentiated proliferation of human ES cells, *Nat Methods* 2:185–190.

Yamamoto, K., Sokabe, T., Watabe, T., Miyazono, K., Yamashita, J. K., Obi, S., Ohura, N., Matsushita, A., Kamiya, A. and Ando, J. 2005. Fluid shear stress induces differentiation of Flk-1-positive embryonic stem cells into vascular endothelial cells *in vitro*, *Am J Physiol Heart Circ Physiol* 288:H1915–1924.

Yao, S., Chen, S., Clark, J., Hao, E., Beattie, G. M., Hayek, A. and Ding, S. 2006. Long-term self-renewal and directed differentiation of human embryonic stem cells in chemically defined conditions, *Proc Natl Acad Sci U S A* 103:6907–6912.

Yener, B., Acar, E., Aguis, P., Bennett, K., Vandenberg, S. L. and Plopper, G. E. 2008. Multiway modeling and analysis in stem cell systems biology, *BMC Syst Biol* 2:63.

Yin, C. H., Chen, W., Hsiao, C. C., Kuo, C. Y., Chen, C. L. and Wu, W. T. 2007. Production of mouse embryoid bodies with hepatic differentiation potential by stirred tank bioreactor, *Biosci Biotechnol Biochem* 71:728–734.

Ying, Q. L., Nichols, J., Chambers, I. and Smith, A. 2003. BMP induction of Id proteins suppresses differentiation and sustains embryonic stem cell self-renewal in collaboration with STAT3, *Cell* 115:281–292.

Yirme, G., Amit, M., Laevsky, I., Osenberg, S. and Itskovitz-Eldor, J. 2008. Establishing a dynamic process for the formation, propagation, and differentiation of human embryoid bodies, *Stem Cells Dev* 17:1227–1241.

Yu, J., Vodyanik, M. A., Smuga-Otto, K., Antosiewicz-Bourget, J., Frane, J. L., Tian, S., Nie, J., Jonsdottir, G. A., Ruotti, V., Stewart, R., Slukvin, II, and Thomson, J. A. 2007. Induced pluripotent stem cell lines derived from human somatic cells, *Science* 318:1917–1920.

Zandstra, P. W., Bauwens, C., Yin, T., Liu, Q., Schiller, H., Zweigerdt, R., Pasumarthi, K. B. and Field, L. J. 2003. Scalable production of embryonic stem cell-derived cardiomyocytes, *Tissue Eng* 9:767–778.

Zhao, F. and Ma, T. 2005. Perfusion bioreactor system for human mesenchymal stem cell tissue engineering: dynamic cell seeding and construct development, *Biotechnol Bioeng* 91:482–493.

Zhu, S., Wurdak, H., Wang, J., Lyssiotis, C. A., Peters, E. C., Cho, C. Y., Wu, X. and Schultz, P. G. 2009. A small molecule primes embryonic stem cells for differentiation, *Cell Stem Cell* 4:416–426.

zur Nieden, N. I., Cormier, J. T., Rancourt, D. E. and Kallos, M. S. 2007. Embryonic stem cells remain highly pluripotent following long term expansion as aggregates in suspension bioreactors, *J Biotechnol* 129:421–432.

Engineering Bioactive Scaffolds for Stem Cell Vascular Therapy 8

Abigail C. Cuddy, Yu-I Shen, and Sharon Gerecht

8.1. Introduction

8.1.1. *A general history of tissue engineering and regenerative medicine and the need for vascular engineering*

Tissue engineering (TE) is defined as the use of organ-specific cells for seeding a scaffold *ex vivo* (Bianco and Robey, 2001), while regenerative medicine (RM) encompasses research into treatments which aim to restore body tissues and organs. The concept of tissue regeneration and engineering emerged when Robert Hooke (1635–1703) identified and described the simplest unit of life, the cell, in his book *Micrographia* (Hooke, 1665). It wasn't until 1858, however, that Rudolph Virchow (1821–1902) claimed that new cells only arise from pre-existing cells, a proclamation that spurned a great deal of excitement into the realm of cellular proliferation. Applying the concepts of cellular proliferation to wound healing, German surgeon Carl Thiersch (1822–1895) introduced the technique of split skin grafting, a procedure involving the transplantation of a portion of epidermis and dermis to a wound for the purpose of expediting the healing process. His pioneering work in 1874 is renowned amongst surgeons and tissue engineers and to Thiersch's credit, the technique is known as the Thiersch graft. Once a dream, TE and RM became a reality as a result of the revolutionary work of Thiersch and countless others. However, like anything in science and medicine, new discoveries are always being made and what was considered the best of the best is always made better. These days, TE and RM are dynamic fields in which new ideas, better biomaterials, and novel techniques are tested for their clinical relevance. In order to fully appreciate these breakthrough discoveries, it is worth discussing the mix of both biological and artificial components which must work intricately with one another so that they may together accomplish their intended purpose.

8.1.1.1. *Cell transplantation for TE and RM*

The cell is the biological force driving tissue regeneration. Therefore, the success of any transplantation depends upon cells, which not only engraft in the desired location but also proliferate, promoting tissue regeneration. Introducing cells alone has proven problematic, because they have limited ability to survive in the host. For instance, the cells now being tested clinically for TE and RM are adult stem cells — tissue-derived stem cells which produce all the cells within a given tissue (e.g. blood, skin, gut, and nervous system). Within an organism, these stem cell populations live in an environment which not only promotes their survival but governs when they grow and divide. Thus, the introduction of such stem cells alone affects their survival, proliferation, and differentiation in the host, since these cells do not have the required environmental cues to support their growth. Therefore, in order to hasten the tissue regeneration process, engineers continue to develop better biomaterials to mimic the natural environments of such stem cells. Equally important is the design of biomaterials, promoting the development of blood vessels to expedite the tissue regeneration process.

8.1.1.2. *Introduction to vascular RM*

While the cellular environment plays a key role in directing stem cell fate, the ubiquitous availability of oxygen and nutrients to drive sustained cellular proliferation and tissue growth is essential. Indeed, without the availability of oxygen and nutrients carried by red blood cells circulating within patent blood vessels, all attempts to regenerate a tissue of interest will be futile. Thus, researchers have focused on designing biomaterials which facilitate the formation of blood vessels both *in vivo* and *ex vivo*. Appreciating the cellular components that comprise blood vessels must precede understanding what cell populations produce blood vessels and which biomaterials favor their growth. While variations exist depending on the type of vessel (i.e. artery, capillary or vein), all blood vessels have the same general structure: the tunica intima, the innermost layer of the blood vessel, is composed of endothelial cells (ECs) which line the lumen of the vessel; the tunica media, the middle layer of the blood vessel, comprises of mural cells [pericytes and smooth muscle cells (SMCs)] which provide the vessel with its contractile and mechanical properties; and the tunica adventitia, the outermost layer of the blood vessel, comprises a variety of connective tissues and fibroblasts which provide the vessel with structural stability. Along these lines, biomaterials must be designed so as to favor the migration, proliferation, and tube-like formation of vascular ECs. In order to design such a scaffold, these materials must not only provide favorable attachment sites for ECs but must additionally possess pores of an appropriate size, quantity, and distribution to allow both the three-dimensional (3D) migration and tube formation from ECs and the subsequent diffusion of oxygen and nutrients necessary to facilitate tissue regeneration (Sachlos and Czernuszka, 2003; Botchwey *et al.*, 2003). Such biomaterials will be discussed in greater detail in subsequent sections. Along with the need for biomimetic scaffolds, which allow the assembly of patent blood vessels, the cell population, which will eventually regenerate the tissue of interest, must also be introduced into the scaffold. These populations of stem cells have self-renewal and differentiation capabilities. Next, we will explore the available stem cell sources and their potential applications for tissue regeneration and for vascular engineering, which is crucial to the sustainability of the rapidly regeneration tissue.

8.2. Stem Cells for TE and RM

8.2.1. *Introduction to the use of stem cells for TE and RM*

Stem cells are defined as clonogenic populations of cells, capable of self-renewal into additional stem cells and of differentiation into the various cell lineages (Slack, 2000). Not all stem cells are alike, differing in their degree of plasticity. For example, totipotent stem cells (from fertilized embryos) can give rise to a functional organism; pluripotent stem cells (from the inner cell mass of the embryo) can produce all tissue types; and multi-potent stem cells (from adult tissue stem cells) can generate a limited range of cells within a given tissue (e.g. blood and lymphoid cells in the hematopoietic system). As our understanding of these stem cell populations (briefly explored below) continue to grow, scientists are harnessing the seemingly unlimited potential of these once ambiguous cells for a wide array of therapeutic modalities.

8.2.1.1. *Human embryonic stem cells*

Human embryonic stem cells (hESCs) are pluripotent cells isolated from the inner cell mass of the developing blastocyst. *In vitro*, these cells can be maintained as undifferentiated cells by culturing them on mouse embryonic fibroblasts (MEF) or in feeder-free conditions with a defined medium composition to supply essential cytokines and nutrients, which promote the undifferentiated phenotype. These stem cells are highly suited for vascular regeneration since they renew indefinitely and provide an inexhaustible source of vascular progenitor cells. Human ESCs have been reported to give rise to vascular progenitor cells which, in turn, are capable of differentiating into ECs and SMCs, depending on the cell culture conditions (Leor *et al.*, 2007; Jin *et al.*, 0000; Kim *et al.*, 2008; Li *et al.*, 2009; Ferreira *et al.*, 2007). When injected into nude mice, both ECs and SMCs integrated into the murine vasculature (Ferreira *et al.*, 2007), indicating the potential use of hESC-derived vascular progenitor cells as a source for the generation of vascular structures for TE and RM.

Indeed, hESCs are an invaluable source from which any tissue population of interest can be derived; however, clinical and ethical limitations still exist, preventing the translation of hESCs from bench to bedside. Additional research is warranted to ensure the terminal differentiation, maturation, and functionality of the derivatives as well as their immunocompatibility with the host.

8.2.1.2. *Induced pluripotent stem cells*

The field of stem cell technology was revolutionized when the discovery was announced that a patient's terminally differentiated cells (somatic cells) could be reprogrammed to produce any cell type of interest (Yamanaka, 2007). Reportedly, such cells not only offer the advantage of supplying unlimited numbers of cells for patient needs, but perhaps more importantly, pose no threat to immune-mediated graft rejection since the cells are derived directly from the patient. Such cells have been termed induced pluripotent stem cells (iPS) — adult somatic cells which have been reprogrammed to assume an ESC-like phenotype via introduction of either of two transcription factor sets: Oct4, Nanog, Sox2, Lin28 or Oct4, cMyc, Sox2, Klf4 (Yu *et al.*, 2007). The breakthrough discovery that terminally differentiated adult cells could be induced to de-differentiate into pluripotent stem cells intensified research in the use of such cells for the treatment of a wide array of diseases. Using animal models, iPS cells have been demonstrated to regenerate cardiac tissue and restore contractile function after experimentally induced ischemia (Nelson *et al.*, 2009). Additionally, iPS cells have shown promise for the treatment of sickle cell anemia, a debilitating disease in which the compromised structure of red blood cells limits their oxygen-carrying capacity and thus their ability to nourish tissues. In this report, the gene-corrected iPS cells reversed the systemic failure as evidenced through increased respiratory function, increased body weight and enhanced kidney function (Hanna *et al.*, 2007). Furthermore, iPS cells have demonstrated therapeutic efficacy in animal models for the treatment of Parkinson's disease (Wernig *et al.*, 2008), a neurodegenerative disorder characterized by resting tremors and limited mobility, and for hemophilia A (Xu *et al.*,, 2009), a life-threatening disorder characterized by uncontrolled bleeding due to defects in a protein responsible for clotting. In addition to their tissue regenerative properties, iPS cells have been successfully differentiated into vascular cells (Taura *et al.*, 2009). For instance, researchers discovered that iPS cell lines could be guided to form vascular structures through the use of methodology previously validated to direct the differentiation of hESCs into vascular cells (Taura *et al.*, 2009). The authors reported that the iPS-derived vascular cells expressed EC-specific markers and organize themselves into capillary-like structures (CLS) when cultured on Matrigel, an extracellular matrix (ECM)- rich material known to support the formation of CLS (Taura *et al.*, 2009). Interestingly, a fraction of the vascular lineage-differentiated iPS cells were further differentiated into perivascular-like cells (e.g. SMCs) as evidenced by their expression of SMC-specific markers α-smooth muscle actin (α-SMA) and calponin (Taura *et al.*, 2009).

While iPS cells have the potential for treating a wide range of chronic disorders, clinical limitations still exist. Similar to hESCs, iPS cells are predisposed to form tumors, which occur as a result of the use of retro- and lenti-viral vectors for the introduction of "stem-ness" genes. Therefore, efforts are underway to circumvent the use of viral vectors. Indeed, the use of viral vectors that do not integrate into the genome and vectors that are efficiently capable of snipping themselves out of the genome following completion of genetic reprogramming, have been described (Kaji *et al.*, 2009; Woltjen *et al.*, 2009; Yu *et al.*, 2009). While the technology for reprogramming continues to improve in this field, other stem cell populations are concurrently being investigated for their curative potential.

8.2.1.3. *Human umbilical cord blood cells*

Human umbilical cord blood (hUCB) cells, another source for stem cell therapies, contain a highly enriched collection of both early and committed endothelial progenitor cells (EPCs), as compared to adult peripheral blood (Broxmeyer *et al.*, 1990). Upon investigating the use of hUCB cells for the regeneration of bone, researchers discovered that these cells, implanted in a collagen-containing scaffold, promoted local bone formation in a rat model of critical size femoral defect (Jager *et al.*, 2007). Additionally, these cells have the capacity to treat myocardial infarcts, as evidenced by observations in which they not only localized to the infarct area, but promoted neoangiogenesis, thus expediting the healing process (Ma *et al.*, 2005). Specifically, hUCB cells have been investigated for their ability to differentiate into vascular cells. In an effort to understand their vasculogenic capability, researchers discovered that hUCB cells differentiated into ECs, which retained their EC phenotype following *in vitro* expansion (Wu *et al.*, 2004). Similar reports have shown that hUCB-derived ECs express the EC-associated markers CD31 and VE-cadherin (Melero-Martin *et al.*, 2008). hUCB cells have additionally been documented to produce mesenchymal stem cells (MSCs), a bone marrow-derived cell which gives rise to fat, cartilage, bone, and the perivascular cells of the blood vessels (Melero-Martin *et al.*, 2008). These *in vitro* hUCB-derived cells were confirmed to be MSCs by their expression of the MSC marker CD90, the absence of EC-specific markers, and the presence of the SMC-associated markers calponin and α-SMA (Melero-Martin *et al.*, 2008).

These and many other studies have demonstrated the potency of hUCB cells as a novel therapeutic modality; however, their use has drawbacks. For instance, hUCB cells represent a limited pool of available EPCs, resulting in efforts to expand them *in vitro* without inducing differentiation. In any case, they have immense therapeutic potency and with continued research, hUCB cells will no doubt revolutionize the fields of TE, RM, and vascular engineering.

8.2.1.4. *Adult mesenchymal stem cells*

As discussed up to this point, numerous criteria must be fulfilled in order for the therapeutic applications of stem cell populations to move from the bench to the bedside. For example, the cell population must be readily available in sufficient quantities, immunocompatible and should pose no threat to the development of secondary diseases. One population, mentioned above, meets almost all of these criteria and has been successfully employed in the clinic as a novel therapeutic strategy for the treatment of a variety of disorders. This population, known as adult mesenchymal stem cells (MSCs) (i.e. multipotent stem cells), encompasses a wide range of tissue-specific stem cells. MSCs can differentiate into many different cell types of the mesodermal lineage, including but not limited to, osteoblasts, chondrocytes, adipocytes, and SMCs. In two separate clinical studies, for example, investigators demonstrated that human MSCs (hMSCs) promoted bone growth and reduced the fracture rate in patients suffering from osteogenesis imperfecta (Horwitz *et al.*, 1999; 2002). Human MSCs have additionally been successfully utilized to treat graft versus host disease (Ning *et al.*, 2008) and cardiac ischemia in which left ventricular function was reported to have improved following autologous MSC injection (Chen *et al.*, 2004). Utilizing synthetic scaffolds, experimental studies have demonstrated that murine MSCs (mMSCs) successfully expedited wound healing in mice (Perng *et al.*, 2008) and in a rodent model of cardiac ischemia, improved cardiac function and increased cardiac myogenesis (Jin *et al.*, 2009). In addition, hMSCs have been reported to give rise to vascular cells. For instance, it was shown that hMSCs could differentiate *in vitro* into ECs (Oswald *et al.*, 2004). In this report, hMSCs were initially cultivated in endothelial media supplemented with 50 ng/ml of VEGF, a pro-angiogenic growth factor (Oswald *et al.*, 2004). Following growth of hMSCs under these cell culture conditions, it was found that the resulting cells expressed VEGFR1 and VEGFR2 (refer to Table 8.1), cell surface markers present on ECs, and formed CLS *in vitro* (Oswald *et al.*, 2004). Investigating the tissue restorative capabilities of MSCs, canine MSCs were introduced into experimentally induced ischemic myocardium of canine models (Silva *et al.*, 2005). Following the introduction of canine MSCs, the MSCs gave rise to ECs and SMCs, which integrated into the host vasculature, thus improving its capacity to restore cardiac function (Silva *et al.*, 2005).

While hMSCs have the capacity to treat and regenerate numerous tissues and even generate vascular cells, they have limitations. Like HUCB cells, hMSCs are limited in number, restricting their ameliorative capabilities. Regardless, hMSCs — along with hESCs, iPS cells, and HUCB cells — may represent the future of TE and RM and with further investigation, could advance the realization of practical clinical applications.

8.2.1.5. *The use of stem cells for the generation of vascular structures*

Previous sections have discussed the ability of stem cell populations to regenerate new tissues and to produce vascular cells. At this point, it is worthwhile to discuss how these stem cell populations are presently being utilized for their ability to generate functional vascular structures. Subsequent sections will address the biomaterials presently utilized to promote the assembly of patent vascular conduits.

In addition to the tissue-regenerating capabilities of the stem cell populations mentioned above, these stem cell sources have been investigated for their ability to organize into functional, long-lasting blood vessels. Utilizing ECs derived from hESCs, investigators demonstrated that such hESC-derived ECs, when coimplanted in the presence of mMSCs, organized into CLS 2 days after implantation into murine hosts. Additionally, a portion of the hESC-derived ECs formed vascular conduits with the host vascular network, suggesting that hESC-derived ECs may offer a renewable source of vascular cells for future vascular engineering initiatives (Wang *et al.*, 2007). A group investigating the vascular potential of iPS cells reported that iPS-derived ECs, validated by their expression of EC markers VE-cadherin, VEGFR1, and CD31 amongst others, formed CLSs on Matrigel when in the presence of VEGF-containing media (Choi *et al.*, 2009). While this group utilized iPS cells transduced with viral factors, a major limitation to clinical implementation, this report nonetheless demonstrates the successful derivation of ECs and their subsequent ability to successfully organize into vascular structures *in vivo*. Additionally, the vascular-like capabilities of the EPC-rich hUCB,

Table 8.1. Growth factors promoting angiogenesis.

Growth Factor	Receptor Interaction	Effect
Ang1	Tie2 (Xu and Yu, 2001)	EC migration, adhesion, and spreading (Carlson *et al.*, 2001; Harfouche *et al.*, 2002)
Ang2	Tie2 (Maisonpierre *et al.*, 1997)	Angiogenic in the presence of VEGF (Hanahan, 1997)
Angiogenin	Cell surface actin (Hu *et al.*, 1993)	EC migration, proliferation and tube formation (Hu *et al.*, 1994; Jimi *et al.*, 1995; Hu *et al.*, 1997)
EGF	EGFR (Toby *et al.*, 2007; Schreiber *et al.*, 1986)	EC migration and proliferation (Schreiber *et al.*, 1986; McAuslan *et al.*, 1985)
Ephrins	EphA, EphB (Gale and Yancopoulos, 1999)	EC migration, tube formation and vessel sprouting (Pandey *et al.*, 1995; Daniel *et al.*, 1996; Adams *et al.*, 1999)
aFGF	FGFR1-4 (Ornitz *et al.*, 1996)	EC chemotaxis, growth and vascular SMC proliferation and migration (Baird *et al.*, 1986; Gospodarovicz *et al.*, 1986; Gospodarowicz *et al.*, 1987)
bFGF	FGFR1-4 (Basilico and Moscatelli, 1992; Jaye *et al.*, 1992)	EC motility and mitosis; vascular SMC proliferation and migration (Baird *et al.*, 1986; Gospodarovicz *et al.*, 1986; Gospodarowicz *et al.*, 1987; Basilico and Moscatelli, 1992).
G-CSF	G-CSFR (Lee *et al.*, 2005)	EC proliferation, migration and tube formation (Lee *et al.*, 2005; Bussolino *et al.*, 1989).
GM-CSF	α, β subunits GM-CSFR (Hayashida *et al.*, 1990)	EC proliferation and migration (Bussolino *et al.*, 1989; Mann *et al.*, 2001)
HGF	c-Met (Ding *et al.*, 2003)	EC growth and motility (Morimoto *et al.*, 1991; Bussolino *et al.*, 1992; Xin *et al.*, 2001).
IL-8	CXCR1, CXCR2 (Li *et al.*, 2003)	EC survival, proliferation and tube formation (Li *et al.*, 2003)

(*Continued*)

Table 8.1. (*Continued*)

Growth Factor	Receptor Interaction	Effect
Leptin	Ob-R (Park *et al.*, 2001)	EC migration, proliferation and survival (Sierra-Honigmann *et al.*, 1998; Bouloumie *et al.*, 1998; Goetze *et al.*, 2002)
NGF	trkA, p75(NTR) (Cantarella *et al.*, 2002)	EC mitosis and migration (Cantarella *et al.*, 2002; Moser *et al.*, 2004; Dolle *et al.*, 2005)
PDGFs	α-PDGFR, β-PDGFR (Raines *et al.*, 1990; Heidaran *et al.*, 1990)	EC survival and SMC proliferation and migration (Majack *et al.*, 1990; Au *et al.*, 2009)
PIGF	VEGFR1 (Sawano *et al.*, 1996)	Augments the VEGF-induced effect (Luttun *et al.*, 2002; Park *et al.*, 1994)
TGF-α	EGFR (Schreiber *et al.*, 1986)	With TGF-β1, promotes EC proliferation and migration (Vinals and Pouyssegur, 2001)
TGF-β	ALK1 (Goumans *et al.*, 2002)	EC proliferation, migration, survival and tube formation (Vinals and Pouyssegur, 2001; Goumans *et al.*, 2002)
TNF-α	TNFR-2 (Luo *et al.*, 2006)	EC vascular permeability, survival and migration (Luo *et al.*, 2006; Henning *et al.*, 1987)
VEGF	VEGFR1, VEGFR2 (de Vries *et al.*, 1992; Neufeld *et al.*, 1996)	EC vascular permeability, proliferation, survival and motility (Ferrara, 2002)

*Ang: angiogenin; alk: activin-like kinase; EGF: epidermal growth factor; EGFR: epidermal growth factor receptor; Eph: ephrin; FGF: fibroblast growth factor; FGFR: fibroblast growth factor receptor; G-CSF: granulocyte colony-stimulating factor; G-CSFR: granulocyte colony-stimulating factor receptor; GM-CSF: granulocyte macrophage colony-stimulating factor; GM-CSFR: granulocyte macrophage colony-stimulating factor receptor; HGF: hepatocyte growth factor; IL-8: interleukin 8; CXCR: chemokine CXC motif receptor; NGF: nerve growth factor; trkA: tyrosine kinase A receptor; NTR: neurotrophin receptor; PDGF: platelet-derived growth factor; PDGFR: platelet-derived growth factor receptor; PlGF: placental growth factor; SMC: smooth muscle cell; TGF: transforming growth factor; TNF: tumor necrosis factor; TNFR: tumor necrosis factor receptor; VEGF: vascular endothelial growth factor; VEGFR: vascular endothelial growth factor receptor was investigated (Mukai *et al.*, 2008). Interestingly, EPCs from hUCB both formed CLS on Matrigel and integrated into pre-existing CLS formed by human umbilical vein endothelial cells (HUVECs) (Mukai *et al.*, 2008), mature ECs derived from the neonate. Investigating the vascular potential of hUCB cells *in vivo*, hUCB EPCs with hUCB MSCs were subcutaneously coimplanted in the presence of Matrigel into murine hosts (Melero-Martin *et al.*, 2008). Subsequent histological analyses revealed the presence of vessels containing erythrocytes in the Matrigel coimplants only (Melero-Martin *et al.*, 2008). Further analysis demonstrated the presence of hUCB EPC-derived ECs lining the lumen of the vessel with hUCB MSC-derived perivascular cells (e.g. SMCs) lining the outer rim (Melero-Martin *et al.*, 2008). In an attempt to understand the vascular potential of hMSCs, another group used a collagen/fibronectin-rich matrix to coimplant hMSCs with HUVECs into murine hosts (Au *et al.*, 2008). They demonstrated that the introduction of HUVECs alone led to the formation of vessels which quickly regressed, while coimplantating HUVECs with hMSCs promoted the assembly of functional vessels that anastomosed with the host vasculature and were stable for over 130 days (Au *et al.*, 2008). Subsequent *in vivo* analyses demonstrated that the hMSCs used in this study differentiated into pericytes, as evidenced by their perivascular orientation around HUVEC-derived vessels and their ability to contract in response to endothelin-1, a contractile-promoting agent (Au *et al.*, 2008). This study makes it clear that, when injected *in vivo*, hMSCs differentiate along a SMC-like lineage.

Together, these analyses report the feasibility of using various stem cell populations to generate vascular structures for vascular engineering. However, in order to fully appreciate how stem cell populations and their associated biomaterials form fully functional blood vessels, the process of blood vessel formation requires greater examination. Understanding how environmental cues direct blood vessel formation is an essential step in designing biomaterials that better mimic nature.

8.2.2. *Angiogenesis*

Angiogenesis — the formation of blood vessels from pre-existing blood vessels — is an important natural process whereby development successfully occurs, wound healing is hastened, and the reproductive cycles of female mammals occurs. Apart from these functions, angiogenesis does not occur in healthy mammals, but plays a major role in tumor development, growth and metastasis. The tightly controlled process of angiogenesis is dependent on a variety of chemical and mechanical inputs, including the availability of growth factors, cell-to-extracellular matrix (ECM) interactions and cell-to-cell interactions. Figure 8.1 provides a schematic diagram depicting the process of angiogenesis. As previously stated, the therapeutic efficacy

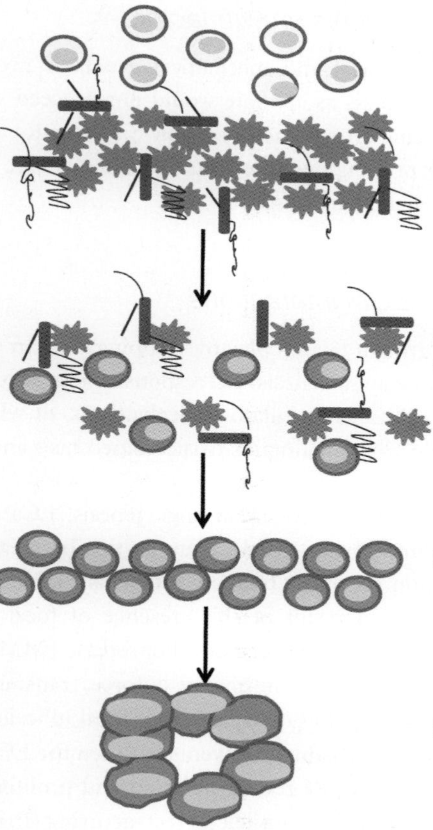

Figure 8.1. Angiogenesis. Angiogenesis, the growth of new vessels from pre-existing vessels, proceeds in a step-wise manner. First, angiogenic growth factors activate ECs to release proteases, enzymes which degrade the vessels underlying basement membrane. Following basement membrane degradation, chemotactic angiogenic molecules, such as VEGF and bFGF, promote the migration of ECs through the dissolved basement membrane. ECs then proliferate behind the leading tip of migration, from lumens and align into tube-like structures. Stabilization of the vessel(s) is accomplished via arrest of EC proliferation and subsequent reconstruction of the basement membrane.

of stem cell therapies depends on the presence of oxygen and nutrients necessary to sustain rapid cellular proliferation, a process solely dependent upon angiogenesis. As a result, the introduction of stem cells must be coupled with the appropriate biomaterials, growth factors and perhaps exogenous ECs so as to promote rapid vascularization and thus expedite the tissue regeneration process. Below, we discuss the biological ingredients essential for promoting angiogenesis.

8.2.2.1. *The role of proangiogenic growth factors*

While the interplay between chemical and mechanical events is necessary for angiogenesis to take place, the process as a whole would not proceed were it not for the availability of proangiogenic growth factors. Table 8.1 provides a list of the known growth factors and their pro-angiogenic effects. Such factors are critical for controlling various steps in the angiogenic cascade.

8.2.2.2. *The role of EC-to-ECM interactions*

As previously seen, numerous growth factors play overlapping roles in controlling one or more steps in the process of angiogenesis. EC responsiveness to such growth factors, however, depends on additional regulatory mechanisms, of which one is the ECM, a non-cellular scaffold which maintains tissue homeostasis and regulates cellular growth (Joshi *et al.*, 1993).

In order for the ECM to exert control upon angiogenesis, ECs must form intimate connections with the underlying ECM. They do this by means of focal adhesions, collections of adhesion receptors (namely integrins) linking the ECM to the cytoskeleton of an EC. As a result of the presence of focal adhesions, mechanical changes in the ECM due to matrix metalloprotease (MMP) activity may be transferred to the EC or vice versa. The change in force, transmitted across the entire EC, is then transferred to the interior of the cell via focal adhesion coupling to the cytoskeleton. These mechanotransduction events between the ECM and the ECs collectively act to change patterns of gene expression and proliferation, and change the migration of ECs as a result of cytoskeletal restructuring (Ingber, 2002). One way that structural changes in the ECM promote the aforementioned changes in EC behavior occurs via coupling of mitogens to the angiogenic response. In an attempt to understand how structural changes in the ECM are linked to extrinsic mitogens, Mammoto *et al.* (2009) discovered that mechanical forces transduced to the EC-modulated EC expression of VEGFR2, a cell surface receptor which binds the proangiogenic growth factor VEGF (Table 8.1). They made this novel discovery when they observed that a relatively compliant ECM substrate of 4000 Pa enhanced VEGFR2 mRNA and protein expression in overlying ECs (Mammoto *et al.*, 2009). Rigid ECM substrates, on the other hand, resulted in lower EC VEGFR2 expression levels (Mammoto *et al.*, 2009). The upregulated VEGFR2 expression on ECs should correlate to enhanced EC responsiveness to VEGF. In fact, this was observed when EC tube formation could be initiated on a compliant gel supplemented with a relatively high (50 ng/ml) concentration of VEGF (Hanjaya-Putra *et al.*, 2009). Indeed, the proangiogenic effect of VEGF in this example can be correlated to the structural changes in the ECM, which not only promote the upregulated expression of growth factor receptors on ECs, but additionally liberate sequestered growth factors. These and many other studies point to the role that structural changes in the ECM play in modulating EC responsiveness to growth factors.

8.2.2.3. *The Role of EC-to-EC interactions*

Like EC interactions with the ECM, EC-to-EC interactions mediated by the actin cytoskeleton provide a means whereby extrinsic changes affecting the behavior of one cell are directly coupled to adjacent cells. Although many adhesion molecules influence angiogenesis, two of the most important ones are the glycoproteins VE-cadherin (VE-Cad) and CD31, located between ECs at adherens junctions and intercellular clefts, respectively. Extracellularly, VE-Cad forms homophilic interactions with VE-Cad on adjacent ECs (Lampugnani *et al.*, 1992) while CD31 forms both homophilic and heterophilic interactions (Muller *et al.*, 1989). Intracellularly, both adhesion molecules are anchored to the actin cytoskeleton (Ayalon *et al.*, 1994). A group investigating the angiogenic potential of CD31 reported that antibody-mediated downregulation of CD31 resulted in severe declines in rat corneal neovascularization *in vitro* and murine vessel growth *in vivo* (DeLisser *et al.*, 1997). Interestingly, the monoclonal antibodies directed against CD31 in this study additionally prevented cytokine-induced vascularization (DeLisser *et al.*, 1997). This was evidenced by the absence of vascular structures in mice and rats, following either the subcutaneous implantation of bFGF-containing Matrigel plugs into murine hosts or the corneal implantation of bFGF-containing Hydron pellets into rat hosts (DeLisser *et al.*, 1997). This study demonstrates the crucial role that CD31 plays in the process of angiogenesis. VE-Cad also plays a direct role in angiogenesis and, similar to CD31, has an apparent role in coordinating with growth factors to regulate the angiogenic response. For instance, investigators found that VEGFR2 clustered to areas with VE-Cad following stimulation of ECs with VEGF (Liebner *et al.*, 2006). The overall response to VEGF, however, has discordant angiogenic effects and seems to be dependent on cellular confluence (Grazia Lampugnani *et al.*, 2003). In addition to its role in facilitating and inhibiting growth factor signaling, VE-Cad has been shown to activate the EC prosurvival pathways Akt and PI3K, whereas interfering with VE-Cad rendered ECs susceptible to proapoptotic stimuli (Yang *et al.*, 1999). While VE-Cad and CD31 apparently exert their proangiogenic effects via growth factors and cytokines, both adhesion molecules have specific overall roles in the angiogenic process. For instance, VE-Cad has been demonstrated to promote vacuolization, or the development of intracellular lumens necessary for the process of forming blood vessel tubes, while CD31 promotes EC migration and elongation (Yang *et al.*, 1999). Taken together, CD31 and VE-Cad, in conjunction with many other cell-to-cell adhesion molecules, are vital components exerting control over the angiogenic cascade.

8.3. Biomaterials for Tissue Engineering and Regenerative Medicine

8.3.1. *Introduction*

TE and RM have the goal of developing a biological substitute that repairs, improves, or maintains the original tissue function. One factor limiting the field is the provision of well-vascularized constructs for nourishment of rapidly proliferating cells. The cell microenvironment consists of the noncellular ECM and the biological molecules which it sequesters, as well as the vascular networks. The ECM provides a living space for cell proliferation and differentiation. Biosignaling molecules sequestered in the ECM serve as messengers to direct cell fate (e.g. growth and differentiation of stem cells and proliferation, differentiation and tube formation from ECs). Vascular networks provide a means of supporting nutrition (e.g. glucose and O_2) and removing waste (e.g. CO_2) in the tissue. Because of limitations of oxygen diffusion, cells cannot survive when they are located greater than 150 μm away from the nearest capillary (Luong and Gerecht, 2009). Without a steady supply of nutrients, cells in the constructs will not survive *in vivo*. Thus, engineering a vascularized construct is an essential step for the successful regeneration of a tissue of interest.

8.3.2. *Design criteria*

Several design criteria must be addressed in order to engineer a vascularized construct. The material should: (1) Be biocompatible to the host (i.e. the material must provide a favorable surface for cell attachment and proliferation). (2) Be biodegradable; the resultant degradation product must be non-toxic and easily eliminated from the body through waste removal systems such as urine. (3) Provide enough mechanical strength in order to withstand surgical implantation and the mechanical stresses imposed by the body itself. (4) Retain a porous structure so that it not only provides sufficient space for cell migration and proliferation, but also allows the encapsulation and delivery of various biosignaling molecules.

8.3.3. *Biocompatible scaffolds for the generation of vascular structures*

Weinberg and Bell were the first to document the design of vascular biomaterials. They formed a vascular conduit initially comprised of SMCs seeded in a collagen gel (Weinberg and Bell, 1986). Fibroblasts were later added to the outer rim of the vessel, ECs were added to the interior of the vessel and a polytetrafluoroethylene (Dacron) graft was introduced to provide additional mechanical strength to the newly synthesized vessel (Weinberg and Bell, 1986). The artificially created vessel possessed differentiated ECs and produced bioactive molecules (Weinberg and Bell, 1986). Following this discovery, other investigations employed a variety of both natural and synthetic materials to generate vascular conduits both to replace diseased arteries and to expedite the generation of vascular structures for tissue regeneration. The biomaterials utilized to design such vascular constructs are discussed below.

8.3.3.1. *Biomaterials inspired by the ECM*

The construction of vascular biomaterials depends on the ability of the cell population of interest to deposit a long-term layer of ECM. Living organisms produce a variety of polymers, which can be categorized into three groups: polysaccharides, proteins, and polyesters. Due to the advances in biotechnology, most of these polymers can be obtained from microorganisms. However, these polymers are still mainly obtained from plants, animals or algae (Mano *et al.*, 2007). Below, we describe the various biomaterials used to mimic the natural environment of ECs.

Collagen, the most abundant protein in the body, belongs to a family consisting of more than 20 genetic splice variants. Collagen gel exhibits shear-thinning behaviors, which means its viscosity decreases as the applied shear increases (Boesel, 2004). Due to collagen's fibrous nature, collagen networks usually require secondary mechanisms to produce scaffolds that can withstand the mechanical stresses induced by encapsulation or TE procedures (Wallace and Rosenblatt, 2003). Thus, collagen can be cross-linked with a PEG-based polymer. One group found that chondrocytes encapsulated in such a scaffold remained viable 3 weeks following encapsulation (Taguchi *et al.*, 2005). Another group copolymerized collagen with dextran in order to mimic tertiary embryonic connective tissue, which demonstrated improved tissue growth into the scaffold (Marston *et al.*, 2005). Lastly, synthetic vascular structures have been designed by embedding collagen and dermatan sulfate into prosthetic scaffolds, followed by the addition of fibroblasts or SMCs and ECs (Ishibashi and Matsuda, 1994). This approach yielded patency rates of 80% for as long as 26 weeks following grafting into animal common carotid arteries (Matsuda and Mina, 1995).

Hyaluronanic acid (HA), a linear glycosaminoglycan which is soluble in water, can readily form hydrogels. Hyaluronidase, which exists in cells and serum, degrades it enzymatically. Much of the attention devoted to HA is due to the following HA chemical properties among others: (1) obtained in a wide range of molecular weights (by hyaluronidase degradation); (2) enzymatically remodeled by selected cell types, such as chondrocytes; (3) functionalized with reactive group to form hydrogels; and (4) modified to adjust its cell adhesive properties (Park *et al.*, 2003). It is possible to control the degradation rate and mechanical behavior of photopolymerized HA networks (Burdick *et al.*, 2005). Moreover, engineers have additionally developed flat sheets of non-woven HA-based scaffolding as vascular grafts (Arrigoni *et al.*, 2006). One group also showed that HA degradation products have proangiogenic activities (Slevin *et al.*, 2002). Finally, HA-based scaffolds correlate positively with the successful treatment of osteochondral defects (Solchaga *et al.*, 2000).

Fibrin is a major component of blood clots and plays an important role in spontaneous tissue repair. It forms a temporary network by polymerization of fibrinogen through the presence of the enzyme thrombin. Clinically, fibrin is used as a glue in surgical procedures. Fibrin can serve as an efficient vehicle for the delivery of growth factors or biomacromolecules (Sahni and Francis, 2000). It has also been used to regenerate skin and is involved in the loading and release of growth factors (Hubbell, 2003). Alternatively, fibrin scaffolds can serve as provisional reservoirs for the delivery of therapeutic proteins or cells for TE (Fasol *et al.*, 1994).

Alginate, a natural hydrophilic polysaccharide derived from algae, can be cross-linked in a gel with the use of divalent cations, such as Ca^{2+} and Mg^{2+}. The shape and strength of the gels in calcium-rich solutions can be altered (Percival, 1990). Because of its hydrophilic nature, it can promote cellular proliferation with a homogeneous cell distribution within the scaffold. One author of this chapter was part of a group that effectively employed capillary action to pull hESCs into an alginate scaffold (Gerecht-Nir *et al.*, 2004). Additional applications include injectable cell delivery vehicles (Bouhadir *et al.*, 2001) and alginate hydrogels used for cartilage TE (Awad *et al.*, 2004). Alginate gels can also serve as delivery vessels for proangiogenic growth factors necessary for the treatment of ischemia (Edelman *et al.*, 1991).

Chitin can be found in animal shells and exoskeletons of insects. It is insoluble in most common solvents but can be chemically modified into soluble derivatives, such as chitosan (Kurita, 2001). The derivatives are enzymatically degradable by lysozyme, which is present in blood serum. The properties of porous chitosan matrices can be varied by altering the chitosan concentration, freezing rate, and molecular weight. It is also utilized for the controlled release of pharmacological agents and as an effective non-viral vector for gene delivery (Erbacher *et al.*, 1998). It can also be engineered into sponge-like scaffolds in bone formation and cartilage TE (Nettles *et al.*, 2002).

Dextran, a bacteria-derived polysaccharide, has been used as a potential biomaterial for TE. It is soluble in both water and organic solvents. Its chemical structure consists of a large number of hydroxyl groups, which are available for chemical modification. Such modifications are necessary since dextran is naturally resistant to cell adhesion and is used as a biomaterial surface coating to limit cell adhesion. Functional group modifications can also alter its chemical and physical properties (Sun *et al.*, 2010). As a result, researchers have observed different vasculature responses. Lastly, by incorporating a cell adhesion peptide, one group showed that dextran, modified with Arg-Gly-Asp (RGD) peptides, promoted EC adhesion and spreading (Massia and Stark, 2001).

Natural polymers are found in the living organism and are appealing for the design of biomimetic scaffolds, since they are inherently biocompatible and can be enzymatically degraded. For example, native, tissue-derived ECM may be harvested from small intestinal submucosa (SIS), umbilical veins, cadaveral arteries (Berglund and Galis, 2003) and even xenogeneic materials such as porcine aortas, common carotid arteries, iliac arteries and bovine ureters (Hoenig *et al.*, 2005). The use of such natural materials is highly desirable since native ECM possesses the appropriate spatiotemporal features (Soucy and Romer, 2009) and acts as a natural reservoir for proangiogenic molecules (Vlodavsky *et al.*, 1991; Schonherr and Hausser, 2000), which together efficiently direct angiogenesis. The major drawbacks associated with the use of natural matrices are their susceptibility to enzymatic digestion, their propensity for thrombus formation, and their decreased mechanical strength. In order to overcome some of these limitations, one group cross-linked decellularized porcine SIS with ethyl carbodiimide to enhance its mechanical strength and treated it with heparin-benzalkonium chloride to reduce the scaffold's natural ability to induce thrombus formation (Huynh *et al.*, 1999). Following implantation of the chemically modified SIS graft into rabbit hosts, they found that the graft material efficiently recruited host cells (Huynh *et al.*, 1999), suggesting that such modification techniques may not limit the potential tissue regenerating capabilities of this biological scaffold. Similarly, heparin-treated decellularized matrices from porcine carotid arteries were introduced into the abdominal arteries of dogs (Tamura *et al.*, 2003). Eighteen weeks following the introduction of the graft material,

ECs aligned over the surface of the entire decellularized graft, fibroblasts and macrophages infiltrated its inner and outer portions, and SMCs made up the graft's neointima, suggesting that this decellularized biological material very effectively promoted native vascular formation (Tamura *et al.*, 2003). Unfortunately, some studies have noted that the use of chemicals to improve the ECM's mechanical strength and enhance its resistance to enzymatic digestion has resulted in deleterious outcomes, including aneurysm formation (Schmitz-Rixen *et al.*, 1991) and tissue calcification (Schmidt and Baier, 2000). Therefore, researchers have evaluated additional measures based on acellular and synthetic materials for their ability to form vascular structures.

8.3.3.2. *Biomaterials from synthetic sources*

Due to present limitations in the application of natural ECM for the fabrication of vascular constructs, other strategies are being developed and investigated. One strategy uses biodegradable (biomimetic) scaffolds to direct the synthesis of vascular structures. Interestingly, when compared to naturally derived polymers, synthetic materials have captured more attention in the recent years. Investigators have increasingly relied on scaffolds made of artificial materials to take advantage of the versatility and functionality they provide (Jay and Saltzman, 2009). Most synthetic polymers have the advantage that they are degraded via chemical hydrolysis and are insensitive to enzymatic processes, which limits variation among batches.

The most frequently utilized biodegradable scaffold is polyglycolic acid (PGA), since it is not only pliable and thus can be formed into a multitude of shapes, but it is also highly porous, allowing the diffusion of oxygen and nutrients necessary to facilitate angiogenesis (Berglund *et al.*, 2003). PGA, which consists of monomers linked by ester bonds, can easily be degraded to glycolic acid by water through hydrolysis, both *in vitro* and *in vivo*. In addition, it is hydrophilic in nature, resulting in the tendency to lose mechanical strength rapidly over a period of weeks and becoming fully absorbed within a few months (Seal *et al.*, 2001). Its controlled rate of degradation allows cells to deposit ECM and, in so doing, permits them to establish themselves within their own microenvironment (Mooney *et al.*, 1996). Utilizing the biomimetic properties of PGA, ovine arterial or venous cells were seeded into a PGA scaffold, the vascular construct was cultured *in vitro* for 7 days, lambs had a 2-cm segment of the pulmonary artery removed, and the hybrid scaffold was then introduced into the avascular region (Shinoka *et al.*, 1998). The authors reported the integration of the hybrid vascular constructs into the host arterial system, concluding that such techniques have the potential to ameliorate diseased and defective cardiac tissues (Shinoka *et al.*, 1998). Another group demonstrated that, with the use of a PGA scaffold, epithelial cells and SMCs could be reconstituted *in vivo* into vaginal tissue (De Filippo *et al.*, 2003). The reconstituted tissue additionally contained a patent microvasculature (De Filippo *et al.*, 2003).

Polylactic acid (PLA) is structurally similar to PGA; however, PLA and PGA have different chemical, physical, and mechanical properties. Due to its chiral structure, PLA can exist in two forms, PLLA or PDLLA. The former is more widely used since L-lactic acid occurs naturally; however, it is more hydrophobic compared to PGA. As a result, its degradation half-time (i.e. from months to a year) is much longer than PGA (Vert *et al.*, 1994). Some studies have indicated that neither PGA nor PLA may actively provoke an immune response since they pack peptide bonds (Kulkarni *et al.*, 1971). The mechanical and degradation profile of the scaffold can be controlled by creating a copolymer of PGA and PLA (PLGA). However, PLGA's characteristic property is not always a linear correlation with the pure polymer. For example, PLGA containing equal ratios of PGA and PLA degrades faster than a pure PGA scaffold (Seal *et al.*, 2001). The hybrid scaffold was developed to differentiate hESCs since it facilitates cellular adhesion, proliferation, and viability (Levenberg *et al.*, 2003).

Polyethylene glycol (PEG) is a hydrophilic polymer, the native structure of which does not permit cellular adhesion. It does, however, contain one hydroxyl group at each end which can be chemically modified so as to facilitate cellular attachment. The reactive group on the PEG polymer can be cross-linked by UV irradiation to form a hydrogel that mimics tissue ECM. Additionally, the chemically modified hydrogel can incorporate bioactive molecules in order to direct cell fate. For example, *in vivo* studies demonstrated that a VEGF-conjugated hydrogel prompted the local release of VEGF, which, in turn, promoted EC survival, growth, and concomitant vessel formation (Zisch *et al.*, 2003). Using PEG polymers has the advantage that they can be easily cleared by the kidneys (Sawhney *et al.*, 1993). Due to their high hydration content, PEG gels can also act as a barrier to macromolecules involved in immune responses (Zalipsky, 1995). One group found that PEG gels provided a biocompatible scaffold in which chondrocytes could readily proliferate and secrete ECM (Elisseeff *et al.*, 1999). Another group also reported that ECs cultured in a modified PEG hydrogel organized into stable, intricate networks of CLS (Moon *et al.*, 2010).

8.3.3.3. *Biomaterials designed using alternative technologies and approaches*

An alternative approach to the use of natural ECM and biodegradable scaffolds is the novel technique of fabricating artificial vessels without the use of a scaffold. L'Heureux *et al.* (1998) were the first to introduce this technique. They discovered that artificial vessels could be engineered using human umbilical vein SMCs and fibroblasts. In this study, the SMCs and fibroblasts, which were cultured separately for 30 days, formed a sheet of cells and ECM which could be manually peeled away from the culture flask (L'Heureux *et al.*, 1998). The SMC sheet (cells and ECM) was first added to a tubular support structure and allowed to mature for 1week, at which time the fibroblast sheet was added on top of the SMC sheet and the structure as a whole was allowed to mature for 7 weeks (L'Heureux *et al.*, 1998). HUVECs were subsequently introduced into the lumen of the structure and allowed to grow for 1 week, after which the newly synthesized vessel was found to efficiently engraft into the femoral artery of mongrel dogs (L'Heureux *et al.*, 1998). Similarly, researchers used a bioreactor to assemble vessels composed of HUVECs and human artery-derived myofibroblasts, which are fibroblasts having contractile capabilities (Kelm *et al.*, 2008). After 7 to 14 days in the bioreactor, they noted accelerated ECM formation in the resulting microvessels, which formed vessel-like tissues with ECs lining the interior of the vessel (Kelm *et al.*, 2008). While they did not perform *in vivo* studies, this model of enhanced cell-derived ECM production offers an attractive means to more efficiently design vascular structures that mimic the natural environment.

Aside from synthetic materials, novel forms of technology have been investigated for their ability to produce synthetic vascular structures. For example, investigators demonstrated a magnetized, synthetic vascular graft that efficiently captured and retained ECs (Pislaru *et al.*, 2006). In this study, porcine-derived ECs endocytosed superparamagnetic iron oxide microspheres and were then introduced into porcine carotid arteries, in the presence of a magnetic sheet held in place by a Dacron graft (Pislaru *et al.*, 2006). The authors concluded that the biophysical interactions were not only sufficient to retain ECs, but additionally allowed the rapid endothelialization of the vascular graft, thus demonstrating the potential for use in coronary artery bypass procedures (Pislaru *et al.*, 2006). Truly remarkable was the description of the fabrication of synthetic vascular structures using electrospinning technology, a technique that allows the construction of scaffolds that structurally mimic the native ECM (Thomas *et al.*, 2007). Of note, blends of synthetic materials, such as PLGA, type I collagen, and elastin, were used to electrospin tubular vascular conduits, which were subsequently seeded with bovine ECs and SMCs and implanted subcutaneously into murine hosts (Stitzel *et al.*, 2006). The hybrid graft persisted *in vivo* for 4 weeks following implantation (Stitzel *et al.*, 2006).

8.3.4. *Biomaterials modification*

8.3.4.1. *Biochemical cue modification*

Previous sections have discussed the importance of cell-matrix interactions in natural ECM. In order to truly mimic the natural ECM, the aforementioned biomaterials require additional surface modifications. Modifying their surfaces with bioactive molecules is a simple way to make biomimetic materials. Earlier studies using two naturally occurring ECM components, fibronectin and laminin, to coat biomaterials, yielded improved cell adhesion and proliferation (Humphries *et al.*, 1986). With recent advances in synthetic biology, active peptide chains on the long ECM proteins have been identified; however, laboratories can produce short peptide chains in larger quantities and more efficiently. In addition, short peptide chains are more stable during the modification process than the original long-chain protein. Moreover, the modified materials can be used in TE and can also serve as an artificial ECM, therefore providing suitable receptor-binding domains for cell attachment (Shin *et al.*, 2003). The biological cue that initiates the cell-signaling cascade as a result of cell-to-matrix interactions can be used to promote viability, to direct cell fate, and to guide 3D tissue formation. One of the most commonly used peptide sequences for the modification of scaffolds is RGD, the active binding site sequence on fibronectin (Table 8.2). For example, to ensure covalent binding of the peptides to the surface, the surface amine reactive moieties were chemically reacted with the carboxyl functional groups presented within the bioactive peptide. A bifunctional cross-linker was used to immobilize the peptide to the surface (Rezania *et al.*, 1999). Utilizing an RGD-modified scaffold containing sphingosine-1-phosphate, one group noted that EC migration speeds were enhanced, thus potentially reducing the time required for ECs to organize into patent blood vessels (Wacker *et al.*, 2008).

Surface modifications provide control over cell behavior; however, these modifications are limited because they are applicable for 2D structures rather than tissue-specific 3D structures. Bulk modifications are preferable in TE and RM applications. Several cell binding peptides have been introduced in 3D scaffolds through chemical (Elbert and Hubbell, 2001), photochemical (Hern and Hubbell, 1998), and ionic cross-linking (Rowley *et al.*, 1999). For example, one group synthesized alginate-based hydrogels covalently coupled with RGD. The peptide-modified hydrogel promoted soft tissue augmentation *in vivo* (Marler *et al.*, 2000).

Table 8.2. ECM sequences and associated functions.

Origin	Synthetic sequence	Function
Fibronectin	RGD	Cell adhesion
	KQAGDV	Smooth muscle cell adhesion (Massia and Stark, 2001)
	REDV	Endothelial cell adhesion (Mann *et al.*, 2001)
Laminin	IKVAV	Neurite extension (Hubbell *et al.*, 1991)
Laminin B1	YIGSR	Cell adhesion (Schense *et al.*, 2000)
Laminin B2	RNIAEIIKDI	Neurite extension (Ranieri *et al.*, 1995)
Cell adhesion molecules	KHIFSDDSSE	Astrocyte adhesion (Kam *et al.*, 2002)
Elastin	VPGIG	Enhance elastic modulus (Panitch *et al.*, 1999)
Heparin binding domain	KRSR	Osteoblast adhesion (Dee *et al.*, 1998)
	AGPL	Collagenase mediated degradation (West and Hubbell, 1999)

Also, another group demonstrated that hydrogels, fabricated with integrin-binding sites and protease cleavage sites, not only promoted CLS formation but also recruited MSCs to mechanically stabilize the newly formed CLSs (Moon *et al.*, 2010).

8.3.4.2. *Biophysical cue incorporation*

The mechanical properties of biomaterial scaffolds are also important. For example, in living skin substitutes used clinically to treat burns, the tension generated by fibroblasts in the collagen scaffolds is essential for the production of constructs with adequate biomechanical integrity. The ability to modify such scaffolds via extrinsic mechanisms is highly desirable, as recent studies have indicated that stem cell differentiation may be modulated by varying the stiffness of the scaffold (Engler *et al.*, 2007). For instance, researchers demonstrated that following the application of cyclic tension, SMCs seeded in scaffolds could be effectively utilized to produce bypass grafts *in vitro* (Niklason *et al.*, 1999). Aside from vascular engineering, scientists have also demonstrated that the growth of fasciculated axons can be sustained via application of continuous mechanical tension (Smith *et al.*, 2001). In addition to the application of mechanical tension, the physical properties of scaffolds may also be modified so as to facilitate cellular retention and proliferation. For instance, one group found that they could alter the local physical properties of scaffolds via exposure to a light source (Kloxin *et al.*, 2009). This study showed that the scaffold degraded following its exposure to light which, in turn, decreased its mechanical strength (Kloxin *et al.*, 2009). As a result of light-induced degradation, the scaffold-encapsulated MSCs exhibited enhanced cell spreading, a prerequisite for cellular survival and proliferation. A vascular engineering group found that vessel diameter was increased and vessel density decreased when EPCs were cultured on a stiff, more concentrated collagen gel (Critser *et al.*, 0000).

8.3.4.3. *Release strategies*

While scaffold architecture is important for promoting cellular attachment, proliferation, and differentiation, the therapeutic efficacy of the cellular source is limited if the mitogens necessary for cellular proliferation are not present. Generally, the local injection of growth factors is not effective, because these biomacromolecules rapidly diffuse away from the injection site (Tabata, 2009). Thus, the local concentration of the injected growth factors cannot effectively facilitate the proliferation of the cells encapsulated in the scaffold. In order to overcome this limitation, scaffolds should be designed so that the biomacromolecules are sequestered in the scaffold, but are also released in a steady state fashion to promote prolonged cell proliferation. Considering that most biosignaling molecules have a short life span and are unstable *in vivo*, the scaffold should be designed to serve as a protective barrier against protease and/or immune degradation. In this way, the scaffold must act as a protective agent for biosignaling molecules, preventing their degradation by the harsh physiologic surroundings. When the scaffold does degrade, whether through hydrolysis or enzymatic degradation, biosignaling molecules may be released, which may recruit surrounding cells, such as ECs, to contribute to the delivery of oxygen and nutrients to the rapidly proliferating cells.

8.4. Concluding Remarks

A great deal of effort has been expended to develop scaffolds which mimic the natural environment in which stem cell populations reside. However, additional work is required in order to obtain clinically successful implants. Therefore, the future of TE and RM requires focusing on design techniques that improve stem cell retention, survival, proliferation, and differentiation. This endeavor will involve many interdisciplinary fields, including medicine, biology, engineering and material science. In addition, such factors as oxygen diffusion, hypoxia, and cell–cell interactions must also be considered when designing a biomimetic scaffold. Continued improvements in biomaterial design and greater understanding of stem cell biology hold the promise of allowing the successful treatment of many diseases. To reach this goal, TE and RM must cooperate to create functional biological substitutes which repair, improve, and maintain the original tissue function.

Bibliography

Adams, R. H., Wilkinson, G. A., Weiss, C., Diella, F., Gale, N. W., Deutsch, U., Risau, W. and Klein, R. 1999. Roles of ephrinB ligands and EphB receptors in cardiovascular development: Demarcation of arterial/venous domains, vascular morphogenesis, and sprouting angiogenesis, *Genes Dev* 13:295–306.

Arrigoni, C., Camozzi, D., Imberti, B., Mantero, S. and Remuzzi, A. 2006. The effect of sodium ascorbate on the mechanical properties of hyaluronan-based vascular constructs, *Biomaterials* 27:623–630.

Au, P., Tam, J., Duda, D. G., Lin, P. C., Munn, L. L., Fukumura, D. and Jain, R. K. 2009. Paradoxical effects of PDGF-BB overexpression in endothelial cells on engineered blood vessels *in vivo*, *Am J Pathol* 175:294–302.

Au, P., Tam, J., Fukumura, D. and Jain, R. K. 2008. Bone marrow-derived mesenchymal stem cells facilitate engineering of long-lasting functional vasculature, *Blood* 111:4551–4558.

Awad, H. A., Wickham, M. Q., Leddy, H. A., Gimble, J. M. and Guilak, F. 2004. Chondrogenic differentiation of adipose-derived adult stem cells in agarose, alginate, and gelatin scaffolds, *Biomaterials* 25:3211–3222.

Ayalon, O., Sabanai, H., Lampugnani, M. G., Dejana, E. and Geiger, B. 1994. Spatial and temporal relationships between cadherins and PECAM-1 in cell-cell junctions of human endothelial cells, *J Cell Biol* 126:247–258.

Baird, A., Esch, F., Mormede, P., Ueno, N., Ling, N., Bohlen, P., Ying, S. Y., Wehrenberg, W. B. and Guillemin, R. 1986. Molecular characterization of fibroblast growth factor: Distribution and biological activities in various tissues, *Recent Prog Horm Res* 42:143–205.

Bard, F., Cannon, C., Barbour, R., Burke, R. L., Games, D., Grajeda, H., Guido, T., Hu, K., Huang, J., Johnson-Wood, K., Khan, K., Kholodenko, D., Lee, M., Lieberburg, I., Motter, R., Nguyen, M., Soriano, F., Vasquez, N., Weiss, K., Welch, B., Seubert, P., Schenk, D. and Yednock, T. 2000. Peripherally administered antibodies against amyloid beta-peptide enter the central nervous system and reduce pathology in a mouse model of Alzheimer disease, *Nat Med* 6:916–919.

Basilico, C. and Moscatelli, D. 1992. The FGF family of growth factors and oncogenes, *Adv Cancer Res* 59:115–165.

Benzing, W. C., Wujek, J. R., Ward, E. K., Shaffer, D., Ashe, K. H., Younkin, S. G. and Brunden, K. R. 1999. Evidence for glial-mediated inflammation in aged APP(SW) transgenic mice, *Neurobiol Aging* 20:581–589.

Berglund, J. D. and Galis, Z. S. 2003. Designer blood vessels and therapeutic revascularization, *Br J Pharmacol* 140:627–636.

Berglund, J. D., Mohseni, M. M., Nerem, R. M. and Sambanis, A. 2003. A biological hybrid model for collagen-based tissue engineered vascular constructs, *Biomaterials* 24:1241–1254.

Bianco, P. A. and Robey, P. G. 2001. Stem cells in tissue engineering, *Nature* 414:118–121.

Boesel, L. F. a. R., R. L. 2004. *Biodegradable Systems in Tissue Engineering and Regnerative Medicine*. CRC Press LLC.

Botchwey, E. A., Dupree, M. A., Pollack, S. R., Levine, E. M. and Laurencin, C. T. 2003. Tissue engineered bone: Measurement of nutrient transport in three-dimensional matrices, *J Biomed Mater Res A* 67:357–367.

Bouhadir, K. H., Lee, K. Y., Alsberg, E., Damm, K. L., Anderson, K. W. and Mooney, D. J. 2001. Degradation of partially oxidized alginate and its potential application for tissue engineering, *Biotechnol Prog* 17:945–950.

Bouloumie, A., Drexler, H. C., Lafontan, M. and Busse, R. 1998. Leptin, the product of Ob gene, promotes angiogenesis, *Circ Res* 83:1059–1066.

Broxmeyer, H. E., Gluckman, E., Auerbach, A., Douglas, G. W., Friedman, H., Cooper, S., Hangoc, G., Kurtzberg, J., Bard, J. and Boyse, E. A. 1990. Human umbilical cord blood: A clinically useful source of transplantable hematopoietic stem/progenitor cells, *Int J Cell Cloning* 8 Suppl 1:76–89; discussion 89–91.

Burdick, J. A., Chung, C., Jia, X., Randolph, M. A. and Langer, R. 2005. Controlled degradation and mechanical behavior of photopolymerized hyaluronic acid networks, *Biomacromolecules* 6:386–391.

Bussolino, F., Di Renzo, M. F., Ziche, M., Bocchietto, E., Olivero, M., Naldini, L., Gaudino, G., Tamagnone, L., Coffer, A. and Comoglio, P. M. 1992. Hepatocyte growth factor is a potent angiogenic factor which stimulates endothelial cell motility and growth, *J Cell Biol* 119:629–641.

Bussolino, F., Wang, J. M., Defilippi, P., Turrini, F., Sanavio, F., Edgell, C. J., Aglietta, M., Arese, P. and Mantovani, A. 1989. Granulocyte- and granulocyte-macrophage-colony stimulating factors induce human endothelial cells to migrate and proliferate, *Nature* 337:471–473.

Cantarella, G., Lempereur, L., Presta, M., Ribatti, D., Lombardo, G., Lazarovici, P., Zappala, G., Pafumi, C. and Bernardini, R. 2002. Nerve growth factor-endothelial cell interaction leads to angiogenesis *in vitro* and *in vivo*, *Faseb J* 16:1307–1309.

Carlson, T. R., Feng, Y., Maisonpierre, P. C., Mrksich, M. and Morla, A. O. 2001. Direct cell adhesion to the angiopoietins mediated by integrins, *J Biol Chem* 276:26516–26525.

Chen, S. L., Fang, W. W., Ye, F., Liu, Y. H., Qian, J., Shan, S. J., Zhang, J. J., Chunhua, R. Z., Liao, L. M., Lin, S. and Sun, J. P. 2004. Effect on left ventricular function of intracoronary transplantation of autologous bone marrow mesenchymal stem cell in patients with acute myocardial infarction, *Am J Cardiol* 94:92–95.

Choi, K. D., Yu, J., Smuga-Otto, K., Salvagiotto, G., Rehrauer, W., Vodyanik, M., Thomson, J. and Slukvin, I. 2009. Hematopoietic and endothelial differentiation of human induced pluripotent stem cells, *Stem Cells* 27:559–567.

Critser, P. J., Kreger, S. T., Voytik-Harbin, S. L. and Yoder, M. C. Collagen matrix physical properties modulate endothelial colony forming cell-derived vessels *in vivo*, *Microvasc Res* 80:23–30.

Daniel, T. O., Stein, E., Cerretti, D. P., St John, P. L., Robert, B. and Abrahamson, D. R. 1996. ELK and LERK-2 in developing kidney and microvascular endothelial assembly, *Kidney Int Suppl* 57:S73–S81.

De Filippo, R. E., Yoo, J. J. and Atala, A. 2003. Engineering of vaginal tissue *in vivo*, *Tissue Eng* 9:301–306.

de Vries, C., Escobedo, J. A., Ueno, H., Houck, K., Ferrara, N. and Williams, L. T. 1992. The fms-like tyrosine kinase, a receptor for vascular endothelial growth factor, *Science* 255:989–991.

Dee, K. C., Andersen, T. T. and Bizios, R. 1998. Design and function of novel osteoblast-adhesive peptides for chemical modification of biomaterials, *J Biomed Mater Res* 40:371–377.

DeLisser, H. M., Christofidou-Solomidou, M., Strieter, R. M., Burdick, M. D., Robinson, C. S., Wexler, R. S., Kerr, J. S., Garlanda, C., Merwin, J. R., Madri, J. A. and Albelda, S. M. 1997. Involvement of endothelial PECAM-1/CD31 in angiogenesis, *Am J Pathol* 151:671–677.

Ding, S., Merkulova-Rainon, T., Han, Z. C. and Tobelem, G. 2003. HGF receptor up-regulation contributes to the angiogenic phenotype of human endothelial cells and promotes angiogenesis *in vitro*, *Blood* 101:4816–4822.

Dolle, J. P., Rezvan, A., Allen, F. D., Lazarovici, P. and Lelkes, P. I. 2005. Nerve growth factor-induced migration of endothelial cells, *J Pharmacol Exp Ther* 315:1220–1227.

Edelman, E. R., Mathiowitz, E., Langer, R. and Klagsbrun, M. 1991. Controlled and modulated release of basic fibroblast growth factor, *Biomaterials* 12:619–626.

Elbert, D. L. and Hubbell, J. A. 2001. Conjugate addition reactions combined with free-radical cross-linking for the design of materials for tissue engineering, *Biomacromolecules* 2:430–441.

Elisseeff, J., Anseth, K., Sims, D., McIntosh, W., Randolph, M. and Langer, R. 1999. Transdermal photopolymerization for minimally invasive implantation, *Proc Natl Acad Sci USA* 96:3104–3107.

Engler, A. J., Rehfeldt, F., Sen, S. and Discher, D. E. 2007. Microtissue elasticity: Measurements by atomic force microscopy and its influence on cell differentiation, *Methods Cell Biol* 83:521–545.

Erbacher, P., Zou, S., Bettinger, T., Steffan, A. M. and Remy, J. S. 1998. Chitosan-based vector/DNA complexes for gene delivery: Biophysical characteristics and transfection ability, *Pharm Res* 15:1332–1339.

Fasol, R., Schumacher, B., Schlaudraff, K., Hauesnstein, K. H. and Seitelberger, R. 1994. Experimental use of a modified fibrin glue to induce site-directed angiogenesis from the aorta to the heart, *J Thorac Cardiovasc Surg* 107:1432–1439.

Ferrara, N. 2002. Role of vascular endothelial growth factor in physiologic and pathologic angiogenesis: Therapeutic implications, *Semin Oncol* 29:10–14.

Ferreira, L. S., Gerecht, S., Shieh, H. F., Watson, N., Rupnick, M. A., Dallabrida, S. M., Vunjak-Novakovic, G. and Langer, R. 2007. Vascular progenitor cells isolated from human embryonic stem cells give rise to endothelial and smooth muscle like cells and form vascular networks *in vivo*, *Circ Res* 101:286–294.

Gale, N. W. and Yancopoulos, G. D. 1999. Growth factors acting via endothelial cell-specific receptor tyrosine kinases: VEGFs, angiopoietins, and ephrins in vascular development, *Genes Dev* 13:1055–1066.

Gerecht-Nir, S., Cohen, S., Ziskind, A. and Itskovitz-Eldor, J. 2004. Three-dimensional porous alginate scaffolds provide a conducive environment for generation of well-vascularized embryoid bodies from human embryonic stem cells, *Biotechnol Bioeng* 88:313–320.

Goetze, S., Bungenstock, A., Czupalla, C., Eilers, F., Stawowy, P., Kintscher, U., Spencer-Hansch, C., Graf, K., Nurnberg, B., Law, R. E., Fleck, E. and Grafe, M. 2002. Leptin induces endothelial cell migration through Akt, which is inhibited by PPARgamma-ligands, *Hypertension* 40:748–754.

Gospodarovicz, D., Neufeld, G. and Schweigerer, L. 1986. Molecular and biological characterization of FGF, an angiogenic factor which also controls the proliferation and differentiation of mesoderm-derived cells, *Cell Differ Dev* 91:1–17.

Gospodarowicz, D., Ferrara, N., Schweigerer, L. and Neufeld, G. 1987. Structural characterization and biological functions of fibroblast growth factor, *Endocr Rev* 8:95–114.

Goumans, M. J., Valdimarsdottir, G., Itoh, S., Rosendahl, A., Sideras, P. and ten Dijke, P. 2002. Balancing the activation state of the endothelium via two distinct TGF-beta type I receptors, *Embo J* 21:1743–1753.

Grazia Lampugnani, M., Zanetti, A., Corada, M., Takahashi, T., Balconi, G., Breviario, F., Orsenigo, F., Cattelino, A., Kemler, R., Daniel, T. O. and Dejana, E. 2003. Contact inhibition of VEGF-induced proliferation requires vascular endothelial cadherin, beta-catenin, and the phosphatase DEP-1/CD148, *J Cell Biol* 161:793–804.

Hanahan, D. 1997. Signaling vascular morphogenesis and maintenance, *Science* 277: 48–50.

Hanjaya-Putra, D., Yee, J., Ceci, D., Truitt, R., Yee, D. and Gerecht, S. 2009. Vascular endothelial growth factor and substrate mechanics regulate *in vitro* tubulogenesis of endothelial progenitor cells, *J Cell Mol Med* 14:2436–2447.

Hanna, J., Wernig, M., Markoulaki, S., Sun, C. W., Meissner, A., Cassady, J. P., Beard, C., Brambrink, T., Wu, L. C., Townes, T. M. and Jaenisch, R. 2007. Treatment of sickle cell anemia mouse model with iPS cells generated from autologous skin, *Science* 318:1920–1923.

Harfouche, R., Hassessian, H. M., Guo, Y., Faivre, V., Srikant, C. B., Yancopoulos, G. D. and Hussain, S. N. 2002. Mechanisms which mediate the antiapoptotic effects of angiopoietin-1 on endothelial cells, *Microvasc Res* 64:135–147.

Hayashida, K., Kitamura, T., Gorman, D. M., Arai, K., Yokota, T. and Miyajima, A. 1990. Molecular cloning of a second subunit of the receptor for human granulocyte-macrophage colony-stimulating factor (GM-CSF): Reconstitution of a high-affinity GM-CSF receptor, *Proc Natl Acad Sci U S A* 87:9655–9659.

Heidaran, M. A., Pierce, J. H., Jensen, R. A., Matsui, T. and Aaronson, S. A. 1990. Chimeric alpha- and beta-platelet-derived growth factor (PDGF) receptors define three immunoglobulin-like domains of the alpha-PDGF receptor that determine PDGF-AA binding specificity, *J Biol Chem* 265:18741–18744.

Henning, B., Goldblum, S. E. and McClain, C. J. 1987. Interleukin-1 (IL-1) and tumor necrosis factor/cachectin (TNF) increase endothelial permeability *in vitro*, *Leukocyte Biol* 42:551.

Hern, D. L. and Hubbell, J. A. 1998. Incorporation of adhesion peptides into nonadhesive hydrogels useful for tissue resurfacing, *J Biomed Mater Res* 39:266–276.

Hoenig, M. R., Campbell, G. R., Rolfe, B. E. and Campbell, J. H. 2005. Tissue-engineered blood vessels: Alternative to autologous grafts? *Arterioscler Thromb Vasc Biol* 25:1128–1134.

Hooke, R. 1665. *Micrographia: Or Some Physiological Descriptions of some Minute Bodies made by Magnifying Glasses with Observations and Inquiries made Thereupon.* J. Martyn and J. Allestry, London.

Horwitz, E. M., Gordon, P. L., Koo, W. K., Marx, J. C., Neel, M. D., McNall, R. Y., Muul, L. and Hofmann, T. 2002. Isolated allogeneic bone marrow-derived mesenchymal cells engraft and stimulate growth in children with osteogenesis imperfecta: Implications for cell therapy of bone, *Proc Natl Acad Sci U S A* 99:8932–8937.

Horwitz, E. M., Prockop, D. J., Fitzpatrick, L. A., Koo, W. W., Gordon, P. L., Neel, M., Sussman, M., Orchard, P., Marx, J. C., Pyeritz, R. E. and Brenner, M. K. 1999. Transplantability and therapeutic effects of bone marrow-derived mesenchymal cells in children with osteogenesis imperfecta, *Nat Med* 5:309–313.

Hu, G., Riordan, J. F. and Vallee, B. L. 1994. Angiogenin promotes invasiveness of cultured endothelial cells by stimulation of cell-associated proteolytic activities, *Proc Natl Acad Sci U S A* 91:12096–12100.

Hu, G. F., Riordan, J. F. and Vallee, B. L. 1997. A putative angiogenin receptor in angiogenin-responsive human endothelial cells, *Proc Natl Acad Sci U S A* 94: 2204–2209.

Hu, G. F., Strydom, D. J., Fett, J. W., Riordan, J. F. and Vallee, B. L. 1993. Actin is a binding protein for angiogenin, *Proc Natl Acad Sci U S A* 90:1217–1221.

Hubbell, J. A. 2003. Materials as morphogenetic guides in tissue engineering, *Curr Opin Biotechnol* 14:551–558.

Hubbell, J. A., Massia, S. P., Desai, N. P. and Drumheller, P. D. 1991. Endothelial cell-selective materials for tissue engineering in the vascular graft via a new receptor, *Biotechnology (NY)* 9:568–572.

Humphries M. J., Akiyama, S. K., Komoriya, A., Olden, K. and Yamada, K. M. 1986. Identification of an alternatively spliced site in human-plasma fibronectin that mediates cell type-specific adhesion. *J Cell Biol* 103: 2637–2647.

Huynh, T., Abraham, G., Murray, J., Brockbank, K., Hagen, P. O. and Sullivan, S. 1999. Remodeling of an acellular collagen graft into a physiologically responsive neovessel, *Nat Biotechnol* 17:1083–1086.

Ingber, D. E. 2002. Mechanical signaling and the cellular response to extracellular matrix in angiogenesis and cardiovascular physiology, *Circ Res* 91:877–887.

Ishibashi, K. and Matsuda, T. 1994. Reconstruction of a hybrid vascular graft hierarchically layered with three cell types, *Asaio J* 40:M284–M290.

Jager, M., Degistirici, O., Knipper, A., Fischer, J., Sager, M. and Krauspe, R. 2007. Bone healing and migration of cord blood-derived stem cells into a critical size femoral defect after xenotransplantation, *J Bone Miner Res* 22:1224–1233.

Jay, S. M. and Saltzman, W. M. 2009. Shining light on a new class of hydrogels, *Nat Biotechnol* 27:543–544.

Jaye, M., Schlessinger, J. and Dionne, C. 1992. Fibroblast growth factor receptor for acidic and basic fibroblast growth factors, *Biochem Biophys Acta* 1135:185–199.

Jimi, S., Ito, K., Kohno, K., Ono, M., Kuwano, M., Itagaki, Y. and Ishikawa, H. 1995. Modulation by bovine angiogenin of tubular morphogenesis and expression of plasminogen activator in bovine endothelial cells, *Biochem Biophys Res Commun* 211: 476–483.

Jin, J., Jeong, S. I., Shin, Y. M., Lim, K. S., Shin, H., Lee, Y. M., Koh, H. C. and Kim, K. S. 2009. Transplantation of mesenchymal stem cells within a poly(lactide-co-epsilon-caprolactone) scaffold improves cardiac function in a rat myocardial infarction model, *Eur J Heart Fail* 11:147–153.

Jin, K., Mao, X., Xie, L., Galvan, V., Lai, B., Wang, Y., Gorostiza, O., Wang, X. and Greenberg, D. A. Transplantation of human neural precursor cells in Matrigel scaffolding improves outcome from focal cerebral ischemia after delayed postischemic treatment in rats, *J Cereb Blood Flow Metab* 30:534–544.

Joshi, P., Chung, C. Y., Aukhil, I. and Erickson, H. P. 1993. Endothelial cells adhere to the RGD domain and the fibrinogen-like terminal knob of tenascin, *J Cell Sci* 106 (Pt 1):389–400.

Kaji, K., Norrby, K., Paca, A., Mileikovsky, M., Mohseni, P. and Woltjen, K. 2009. Virus-free induction of pluripotency and subsequent excision of reprogramming factors, *Nature* 458:771–775.

Kam, L., Shain, W., Turner, J. N. and Bizios, R. 2002. Selective adhesion of astrocytes to surfaces modified with immobilized peptides, *Biomaterials* 23:511–515.

Kelm, J. S., Lorber, V. and Schmidt, D. 2008. Tissue engineering of living, small diameter vascular grafts: A novel concept based exclusively on self assembled microtissue building blocks, *Circulation* 118:S_1018.

Kim, S., Kim, S. S., Lee, S. H., Eun Ahn, S., Gwak, S. J., Song, J. H., Kim, B. S. and Chung, H. M. 2008. In vivo bone formation from human embryonic stem cell-derived osteogenic cells in poly(d,l-lactic-co-glycolic acid)/hydroxyapatite composite scaffolds, *Biomaterials* 29:1043–1053.

Kloxin, A. M., Kasko, A. M., Salinas, C. N. and Anseth, K. S. 2009. Photodegradable hydrogels for dynamic tuning of physical and chemical properties, *Science* 324:59–63.

Kulkarni, R. K., Moore, E. G., Hegyeli, A. F. and Leonard, F. 1971. Biodegradable poly(lactic acid) polymers, *J Biomed Mater Res* 5:169–181.

Kurita, K. 2001. Controlled functionalization of the polysaccharide chitin, *Prog Polym Sci* 26:1921–1971.

L'Heureux, N., Paquet, S., Labbe, R., Germain, L. and Auger, F. A. 1998. A completely biological tissue-engineered human blood vessel, *Faseb J* 12:47–56.

Lampugnani, M. G., Resnati, M., Raiteri, M., Pigott, R., Pisacane, A., Houen, G., Ruco, L. P. and Dejana, E. 1992. A novel endothelial-specific membrane protein is a marker of cell-cell contacts, *J Cell Biol* 118:1511–1522.

Laporte, V., Ait-Ghezala, G., Volmar, C. H. and Mullan, M. 2006. CD40 deficiency mitigates Alzheimer's disease pathology in transgenic mouse models, *J Neuroinflammation* 3:3.

Lee, M., Aoki, M., Kondo, T., Kobayashi, K., Okumura, K., Komori, K. and Murohara, T. 2005. Therapeutic angiogenesis with intramuscular injection of low-dose recombinant granulocyte-colony stimulating factor, *Arterioscler Thromb Vasc Biol* 25:2535–2541.

Leor, J., Gerecht, S., Cohen, S., Miller, L., Holbova, R., Ziskind, A., Shachar, M., Feinberg, M. S., Guetta, E. and Itskovitz-Eldor, J. 2007. Human embryonic stem cell transplantation to repair the infarcted myocardium, *Heart* 93:1278–1284.

Levenberg, S., Huang, N. F., Lavik, E., Rogers, A. B., Itskovitz-Eldor, J. and Langer, R. 2003. Differentiation of human embryonic stem cells on three-dimensional polymer scaffolds, *Proc Natl Acad Sci U S A* 100:12741–12746.

Li, A., Dubey, S., Varney, M. L., Dave, B. J. and Singh, R. K. 2003. IL-8 directly enhanced endothelial cell survival, proliferation, and matrix metalloproteinases production and regulated angiogenesis, *J Immunol* 170:3369–3376.

Li, Y., Zeng, H., Xu, R. H., Liu, B. and Li, Z. 2009. Vaccination with human pluripotent stem cells generates a broad spectrum of immunological and clinical responses against colon cancer, *Stem Cells* 27:3103–3111.

Liebner, S., Cavallaro, U. and Dejana, E. 2006. The multiple languages of endothelial cell-to-cell communication, *Arterioscler Thromb Vasc Biol* 26:1431–1438.

Lim, G. P., Yang, F., Chu, T., Gahtan, E., Ubeda, O., Beech, W., Overmier, J. B., Hsiao-Ashec, K., Frautschy, S. A. and Cole, G. M. 2001. Ibuprofen effects on Alzheimer pathology and open field activity in APPsw transgenic mice, *Neurobiol Aging* 22:983–991.

Luo, D., Luo, Y., He, Y., Zhang, H., Zhang, R., Li, X., Dobrucki, W. L., Sinusas, A. J., Sessa, W. C. and Min, W. 2006. Differential functions of tumor necrosis factor receptor 1 and 2 signaling in ischemia-mediated arteriogenesis and angiogenesis, *Am J Pathol* 169:1886–1898.

Luong, E. and Gerecht, S. 2009. Stem cells and scaffolds for vascularizing engineered tissue constructs, *Adv Biochem Eng Biotechnol* 114:129–172.

Luttun, A., Tjwa, M., Moons, L., Wu, Y., Angelillo-Scherrer, A., Liao, F., Nagy, J. A., Hooper, A., Priller, J., De Klerck, B., Compernolle, V., Daci, E., Bohlen, P., Dewerchin, M., Herbert, J. M., Fava, R., Matthys, P., Carmeliet, G., Collen, D., Dvorak, H. F., Hicklin, D. J. and Carmeliet, P. 2002. Revascularization of ischemic tissues by PlGF treatment, and inhibition of tumor angiogenesis, arthritis and atherosclerosis by anti-Flt1, *Nat Med* 8:831–840.

Ma, N., Stamm, C., Kaminski, A., Li, W., Kleine, H. D., Muller-Hilke, B., Zhang, L., Ladilov, Y., Egger, D. and Steinhoff, G. 2005. Human cord blood cells induce angiogenesis following myocardial infarction in NOD/scid-mice, *Cardiovasc Res* 66:45–54.

Maisonpierre, P. C., Suri, C., Jones, P. F., Bartunkova, S., Wiegand, S. J., Radziejewski, C., Compton, D., McClain, J., Aldrich, T. H., Papadopoulos, N., Daly, T. J., Davis, S., Sato, T. N. and Yancopoulos, G. D. 1997. Angiopoietin-2, a natural antagonist for Tie2 that disrupts *in vivo* angiogenesis, *Science* 277:55–60.

Majack, R. A., Majesky, M. W. and Goodman, L. V. 1990. Role of PDGF-A expression in the control of vascular smooth muscle cell growth by transforming growth factor-beta, *J Cell Biol* 111:239–247.

Mammoto, A., Connor, K. M., Mammoto, T., Yung, C. W., Huh, D., Aderman, C. M., Mostoslavsky, G., Smith, L. E. and Ingber, D. E. 2009. A mechanosensitive transcriptional mechanism that controls angiogenesis, *Nature* 457:1103–1108.

Mann, A., Breuhahn, K., Schirmacher, P. and Blessing, M. 2001. Keratinocyte-derived granulocyte-macrophage colony stimulating factor accelerates wound healing: Stimulation of keratinocyte proliferation, granulation tissue formation, and vascularization, *J Invest Dermatol* 117:1382–1390.

Mann, B. K., Schmedlen, R. H. and West, J. L. 2001. Tethered-TGF-beta increases extracellular matrix production of vascular smooth muscle cells, *Biomaterials* 22:439–444.

Mano, J. F., Silva, G. A., Azevedo, H. S., Malafaya, P. B., Sousa, R. A., Silva, S. S., Boesel, L. F., Oliveira, J. M., Santos, T. C., Marques, A. P., Neves, N. M. and Reis, R. L. 2007. Natural origin biodegradable systems in tissue engineering and regenerative medicine: Present status and some moving trends, *J R Soc Interface* 4:999–1030.

Marler, J. J., Guha, A., Rowley, J., Koka, R., Mooney, D., Upton, J. and Vacanti, J. P. 2000. Soft-tissue augmentation with injectable alginate and syngeneic fibroblasts, *Plast Reconstr Surg* 105:2049–2058.

Marston, W. A., Usala, A., Hill, R. S., Mendes, R. and Minsley, M. A. 2005. Initial report of the use of an injectable porcine collagen-derived matrix to stimulate healing of diabetic foot wounds in humans, *Wound Repair Regen* 13:243–247.

Massia, S. P. and Stark, J. 2001. Immobilized RGD peptides on surface-grafted dextran promote biospecific cell attachment, *J Biomed Mater Res* 56:390–399.

Matsuda, T. and Miwa, H. 1995. A hybrid vascular model biomimicking the hierarchic structure of arterial wall: Neointimal stability and neoarterial regeneration process under arterial circulation, *J Thorac Cardiovasc Surg* 110:988–997.

McAuslan, B. R., Bender, V., Reilly, W. and Moss, B. A. 1985. New functions of epidermal growth factor: Stimulation of capillary endothelial cell migration and matrix dependent proliferation, *Cell Biol Int Rep* 9:175–182.

Melero-Martin, J. M., De Obaldia, M. E., Kang, S. Y., Khan, Z. A., Yuan, L., Oettgen, P. and Bischoff, J. 2008. Engineering robust and functional vascular networks *in vivo* with human adult and cord blood-derived progenitor cells, *Circ Res* 103:194–202.

Moon, J. J., Saik, J. E., Poche, R. A., Leslie-Barbick, J. E., Lee, S. H., Smith, A. A., Dickinson, M. E. and West, J. L. 2010. Biomimetic hydrogels with pro-angiogenic properties, *Biomaterials* 31:3840–3847.

Mooney, D. J., Mazzoni, C. L., Breuer, C., McNamara, K., Hern, D., Vacanti, J. P. and Langer, R. 1996. Stabilized polyglycolic acid fibre-based tubes for tissue engineering, *Biomaterials* 17:115–124.

Morimoto, A., Okamura, K., Hamanaka, R., Sato, Y., Shima, N., Higashio, K. and Kuwano, M. 1991. Hepatocyte growth factor modulates migration and proliferation of human microvascular endothelial cells in culture, *Biochem Biophys Res Commun* 179:1042–1049.

Moser, K. V., Reindl, M., Blasig, I. and Humpel, C. 2004. Brain capillary endothelial cells proliferate in response to NGF, express NGF receptors and secrete NGF after inflammation, *Brain Res* 1017:53–60.

Mukai, N., Akahori, T., Komaki, M., Li, Q., Kanayasu-Toyoda, T., Ishii-Watabe, A., Kobayashi, A., Yamaguchi, T., Abe, M., Amagasa, T. and Morita, I. 2008. A comparison of the tube forming potentials of early and late endothelial progenitor cells, *Exp Cell Res* 314:430–440.

Muller, W. A., Ratti, C. M., McDonnell, S. L. and Cohn, Z. A. 1989. A human endothelial cell-restricted, externally disposed plasmalemmal protein enriched in intercellular junctions, *J Exp Med* 170:399–414.

Nelson, T. J., Martinez-Fernandez, A., Yamada, S., Perez-Terzic, C., Ikeda, Y. and Terzic, A. 2009. Repair of acute myocardial infarction by human stemness factors induced pluripotent stem cells, *Circulation* 120:408–416.

Nettles, D. L., Elder, S. H. and Gilbert, J. A. 2002. Potential use of chitosan as a cell scaffold material for cartilage tissue engineering, *Tissue Eng* 8:1009–1016.

Neufeld, G., Cohen, T., Gitay-Goren, H., Poltorak, Z., Tessler, S., Sharon, R., Gengrinovitch, S. and Levi, B. Z. 1996. Similarities and differences between the vascular endothelial growth factor (VEGF) splice variants, *Cancer Metastasis Rev* 15: 153–158.

Niklason, L. E., Gao, J., Abbott, W. M., Hirschi, K. K., Houser, S., Marini, R. and Langer, R. 1999. Functional arteries grown *in vitro*, *Science* 284:489–493.

Nikolic, W. V., Hou, H., Town, T., Zhu, Y., Giunta, B., Sanberg, C. D., Zeng, J., Luo, D., Ehrhart, J., Mori, T., Sanberg, P. R. and Tan, J. 2008. Peripherally administered human umbilical cord blood cells reduce parenchymal and vascular beta-amyloid deposits in Alzheimer mice, *Stem Cells Dev* 17:423–439.

Ning, H., Yang, F., Jiang, M., Hu, L., Feng, K., Zhang, J., Yu, Z., Li, B., Xu, C., Li, Y., Wang, J., Hu, J., Lou, X. and Chen, H. 2008. The correlation between cotransplantation of mesenchymal stem cells and higher recurrence rate in hematologic malignancy patients: outcome of a pilot clinical study, *Leukemia* 22:593–599.

Ornitz, D. M., Xu, J., Colvin, J. S., McEwen, D. G., MacArthur, C. A., Coulier, F., Gao, G. and Goldfarb, M. 1996. Receptor specificity of the fibroblast growth factor family, *J Biol Chem* 271:15292–15297.

Oswald, J., Boxberger S., Jorgensen B., Feldman S., Ehninger G., Bornhauser M. and Werner C. 2004. Mesenchymal stem cells can be differentiated into endothelial cells *in vitro*, *Stem Cells* 22:377–384.

Pandey, A., Shao, H., Marks, R. M., Polverini, P. J. and Dixit, V. M. 1995. Role of B61, the ligand for the Eck receptor tyrosine kinase, in TNF-alpha-induced angiogenesis, *Science* 268:567–569.

Panitch, A., Yamaoka, T.,Fournier, M. J., Mason, T. L. and Tirrell, D. A. 1999. Design and biosynthesis of elastin-like artificial extracellular matrix proteins containing periodically spaced fibronectin CS5 domains, *Macromolecules* 32:1701–1703.

Park, H. Y., Kwon, H. M., Lim, H. J., Hong, B. K., Lee, J. Y., Park, B. E., Jang, Y., Cho, S. Y. and Kim, H. S. 2001. Potential role of leptin in angiogenesis: Leptin induces endothelial cell proliferation and expression of matrix metalloproteinases *in vivo* and *in vitro*, *Exp Mol Med* 33:95–102.

Park, J. E., Chen, H. H., Winer, J., Houck, K. A. and Ferrara, N. 1994. Placenta growth factor. Potentiation of vascular endothelial growth factor bioactivity, *in vitro* and *in vivo*, and high affinity binding to Flt-1 but not to Flk-1/KDR, *J Biol Chem* 269:25646–25654.

Park, Y. D., Tirelli, N. and Hubbell, J. A. 2003. Photopolymerized hyaluronic acid-based hydrogels and interpenetrating networks, *Biomaterials* 24:893–900.

Percival, E. M. a. M., R. H. 1990. *Methods in Plant Biochemistry*. Carbohydrates, in Algal polysaccharides. Academic Press.

Perng, C. K., Kao, C. L., Yang, Y. P., Lin, H. T., Lin, W. B., Chu, Y. R., Wang, H. J., Ma, H., Ku, H. H. and Chiou, S. H. 2008. Culturing adult human bone marrow stem cells on gelatin scaffold with pNIPAAm as transplanted grafts for skin regeneration, *J Biomed Mater Res A* 84:622–630.

Pislaru, S. V., Harbuzariu, A., Agarwal, G., Witt, T., Gulati, R., Sandhu, N. P., Mueske, C., Kalra, M., Simari, R. D. and Sandhu, G. S. 2006. Magnetic forces enable rapid endothelialization of synthetic vascular grafts, *Circulation* 114:I314–I318.

Raines, E. W., Bowen-Pope, D. F. and Ross, R. 1990. *Platelet-Derived Growth Factor*. Springer-Verlad.

Ranieri, J. P., Bellamkonda, R., Bekos, E. J., Vargo, T. G., Gardella, J. A. Jr. and Aebischer, P. 1995. Neuronal cell attachment to fluorinated ethylene propylene films with covalently immobilized laminin oligopeptides YIGSR and IKVAV. II, *J Biomed Mater Res* 29:779–785.

Rezania, A., Johnson, R., Lefkow, A. R. and Healy, K. E. 1999. Bioactivation of metal oxide surfaces. 1. Surface characterization and cell response, *Langmuir* 15:6931–6939.

Rowley, J. A., Madlambayan, G. and Mooney, D. J. 1999. Alginate hydrogels as synthetic extracellular matrix materials, *Biomaterials* 20:45–53.

Sachlos, E. and Czernuszka, J. T. 2003. Making tissue engineering scaffolds work. Review: The application of solid freeform fabrication technology to the production of tissue engineering scaffolds, *Eur Cell Mater* 5:29–39; discussion 39–40.

Sahni, A. and Francis, C. W. 2000. Vascular endothelial growth factor binds to fibrinogen and fibrin and stimulates endothelial cell proliferation, *Blood* 96:3772–3778.

Sawano, A., Takahashi, T., Yamaguchi, S., Aonuma, M. and Shibuya, M. 1996. Flt-1 but not KDR/Flk-1 tyrosine kinase is a receptor for placenta growth factor, which is related to vascular endothelial growth factor, *Cell Growth Differ* 7:213–221.

Sawhney, A. S., Pathak, C. P. and Hubbell, J. A. 1993. Bioerodible hydrogels based on photopolymerized poly(ethylene glycol)-co-poly(.alpha.-hydroxy acid) diacrylate macromers, *Macromolecules* 26:581–587.

Schense, J. C., Bloch, J., Aebischer, P. and Hubbell, J. A. 2000. Enzymatic incorporation of bioactive peptides into fibrin matrices enhances neurite extension, *Nat Biotechnol* 18:415–419.

Schmidt, C. E. and Baier, J. M. 2000. Acellular vascular tissues: Natural biomaterials for tissue repair and tissue engineering, *Biomaterials* 21:2215–2231.

Schmitz-Rixen, T., Megerman, J., Anderson, J. M., Warnock, D. F., L'Italien, G. J., Erasmi, H., Horsch, S. and Abbott, W. M. 1991. Longterm study of a compliant biological vascular graft, *Eur J Vasc Surg* 5:149–158.

Schonherr, E. and Hausser, H. J. 2000. Extracellular matrix and cytokines: A functional unit, *Dev Immunol* 7:89–101.

Schreiber, A. B., Winkler, M. E. and Derynck, R. 1986. Transforming growth factor-alpha: A more potent angiogenic mediator than epidermal growth factor, *Science* 232: 1250–1253.

Seal, B. L., Otero, T. C. and Panitch, A. 2001. Polymeric biomaterials for tissue and organ regeneration, *Mat Sci Eng R-Reports* 34:147–230.

Shin, H., Jo, S. and Mikos, A. G. 2003. Biomimetic materials for tissue engineering, *Biomaterials* 24:4353–4364.

Shinoka, T., Shum-Tim, D., Ma, P. X., Tanel, R. E., Isogai, N., Langer, R., Vacanti, J. P. and Mayer, J. E., Jr. 1998. Creation of viable pulmonary artery autografts through tissue engineering, *J Thorac Cardiovasc Surg* 115:536–545; discussion 545–546.

Sierra-Honigmann, M. R., Nath, A. K., Murakami, C., Garcia-Cardena, G., Papapetropoulos, A., Sessa, W. C., Madge, L. A., Schechner, J. S., Schwabb, M. B., Polverini, P. J. and Flores-Riveros, J. R. 1998. Biological action of leptin as an angiogenic factor, *Science* 281:1683–1686.

Silva, G. V., Litovsky, S., Assad, J. A., Sousa, A. L., Martin, B. J., Vela, D., Coulter, S. C., Lin, J., Ober, J., Vaughn, W. K., Branco, R. V., Oliveira, E. M., He, R., Geng, Y. J., Willerson, J. T. and Perin, E. C. 2005. Mesenchymal stem cells differentiate into an endothelial phenotype, enhance vascular density, and improve heart function in a canine chronic ischemia model, *Circulation* 111:150–156.

Slack, J. M. 2000. Stem cells in epithelial tissues, *Science* 287:1431–1433.

Slevin, M., Kumar, S. and Gaffney, J. 2002. Angiogenic oligosaccharides of hyaluronan induce multiple signaling pathways affecting vascular endothelial cell mitogenic and wound healing responses, *J Biol Chem* 277:41046–41059.

Smith, D. H., Wolf, J. A. and Meaney, D. F. 2001. A new strategy to produce sustained growth of central nervous system axons: Continuous mechanical tension, *Tissue Eng* 7:131–139.

Solchaga, L. A., Yoo, J. U., Lundberg, M., Dennis, J. E., Huibregtse, B. A., Goldberg, V. M. and Caplan, A. I. 2000. Hyaluronan-based polymers in the treatment of osteochondral defects, *J Orthop Res* 18:773–780.

Soucy, P. A. and Romer, L. H. 2009. Endothelial cell adhesion, signaling, and morphogenesis in fibroblast-derived matrix, *Matrix Biol* 28:273–283.

Stitzel, J., Liu, J., Lee, S. J., Komura, M., Berry, J., Soker, S., Lim, G., Van Dyke, M., Czerw, R., Yoo, J. J. and Atala, A. 2006. Controlled fabrication of a biological vascular substitute, *Biomaterials* 27:1088–1094.

Sun, G., Shen, Y. I., Ho, C. C., Kusuma, S. and Gerecht, S. 2010. Functional groups affect physical and biological properties of dextran-based hydrogels, *J Biomed Mater Res A* 93:1080–1090.

Tabata, Y. 2009. Biomaterial technology for tissue engineering applications, *J R Soc Interface* 6 Suppl 3:S311–S324.

Taguchi, T., Xu, L., Kobayashi, H., Taniguchi, A., Kataoka, K. and Tanaka, J. 2005. Encapsulation of chondrocytes in injectable alkali-treated collagen gels prepared using poly(ethylene glycol)-based 4-armed star polymer, *Biomaterials* 26:1247–1252.

Tamura, N., Nakamura, T., Terai, H., Iwakura, A., Nomura, S., Shimizu, Y. and Komeda, M. 2003. A new acellular vascular prosthesis as a scaffold for host tissue regeneration, *Int J Artif Organs* 26:783–792.

Taura, D., Sone, M., Homma, K., Oyamada, N., Takahashi, K., Tamura, N., Yamanaka, S. and Nakao, K. 2009. Induction and isolation of vascular cells from human induced pluripotent stem cells–brief report, *Arterioscler Thromb Vasc Biol* 29:1100–1103.

Thomas, V., Zhang, X., Catledge, S. A. and Vohra, Y. K. 2007. Functionally graded electrospun scaffolds with tunable mechanical properties for vascular tissue regeneration, *Biomed Mater* 2:224–232.

Toby, I., Chicoine, L. G. and Nelin, L. D. 2007. EGFR stimulation promotes proliferation in human pulmonary microvascular endothelial cells, *Faseb J* 21:977–978.

Vert, M., Li, S. M. and Garreau, H. 1994. Attempts to map the structure and degradation characteristics of aliphatic polyesters derived from lactic and glycolic acids, *J Biomater Sci Polym Ed* 6:639–649.

Vinals, F. and Pouyssegur, J. 2001. Transforming growth factor beta1 (TGF-beta1) promotes endothelial cell survival during *in vitro* angiogenesis via an autocrine mechanism implicating TGF-alpha signaling, *Mol Cell Biol* 21:7218–7230.

Vlodavsky, I., Fuks, Z., Ishai-Michaeli, R., Bashkin, P., Levi, E., Korner, G., Bar-Shavit, R. and Klagsbrun, M. 1991. Extracellular matrix-resident basic fibroblast growth factor: Implication for the control of angiogenesis, *J Cell Biochem* 45:167–176.

Wacker, B. K., Alford, S. K., Scott, E. A., Das Thakur, M., Longmore, G. D. and Elbert, D. L. 2008. Endothelial cell migration on RGD-peptide-containing PEG hydrogels in the presence of sphingosine 1-phosphate, *Biophys J* 94:273–285.

Wallace, D. G. and Rosenblatt, J. 2003. Collagen gel systems for sustained delivery and tissue engineering, *Adv Drug Deliv Rev* 55:1631–1649.

Wang, Z. Z., Au, P., Chen, T., Shao, Y., Daheron, L. M., Bai, H., Arzigian, M., Fukumura, D., Jain, R. K. and Scadden, D. T. 2007. Endothelial cells derived from human embryonic stem cells form durable blood vessels *in vivo*, *Nat Biotechnol* 25:317–318.

Weinberg, C. B. and Bell, E. 1986. A blood vessel model constructed from collagen and cultured vascular cells, *Science* 231:397–400.

Wernig, M., Zhao, J. P., Pruszak, J., Hedlund, E., Fu, D., Soldner, F., Broccoli, V., Constantine-Paton, M., Isacson, O. and Jaenisch, R. 2008. Neurons derived from reprogrammed fibroblasts functionally integrate into the fetal brain and improve symptoms of rats with Parkinson's disease, *Proc Natl Acad Sci U S A* 105:5856–5861.

West, J. L. and Hubbell, J. A. 1999. Polymeric biomaterials with degradation sites for proteases involved in cell migration, *Macromolecules* 32:241–244.

Woltjen, K., Michael, I. P., Mohseni, P., Desai, R., Mileikovsky, M., Hamalainen, R., Cowling, R., Wang, W., Liu, P., Gertsenstein, M., Kaji, K., Sung, H. K. and Nagy, A. 2009. piggyBac transposition reprograms fibroblasts to induced pluripotent stem cells, *Nature* 458:766–770.

Wu, X., Rabkin-Aikawa, E., Guleserian, K. J., Perry, T. E., Masuda, Y., Sutherland, F. W., Schoen, F. J., Mayer, J. E., Jr. and Bischoff, J. 2004. Tissue-engineered microvessels on three-dimensional biodegradable scaffolds using human endothelial progenitor cells, *Am J Physiol Heart Circ Physiol* 287:H480–H487.

Xin, X., Yang, S., Ingle, G., Zlot, C., Rangell, L., Kowalski, J., Schwall, R., Ferrara, N. and Gerritsen, M. E. 2001. Hepatocyte growth factor enhances vascular endothelial growth factor-induced angiogenesis *in vitro* and *in vivo*, *Am J Pathol* 158:1111–1120.

Xu, D., Alipio, Z., Fink, L. M., Adcock, D. M., Yang, J., Ward, D. C. and Ma, Y. 2009. Phenotypic correction of murine hemophilia A using an iPS cell-based therapy, *Proc Natl Acad Sci U S A* 106:808–813.

Xu, Y. and Yu, Q. 2001. Angiopoietin-1, unlike angiopoietin-2, is incorporated into the extracellular matrix via its linker peptide region, *J Biol Chem* 276:34990–34998.

Yamanaka, S. 2007. Strategies and new developments in the generation of patient-specific pluripotent stem cells, *Cell Stem Cell* 1:39–49.

Yang, S., Graham, J., Kahn, J. W., Schwartz, E. A. and Gerritsen, M. E. 1999. Functional roles for PECAM-1 (CD31) and VE-cadherin (CD144) in tube assembly and lumen formation in three-dimensional collagen gels, *Am J Pathol* 155:887–895.

Yu, J., Hu, K., Smuga-Otto, K., Tian, S., Stewart, R., Slukvin, II and Thomson, J. A. 2009. Human induced pluripotent stem cells free of vector and transgene sequences, *Science* 324:797–801.

Yu, J., Vodyanik, M. A., Smuga-Otto, K., Antosiewicz-Bourget, J., Frane, J. L., Tian, S., Nie, J., Jonsdottir, G. A., Ruotti, V., Stewart, R., Slukvin, II, and Thomson, J. A. 2007. Induced pluripotent stem cell lines derived from human somatic cells, *Science* 318:1917–1920.

Zalipsky, S. 1995. Functionalized poly(ethylene glycol) for preparation of biologically relevant conjugates, *Bioconjug Chem* 6:150–165.

Zisch, A. H., Lutolf, M. P. and Hubbell, J. A. 2003. Biopolymeric delivery matrices for angiogenic growth factors, *Cardiovasc Pathol* 12:295–310.

Biofunctionalized Surfaces Controlling Stem Cell Fate Decisions

9

Leonora Buzanska, Ana Ruiz, and Laura Ceriotti

How is stem cell fate regulated by the surrounding microenvironment? The answer to this question is crucial towards our understanding of the unique properties and developmental potential of stem cells and their application in clinical and diagnostic settings. An appreciable effort has been made to identify conditions that exist in stem cell niches *in vivo*. Biotechnology aims to recreate these microenvironmental cues using artificial cell growth platforms with controlled chemistry, cell content, 2D and 3D geometry, spatial and temporal distribution of bioactive areas. Micro/nanofabrication techniques allow the controlled deposition of biomolecules on a cell repellent surface to generate microscale functional patterns for the investigation of crucial stem cell developmental processes such as proliferation, migration and differentiation. Here, different sources and models of stem cells are introduced to describe stem cell interactions with bioactive surfaces and their responses to microenvironmental cues. Effective control of developmental fate decision can be achieved by tailoring the composition and geometry of bioactive patterns and molecular targeting of stem cell signaling pathways. Miniaturization of bioengineered active domains may allow the application of stem cell growth platforms for multiparameter biotests and provide important additional information about the sensitivity of certain stem cell molecular pathways to natural environmental and industrial factors. This chapter overviews the technological progress towards recreating stem cell interactions with biofunctional surface domains and enabling the application of stem cells for biotesting and tissue engineering.

9.1. Introduction

Stem cells are capable of generating, in a highly organized and patterned form, the required number and diversity of functional cell types. This has been demonstrated in the developing organism, and in adulthood, where stem cells are responsible for the physiological turnover and regenerative processes in the organized tissues they reside in. These processes are thought to be orchestrated spatially and temporarily by external and internal stimuli from the surrounding microenvironment called the stem cell niche (Scadden, 2006).

What are the driving forces behind the determination of stem cell behavior and fate? The fate of stem cells depends on a complex plethora of signals emanating from their intracellular and extracellular microenvironment. Stem cells coexist and interact with other lineage-related cells, soluble factors, mechanical and ionic forces, and components of the extracellular matrix (ECM) (see also Chapter 4). The ECM consists of matrix proteins and a class of negatively charged polysaccharides (called glycosaminoglycans), of which the most abundant type is hyaluronic acid. During embryogenesis, the onset of the development of ECM components is temporally ordered: laminin is the first ECM protein to be excreted, followed by fibronectin, heparan sulfate and collagen IV (Ingber, 2006). The adhesion of stem cells to ECM proteins is mediated through integrins — transmembrane receptors that bind mainly to the RGD domain (specific amino acid sequence arginine-glycine-aspartic acid) and other domains of ECM proteins (Ffrench-Constant and Colognato, 2004). Human embryonic stem cells and cord blood-derived stem cells have been shown to express mRNA for hyaluronic acid receptors and a range of subunits of the integrin family (Gerecht *et al.*, 2007; Buzanska *et al.*, 2006). It has also been shown that the surface chemistry has profound consequences on cellular responses by modulating the binding of integrin to the ECM ligands (Keselowsky *et al.*, 2005). ECM–stem cell interactions activate intracellular signaling mechanisms, which in turn regulate the expression of genes vital for developmental processes. In response to environmental signals, stem cells proliferate (with the ability to self-renew), migrate, undergo apoptosis, or differentiate to the required phenotype of the cells which assemble into organized tissues and organs. The spatial nanotopography provided by the surrounding matrix milieu is also important. For example, in addition to the specific receptor-mediated interactions, cellular responses are sensitive to protein folding and the nanostructure of ECM proteins (e.g. fibronectin and collagen have 140- and 67-nm repeat pattern of nanofibers, respectively) (Pompe *et al.*, 2007; Gadegaard *et al.*, 2003). Spatial information is recognized by stem cells via membrane features as small as 10 nm, known as contact guidance spots. Changing the topography of the cellular microenvironment at the nanoscale level has been shown to modulate gene expression and direct the differentiation of mesenchymal stem cells (Dalby *et al.*, 2008).

By isolating and expanding the stem cells in *in vitro* cultures, much has been learned about the nature of molecular signals that influence and direct cell self-renewal, differentiation (Gerecht *et al.*, 2007; Keller, 2005), migration (Kollar *et al.*, 2009; Dittmar *et al.*, 2010), and the spatial patterning of stem cells (Lin *et al.*, 2009).

Yet, it is not entirely known how molecular pathways and cellular events are integrated and coupled to the differentiation and patterning processes at the tissue/organ level. Gaining insights into these mechanisms is essential for the translation of basic research to medical application (Serfozo *et al.*, 2006) and requires recreating, *in vitro*, the conditions from the biological niche.

Recreating these conditions requires nano/microscale control of cellular environments. New tools have become available which enable the development of miniaturized *in vitro* systems that deliver stem cell regulatory signals in a defined manner, thus resembling the stem cell niche. The challenge of this approach is the efficient *in vitro* probing of the spatial and temporal effects of the ECM, and other niche components, on stem cell fate. This may be accomplished through the use of emerging nano/microtechnologies to build up domains which control cell/material interfaces. Surface domains of defined physicochemical properties can be further functionalized by deposition of different biomolecules, as well as optimized for specific cell types and applications. The micropatterning of biomolecules, to generate 2D cell growth platforms or 3D biocompatible scaffolds, has allowed for the development of advanced systems applicable for basic research, multiparameter biotests with biosensing and diagnostic purposes, and cell-based therapies (Flaim *et al.*, 2008; Facca *et al.*, 2008; Fernandes *et al.*, 2009, 2010; Grayson *et al.*, 2009; Kloxin *et al.*, 2009). For the latter, a wide range of biomaterials assembled in 3D scaffolds efficiently improves the impact of transplanted stem cell populations. They are used as carriers for the local delivery of stem cells, display regulatory proteins to support survival and differentiation of transplanted cells, and may also stimulate the function of endogenous stem cells (Lutolf *et al.*, 2009).

The nervous tissue is regarded as being the most difficult to regenerate. However, the dogma established 90 years ago by Ramon Cajal (1928) that "in adult centers the nerve paths are something fixed, ended, immutable" has recently been put into question. Today, we not only have evidence for the presence of stem cells in neurogenic areas of the adult brain, (Eriksson *et al.*, 1998) but also evidence of their activation in response to certain developmental signals and injury (Park *et al.*, 2002; Goldman, 2005) or as a result of direct gene transfer (see Chapter 11). Human neural stem cells (NSCs) have been used in regenerative medicine (Pluchino *et al.*, 2005) and as a model system for *in vitro* screening of putative neurotoxic factors (Buzanska *et al.*, 2009; Moors *et al.*, 2009).

This chapter will provide an overview of the following: (i) emerging technologies in the preparation of bioengineered stem cell growth platforms using 2D and 3D approaches; (ii) stem cell models implemented for fate control studies based on cell membrane/surface interactions; and (iii) possible implications of stem cell micropatterning technologies for biotesting and translational medicine.

9.2. Microtechnology in Stem Cell Research

In the last decade, biotechnologists and stem cell biologists have worked together to identify and understand the parameters modulating stem cell fate. They are now focused on developing well-controlled procedures for *in vitro* stem cell expansion, differentiation and maintenance. Typical strategies for directing stem cell commitment to specific lineages have relied on cell exposure to specific combinations of soluble factors. The properties of surface materials, such as the biomaterial substrate, including surface topography, stiffness, charge, roughness and chemistry have been shown to modulate short-term cell functions (Saltzman, 2000).

It is clear that it is difficult to dissect the individual contribution of each surface parameter to the numerous phenotypic attributes specific to each lineage. How can one discriminate between the effects produced by cell receptor–extra cellular matrix (ECM) protein interaction or cell–cell contact mediated signal exchange? One powerful method to control the chemical properties of the underlying substrate is based on self-assembled monolayers (SAMs) of alkanethiols on gold (Whitesides *et al.*, 2001). Using this approach, it was possible to show that surface chemistry modulates the conformation of the ECM proteins (Keselowsky *et al.*, 2003). Moreover, it was found that depending on the surface chemistry, integrin binding modulates focal adhesion formation and intracellular signal cascades that can result in osteoblast- and myoblast-like cell differentiation (Keselowsky *et al.*, 2004, 2005). The same model was also useful for a systematic assessment of the role of surface chemistry on the phenotypic behavior of adult human mesenchymal stem cells (hMSCs) (Phillips *et al.*, 2010). In particular, four different SAMs were chosen to represent different surface properties: CH_3 (hydrophobic), COOH (negatively charged at pH 7.4), NH_2 (positively charged at pH 7.4) and OH (neutral hydrophilic). Of the four surface chemistries, one ECM protein (fibronectin) was tested at a defined surface density (40 ng/cm^2). The results supported the conclusion that the underlying surface chemistry affects the adsorbed fibronectin conformation, which in turn directs specific integrin–fibronectin interaction in hMSCs and, for instance, induces an osteogenic pathway on NH_2 SAMs. This approach underscores that surface properties are critical parameters to be considered for the design of biomaterials (natural or artificial) aimed at supporting cell growth.

However, in the above experiments, the stem cell microenvironment is far from resembling *in vivo* conditions. Two- and three-dimensional patterns of biomaterials represent a more realistic approach to create conditions mimicking the stem cell niche. Such constructs enable the identification of individual factors that influence stem cell fate and provide spatial and temporal control over cells. The following sections review the most common methods used in fabricating 2D and 3D structures for stem cell biology.

9.2.1. *2D-patterned surfaces*

Cell–cell interactions, consisting of both diffusible and cell-cell contact-mediated signals, are important for many biological processes, including tumor growth, atherosclerotic plaque formation and stem cell self-renewal and differentiation. One way to manipulate cell–cell interactions is to modulate their relative positions. This can be done using cell patterning. Many techniques have been used for cell patterning, including microcontact printing, microspotting and photolithography (Singhvi *et al.*, 1994; Chen *et al.*, 1997; Kuschel *et al.*, 2006; Bretagnol *et al.*, 2007; Falconnet *et al.*, 2006; Liu *et al.*, 2007). All these techniques allow for the formation of microarrays of cells in small areas of a 2D substrate where cell shape, viability, and death can be controlled (Singhvi *et al.*, 1994; Chen *et al.*, 1997). The same cell patterning methods have been recently applied to stem cell research (Ruiz and Chen, 2007; Ruiz *et al.*, 2008; Luo *et al.*, 2008; Koepsel and Murphy, 2009; Flaim *et al.*, 2005; Anderson *et al.*, 2004, 2005; Soen *et al.*, 2006; Ceriotti *et al.*, 2009; Nakajima *et al.*, 2007; El-Ali *et al.*, 2006).

9.2.1.1. *Pattern generation by soft lithography*

Soft lithography is an emerging technology that allows for the fabrication of microscale devices without the use of expensive laboratory equipment. It encompasses a set of microfabrication techniques based on elastomeric polymers, usually polydimethylsiloxane (PDMS), which are used as stamps or moulds, to create 2D and 3D patterns of molecules (Whitesides *et al.*, 2001). In the microcontact printing technique, a hard master (commonly made with silicon) is used to fabricate a replica in soft polymer, typically PDMS. PDMS is a biocompatible material that can be activated and functionalized with multiple molecules. After functionalization, the PDMS is used as a stamp. When applied on the substrate and in contact areas, biomolecules are transferred from the PDMS to the substrate, thus producing the pattern (Figure 9.1). The micropatterning of biomolecules using this technology can achieve patterns that typically range from 100 nm to several μm, and can thus be used to fabricate domains that can contain anywhere from a single cell to large groups of multiple cells with a very high dimensional resolution (Figure 9.2).

Soft lithography is particularly advantageous for stem cell research because it provides the flexibility to create patterns with different geometries. Thus, soluble factors, anchoring molecules, and spatial arrangement can be tested as factors influencing stem cell fate. Many research groups have applied soft lithography, particularly the microcontact printing method, to study stem cell developmental processes (Ruiz and Chen, 2007; Ruiz *et al.*, 2008; Buzanska *et al.*, 2009). They have demonstrated that pattern geometry can greatly influence fate decisions of stem cells. Pattern geometry can be easily engineered by microcontact printing to allow for the formation of simple and complex patterns, where the role of

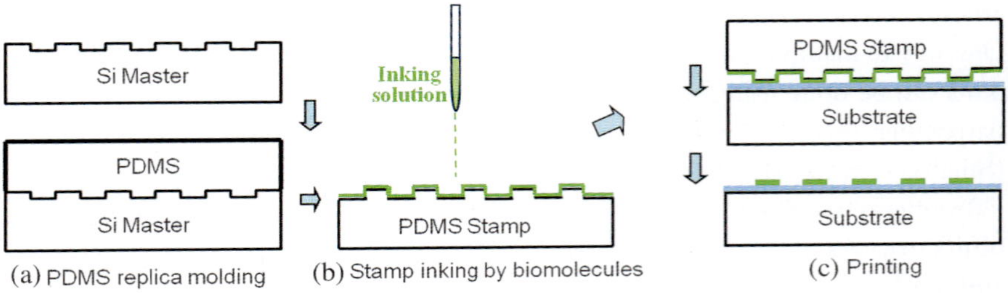

Figure 9.1. Micropatterning of biomolecules by microcontact printing. (a) A PDMS replica is casted against a Silicon master. (b) The PDMS stamp is inked with the solution of interest. (c) Then the stamp is driven into contact with the surface to produce the pattern. Adapted from Colpo P, Ruiz A, Ceriotti L, Rossi F. *Surface functionalisation for protein and cell patterning*. Chapter 2. Advances in Biochemical Engineering/Biotechnology.Springer.DOI:10.1007/10_2009_2.

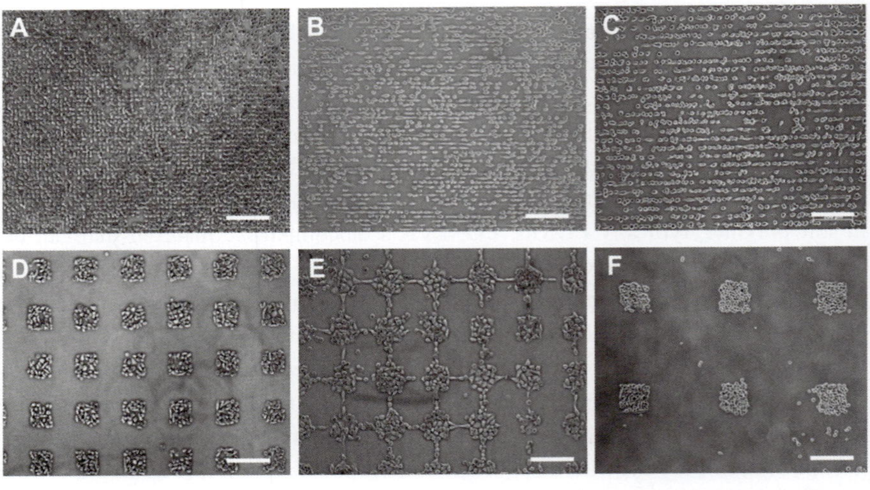

Figure 9.2. Phase contrast images of hUCB-NSC grown on microcontact printed domains with diverse geometries. (a–c) Patterns with single cell resolution. (d–f) Multiple cell patterning. Scale bar = 200 μm. Adapted from Buzanska *et al.*, Toxicology 2010.

cell population density, surface adhesion area, and mechanical force on stem cell differentiation can be investigated. Luo *et al.* (2008) used microcontact printing to pattern alkane thiol self-assembled monolayers (SAMs) on gold-coated glass and studied the adipogenic differentiation of human mesenchymal stem cells. These researchers have identified surface chemistry and geometrical shape as parameters that cause distinct differences in the rate of adipocyte differentiation. Ruiz and Chen (2008) went further and demonstrated that gradients of mechanical forces drive patterning of osteocyte and adipocyte lineages in human mesenchymal stem cells (MSCs). Microcontact printing and PDMS stamps were used to generate 2D patterns of fibronectin in different shapes where MSCs were confined. In regions of high stress, such as convex edges and outer borders, stem cells differentiate into osteocytes while regions of low stress resulted in adipogenesis. Lineage specification, as a function of the multicellular aggregate geometry and cell position within the aggregate, was also observed in 3D structures fabricated by using PDMS molds with wells.

Microscale surface patterns have also been used to determine the factors influencing cell growth and neural differentiation in neural stem cells. Li *et al.* (2009) fabricated patterns of serum on a substrate consisting of hydrophobin II-modified poly(lactic-co-glycolic acid) to study the adhesion of neural stem cells obtained from the brain of mice embryos. The neural differentiation of cortical precursor cells has been investigated using micropatterns of ECM gels on Petri dishes and glass (Magnusson *et al.*, 2008). More mature morphologies and greater expression of neuronal markers were observed when the pattern shape encouraged cells to extend processes along lines. However, electrophysiological measurements showed that the functional properties of these cells were not fully achieved regardless of their neural-like morphology. Buzanska *et al.* (Ruiz *et al.*, 2008; Buzanska *et al.*, 2009; Ruiz *et al.*, 2009; Buzanska *et al.*, 2010) created patterns of cell adhesion molecules and extracellular matrix proteins on cell repellent PEO-like substrates to study the effects of surface chemistry, pattern size and pattern geometry on the developmental processes of human umbilical cord blood-derived neural stem cells (hUCB-NSCs). Cellular responses were investigated in patterns of small active domains, which created single cell arrays and on patterns of larger active areas. Migration and lineage-specific differentiation of hUCB-NSCs depended on the geometry of the pattern and the type of biomolecule printed. These determined the type of cell/surface interaction, such as specific receptor-mediated adhesion vs. electrostatic interaction.

Soft lithographic methods can also be used to fabricate microfluidic channels and scaffolds for stem cell studies and tissue engineering (Whitesides *et al.*, 2001).

Microfluidic patterning requires forming microchannels between a mould and a substrate. The ligand solution is then flushed and the biomolecules are deposited on the substrate. Koepsel and Murphy (2009) used microfluidic channels to pattern SAMs in order to study cell–material interactions. Arrays of spatially distinct cell culture environments on bioinert SAM substrates were realized by combining microfluids with aqueous chemistries. The aqueous solutions selectively removed bioinert SAMs and locally replaced them with bioactive SAMs which were further functionalized with cell adhesive peptides. Such localized replacement of SAMs allowed for peptide confinement and its density control. The identity and density of peptides strongly influenced human mesenchymal stem cell spreading, focal adhesion density, actin organization, and cell projections that may have an impact on differentiation processes.

9.2.1.2. *Microarray formation by microspotting*

While microcontact printing technology allows for the creation of the cell growth platforms with the pattern prepared from a single type of biomolecule, microspotting permits patterning of different microenvironments on the same slide. For that purpose, robotic spotters, capable of dispensing and immobilizing minute quantities (on the order of nL) of materials, are utilized. With this method, miniaturized growth platforms are obtained where cell–matrix interactions can be followed in a high-throughput manner (Kuschel *et al.*, 2006). Originally, microarray technology was not related to cellular patterning. It was introduced to perform high-throughput RNA analysis in a miniaturized format (Khademhossein, 2005) and to directly investigate protein expression, along with functionality (proteomics). More recently, microarrays were used to help understand how cells interact with different external microenvironments: the effects of extracellular components on cellular behavior, the interaction of new drugs with cell surface polysaccharide molecules (glycomics) and the effects of gene expression or silencing on cell microarrays generated, for instance, by the parallel transfection of cells in an array format (Ziauddin and Sabatini, 2001).

Stem cell researchers used microspotting technology to create extracellular matrix (ECM) microarray platforms. They took advantage of the versatility of this technique, the combinatorial matrix approach and the low amount of proteins required. Different microenvironments were used for probing the differentiation of mouse ESCs into hepatocytes (Flaim *et al.*, 2005) and human neural progenitors into neural lineages (Soen *et al.*, 2006). Our research team has fabricated microarrays of ECM proteins to study the fate decisions of neural stem cells derived from human umbilical cord blood (hUCB-NSCs) (Ceriotti *et al.*, 2009). As shown in

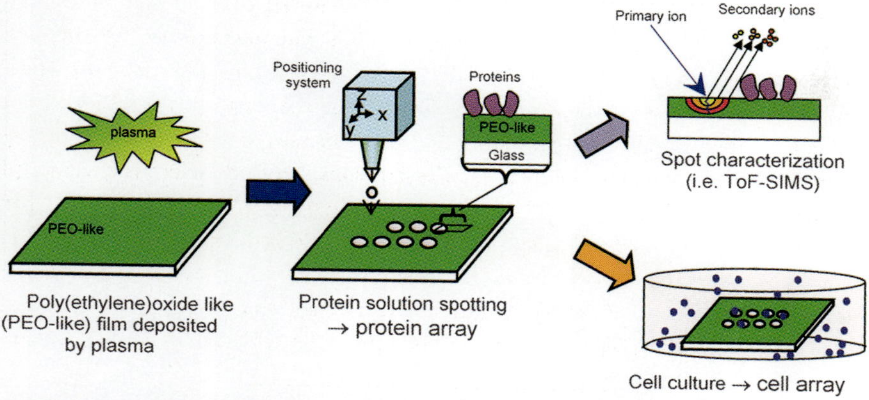

Figure 9.3. Scheme of the steps involved in the fabrication of protein microarrays used for hUCB-NSC patterning. After PEO-like film deposition by plasma, protein solutions are spotted on the coated slidesby piezo electric printing technology. The protein microarrays are then characterized or used as substrate for cell culture. Ceriotti *et al.*, Soft Matter, 2009, 5, 1406–1416.

Figure 9.3, glass slides were coated with a cell repellent PEO-like film deposited by plasma (Bretagnol *et al.*, 2007). Next, ECM protein solutions were spotted by piezo electric printing technology. The protein spots were characterized by time-of-flight secondary ion mass spectrometry (ToF-SIMS), ellipsometry and atomic force microscopy (AFM). Furthermore, the presence and activity of the fibronectin spotted on the PEO-like film was verified by surface plasmon resonance (SPR) imaging, which also provided an estimate of the density of protein available on the surface for cell adhesion. By culturing hUCB-NSCs on such arrays, it was evident that cells behave differently on the spots as a function of protein type and concentration (Figure 9.4). This behavior reflects the expression profile of these cells with respect to the integrin subunits. Cell number on the spots was counted by nuclei staining while immunostaining was introduced to evaluate the cell differentiation stage on such spots. In this preliminary study, only ECM proteins were spotted, but to further investigate the involvement of different cellular signaling pathways in stem cell fate specification, ECM proteins were co-immobilized to the surface with small signaling molecules such as Wnt, Notch, Jagged, CNTF and DKK (Zychowicz, unpublished data; Buzanska *et al.*, 2010). Quantitative single cell analysis by immunostaining against β-tubulin III and GFAP (neuronal and glial differentiation markers, respectively) and multiparameter image analysis revealed complex interactions between some of the signals and the extent and direction of differentiation (Soen *et al.*, 2006; Buzanska *et al.*, 2010).

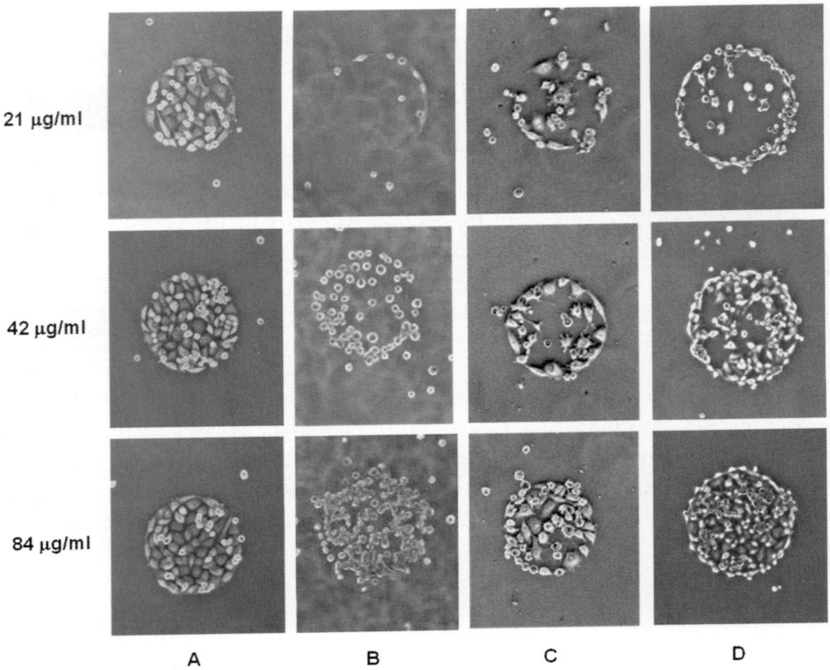

Figure 9.4. HUCB-NSCs on (a) fibronectin, (b) laminin, (c) collagen V, and (d) vitronectin after 4 days of culture on 3-drop protein spots. The effect of different protein types and protein concentrations are visible. Fifty thousand cells were seeded and maintained in standard culture conditions in presence of 2% serum. Scale bar = 100 m. Adapted with modification from Ceriotti *et al.* Soft Matter.

Another approach used to study stem cell differentiation on the microspotted ECM cell growth platforms involves exposure to multiple soluble growth factor environments, while the content of the spots is kept constant (Flaim *et al.*, 2008). Results have confirmed interactions between growth factors and ECM signaling pathways in cell fate decisions (Ffrench-Constant and Colognato, 2004; Nakajima *et al.*, 2007). Furthermore, the results were consistent with the "growth vs. differentiation" paradigm used to describe the switch between self-renewal (proliferation in an undifferentiated state) and differentiation monitored by the changes in gene activity. The expression of Green Fluorescent Protein reporter gene and the cellular DNA content were quantified using fluorescent microscopy as well as laser-based scanner with equivalent results. The latter approach was favored because it effectively tracked cell differentiation in a fraction (~1/30) of the time compared to microscopy.

Why is the fabrication of biomaterial microarrays important for future applications of stem cells in tissue engineering and cell therapy? As already mentioned, surface properties strongly influence cell–material interactions and can affect cell activity and fate (Saltzman, 2000). In general, the development of new biomaterials requires rational design. In addition, their properties when combined with one specific cell type must be individually tested. This has been investigated for ECM microarrays, and more recently for libraries of polymeric biomaterials. Such arrays can be used for the high throughput screening of cell–biomaterial interactions and thus, to identify materials supporting specific cell functions (Yang *et al.*, 2009). Researchers in Langer's group described a platform that enables nanoliter scale synthesis of polymers and cell-based screening of thousands of microarrayed biomaterials using human embryonic stem cells (Anderson *et al.*, 2004). Various polymers that allow cell attachment and spreading have been identified, in addition to biomaterials which, in combination with soluble factors, permit the differentiation of cytokeratin-positive cells. This technology enables the rapid screening of biomaterials, but requires these materials be synthesized on the array in nanoliter droplets. This robotic synthesis strategy limits the choice of usable materials. In a later publication, the same group developed a microarray of polymers synthesized prior to spotting using conventional methods (Anderson *et al.*, 2004). The polymers were spotted on poly(hydroxyethyl methacrylate), which was used as an adhesive layer for the polymers and also as a repellent layer for the cultured cells. Thus, the cells adhere only on the polymer spots and the effects of each material on the cells can be kept independent from the neighboring materials. These polymer microarrays were successfully used to assess material effects on chondrocytes and on neural stem cell growth and differentiation (Anderson *et al.*, 2005).

9.2.1.3. *Other microfabrication methods*

Sasaki *et al.* (2009) have employed a maskless photolithographic method to fabricate a device for the size-controlled adhesion of mouse embryonic cells (mESCs). The device was comprised of arrays of cell-adhesive circular glass domains surrounded by non-cell-adhesive hydrophilic polyacrylamide. mESCs were able to adhere, aggregate and functionally differentiate into cardiac cells on such circular domains. Thus, the use of maskless photolithography enables the rapid production of micropatterned surfaces. It also allows for the investigation and identification of the adhesive domain with the optimal diameter for cardiac differentiation.

Conventional UV photolithography is another microfabrication method currently used to prepare a biofunctional pattern for stem cells. This approach has been used to pattern alkanethiol SAMs to generate an array of composite biomaterials consisting of both natural and synthetic extracellular matrix (ECM) components, and growth factors (Nakajima *et al.*, 2007). Neural stem cells from the rat embryonic striatum have been cultured directly on the array and analyzed for cell adhesion, proliferation and differentiation. Initially, polymers and proteins which enable cell adhesion have been selected. Subsequently, growth factors have been co-immobilized with those biomaterials. The proliferation and differentiation of neuronal and astroglial lineages were measured. This array-based technique identified the most effective matrix-growth factor combinations that allow for parallel culturing, immunostaining and observation of the resulting cells. These studies further demonstrated the crosstalk between growth factors and integrin receptors, revealed earlier by traditional cell biology experiments (Ffrench-Constant and Colognato, 2004).

9.2.2. *3-Dimensional approach*

While cell-based assays in 2D platforms are commonly used to investigate the effects of cell–substrate interactions, there are significant limitations of 2D patterning techniques. For instance, 2D micropatterns deteriorate over time (Nelson *et al.*, 2003) and rarely produce single cell patterns, although such patterns have been reported occasionally (Ruiz *et al.*, 2009). To prevent cell migration, 3D microwells have been generated to confine the cells over time (Ostuni *et al.*, 2001). High-density arrays have been used to analyze, in parallel, the fates of neural stem cells in multiple microwells (Chin *et al.*, 2004). Within such 3D array platforms, it is possible to minimize cell migration between wells, explore cell–cell interactions via contact or paracrine signaling, and monitor cell differentiation by high-magnification immunofluorescence.

Another limitation of 2D cell patterns is that they typically necessitate specific surface agents (i.e. SAMs, ultra-pure gold etc.) which are rare and can prevent researchers from patterning cells onto new substrates (combinatorial chemistry). To overcome these chemical restrictions, a polymeric Bio Flip Chip (BFC) has been microfabricated and used to control the stem cell microenvironment (Rosenthal *et al.*, 2007). The cells are initially trapped in the microwells of the BFC. Subsequently, the chip is flipped upside down onto the desired substrate. The cells fall out of the microwells and adhere to the substrate. This approach allows for positioning of the cells with single cell resolution (using microwells sized to a single stem cell) anywhere on the substrate and onto any substrate, including layers of different cell types. The BFC has been effectively used to monitor cell proliferation and migration on unconfined areas, to track the effects of cell–cell contact on mESC colony formation and to determine molecules involved in these processes.

Following the brief description of the 2D and 3D technological platforms, we will focus on new biomaterials used in 3D networks in order to control stem cell fate. Silva *et al.* described the formation of an artificial 3D scaffold of nanofibers formed by the self-assembly of peptide amphiphile molecules (Silva *et al.*, 2004). Such nanofibers were designed to present neurite-promoting laminin epitope (IKVAV: pentapeptide epitope isolucine-lysine-valine-alanine-valine) to neural progenitor cells. When progenitor cells were encapsulated within the nanofiber network *in vitro*, they displayed greater and faster differentiation into neurons as compared to cells cultured on standard laminin-coated substrates. The peptide density on the nanofibers was shown to be critical in promoting cell differentiation on both 2D and 3D substrates. Another example of 3D scaffolds developed to support the differentiation of stem cells has been described by (Machado *et al.*, 2007). A porous scaffold made of natural chitosan (as adhesive support), gelatine (as differentiation stimulus), and chondroitin (promoting *in vitro* mineralization) ensured the survival of mesenchymal stem cells and promoted their osteogenic differentiation.

The cell material studies described above were limited to one established cell line or type, and only a few cell parameters (usually adhesion and differentiation states) were investigated. Recently, a grid-based platform was proposed to investigate different cell type–biomaterial interactions (Neuss *et al.*, 2008). This approach will be discussed in the subsequent section.

3D array platforms that allow for microscale control of cellular environments are crucial for the systematic screening of stem cell interactions with the niche components and give insight into the molecular mechanisms that drive these interactions. Also, in regenerative medicine, restoration of tissue structure damaged by traumatic insult or disease may be facilitated by delivering stem cells immobilized on 3D scaffolds. This may be particularly important in central nervous system (CNS) therapy in cases where intravenous or intra-arterial infusions of stem cells are ineffective. Scaffolding constructs used for the treatment of CNS injury and the methods used for the fabrication of these constructs have been reviewed by Walker *et al.* (2009). When designing scaffolds applicable to biomedicine, one should take into consideration the characteristics of the scaffold material, especially its cell capacity, biocompatibility, biodegradability, nutrition, uniformity, inertness, size, and the micro/nanostructure. Currently, scaffolds created for biomedical applications are based on either natural or biosynthetic resorbable materials. Different fabrication methods, such as phase separation, leaching, electrospinning, rapid prototyping, stereolithography, or 3D printing are used to prepare the scaffolds. The surface of the scaffold material is further functionalized with ECM proteins or growth factors to obtain effective 3D constructs that can provide structural and nutritive support for stem cells. The shape and spatial arrangement of the scaffold with the biofunctionalized surface allows for cells to become captured inside the scaffold or attached on the surface of the small (50–100 μm) scaffold particles. The latter results in a higher degree of attachment and prevents formation of cell aggregates before grafting into the CNS (Bible *et al.*, 2009).

9.3. Cell-Fate Control by Biomaterials — Studies Utilizing Different Stem Cell Models

Stem cells are characterized by two unique properties: the capacity to self-renew and multi-lineage differentiation potential. These properties make stem cells an ideal material source for cellular therapy, regenerative medicine and pharmaco-toxicological applications. Based on their origination and properties, stem cells are generally classified as embryonic (ESC) or somatic (SSC). ESCs originate from the blastocyst stage of the embryo, while SSCs are isolated from the fetus or postnatal (adult) tissues. The ESCs are pluripotent, meaning they can give rise to any type of somatic cell in the body. In contrast, SSCs are multipotent and believed to differentiate only into specific cell types of the tissue, or organ in which they reside. The SSCs are found in most tissues, including blood, skin, heart, liver, bone and brain, and their function is to maintain cellular homeostasis of the resident tissue. In recent years, it has become increasingly evident that these SSCs may overcome tissue specificity limitation and differentiate into most of the cell types of the body. Both embryonic and somatic stem cell models are widely used in the development of novel nano- and microscale culturing platforms, biological testing and regenerative medicine.

What kind of strategies are used to test cell–biomaterial compatibility and to identify suitable stem cell models for specific applications? The approach used by Neuss *et al.* (2008) involved preliminary screening where (a) a single cell model was tested against different biomaterials, and (b) a single biomaterial was tested with different cell models. To identify the best cell–biomaterial tandem suitable for specific applications, a grid-based platform has been applied that enables testing a variety of models across different biomaterials. In this systematic and parallel approach, 140 combinations of seven different stem cell types and 19 different polymers were assessed for vitality, cytotoxicity, and apoptosis. Several stem cell models were tested in this study: human mesenchymal stem cells, pre-adipocytes, dental pulp, endothelial progenitor cells, mouse mesenchymal, embryonic, and hematopoietic stem cells. However, this study did not include human neural stem cells (hNSCs). Studies of hNSC–biomaterial interactions have been reported by other groups. The biocompatibility of cells derived from developing CNS and a variety of hydrogels were tested by the Thonhoff group, which examined *in vitro* cell survival, migration and differentiation (Thonhoff *et al.*, 2008). Soen *et al.* (2006) applied a combinatory approach to control lineage commitment by the hNSC. Combinations of different ECM proteins with different signaling molecules successfully directed differentiation of these cells. Micro- and nanostructures also influenced the *in vitro* growth and differentiation of rodent neural stem cells. A micropatterned polystyrene surface functionalized with laminin and growth factor cues promoted the differentiation of rodent NSCs into functional neurons (Recknor *et al.*, 2006; Oh *et al.*, 2009). Two different groups reported that fibrous nanostructures can control rodent NSC lineage commitment through the structure itself. Self-assembled nanofibers and electrospun nanofibers of the PCL polymer (Silva *et al.* 2004; Xie *et al.* 2009, respectively) supported neuronal differentiation and inhibited the development of astrocytes.

A number of differences have been found in the molecular regulation of stem cell development between mice and humans (Xu *et al.*, 2002, 2005). Therefore, for both the therapeutic applications in humans and the predictive value of drug screening, it is important that researchers concentrate on human stem cell models. The human systems (ESCs and SSCs, derived from bone marrow and umbilical cord blood) that give rise to cells of different germ layers are introduced further in this section.

9.3.1. *Human embryonic stem cells*

Human (h)ESCs used in research laboratories are derived from the inner cell mass of the blastocyst of 4- to 5-day-old embryos generated by *in vitro* fertilization and donated to research. The ESCs are pluripotent, i.e. they can differentiate into the cell types of all somatic lineages of the three embryonic germ layers: ectoderm, endoderm and mesoderm. An important obstacle that hampers the therapeutic applications of hESCs is the need to expand these cells on a feeder layer of mouse fibroblasts (xeno cells). Also, grafted hESCs have been shown to produce tumors (teratomas). Many research groups are currently working with xeno-free conditions for cell cultures and microenvironmental settings, enabling them to direct ESCs to the desired type of cells without forming teratomas. To address these challenges, the use of 3D scaffolds of synthetic polymer hydrogels, designed to allow independent manipulation of cell adhesion ligands (RGD-containing peptides) and control matrix stiffness has been proposed (Li *et al.*, 2006). The "semi-interpenetrating polymer network" used in this study only allowed for short-term culture of hESCs. In contrast, a 3D hyaluronic acid (HA) hydrogel (Gerecht *et al.*, 2007) supported the long-term controlled self-renewal and differentiation potential of the hESCs. The first step in directing hESC differentiation *in vitro* involves the formation of aggregates, referred to as embryonic bodies (EBs). Therefore, it was necessary to develop a method for generating large quantities of homogenous EBs with defined sizes and shapes. For that purpose, Karp and coworkers have microfabricated cell-repellent poly(ethylene glycol) (PEG) wells, which serve as a template for the controlled growth of EBs. The EBs generated remained viable and maintained their size and shape within the microwells, unlike their suspension counterparts (Karp *et al.*, 2007). Microwells patterned with protein-repellent self-assembled monolayers (SAMs) were suitable to support cultures of undifferentiated hESCs. In addition, they were also used to produce uniform arrays for studying the factor/compound-induced hESC responses (Metallo *et al.*, 2008).

Although substantial advances in hESC culture and differentiation have been made, the problem concerning the formation of teratomas by hESCs remains unresolved. Differentiating cells seeded into 3D poly-(lactic-co-glycolic acid) scaffolds coated with laminin and transplanted between liver lobules were found to produce proteins that are characteristic of several tissues, including the pancreatic, hepatic and neuronal lineages, but they also had teratoma-forming potential (Lees *et al.*, 2007).

9.3.2. *Human bone marrow-derived stem cells*

Human bone marrow is the reservoir of hematopoietic (hHSCs) and mesenchymal (hMSCs) stem cells. hMSCs are isolated from the adherent bone marrow stromal cell fraction typically from patients undergoing routine hip replacement surgery. These cells thus represent a valuable source of cells for possible autologous transplantation. Their adhesive properties, together with their ability to differentiate into many cell types makes hMSCs an attractive candidate for the study of cell–surface interactions and cellular responses to changes in the surrounding microenvironment.

What is the differentiation potential of hMSCs and can one direct their fate using biomaterials? Typically, hMSCs can be differentiated into cells of connective tissues (i.e. bone or cartilage) or adipocyte tissue. Also, neuronal and endothelial differentiations of hMSCs have been reported. The expansion and neuronal differentiation of hMSCs was influenced by biomolecules such as poly-l-lysine, extracellular matrix proteins and matrigel (soluble basement membrane extract) (Qian and Saltzman, 2004).

Markowicz *et al.* (2005), while searching for matrix modifications and cell sources capable of increasing vessel formation, used 3D collagen type 1 scaffolds to differentiate hMSCs into endothelial cells. They showed that hMSCs may differentiate not only into endothelial derivatives, but also to form organized capillary structures throughout the collagen scaffold. The vessel formation capacity depended on matrix modifications (i.e. cross-linking and heparinization), further underscoring the importance of cell membrane–matrix interactions.

Dalby *et al.* (2008) investigated molecular mechanisms underlying hMSCs responses to different surface material properties. In an elegant study, they have shown that nanoscale surface topography influences gene expression and differentiation by the MSCs. The authors used an advanced method of microarray bioinformatics and integrated data analyses to document complex genomic responses to the topographical or chemical stimulation of hMSC differentiation. While differentiation and growth suppression induced by chemical stimuli was accompanied by switching on a vast number of different genes, changes in surface topography allowed for normal cell growth and promoted osteogenic differentiation by targeting only a small number of canonical pathways.

Recently, hMSCs have been used in conjunction with fourth-generation biomaterials, such as photodegradable gels that allow real-time manipulation of the microenvironment (Kloxin *et al.*, 2009). Temporal variations in the structure and biochemical gel composition were induced locally to influence the migration and differentiation of the encapsulated hMSCs. Light-induced changes in the structure (creation of the channels), and chemistry (removal of the RGD ligand for cell membrane integrin receptors) of the hydrogel at specific time-points of the culture allowed for cell migration and chondrocytic differentiation, respectively. Photoinducible real-time manipulation of the physical and chemical properties of the hydrogel results in dynamic environments for the growth and differentiation of stem cells. Such technological advances may constitute an important milestone in the controlled regulation of stem cell fate and have a tremendous impact on the progress of tissue engineering and regenerative medicine (Grayson *et al.*, 2009; Vunjak-Novakovic and Scadden, 2011).

9.3.3. *Human umbilical cord blood (hUCB) and umbilical cord (hUC)-derived stem cells*

Human cord blood contains several types of stem cells including hematopoietic, endothelial, mesenchymal and pluripotent-like stem cells, which have been shown to differentiate into nearly all cell types of the body, including neural cells (McGuckin *et al.*, 2005; Habich *et al.*, 2006; Kucia *et al.*, 2007). In the context of probing cell–biomaterial interactions, it is necessary to concentrate on hUCB-derived cells with adhesive properties, such as mesenchymal, endothelial, or neural progenitor stem cells. The hypoimmunogenic and delayed clamping properties, and the high yield of the progenitor cells make human cord blood an attractive source of the SSCs for biotesting and therapeutic applications (Wang *et al.*, 2009; Walker *et al.*, 2009). Due to the above mentioned features, human cord blood stem cells were named "*Mankind's first natural stem cell transplant*" (Tolosa *et al.*, 2010).

Previously, we showed that stem cells isolated from hUCB attain neural stem-like features (Buzanska *et al.*, 2002) and established a non-transformed, clonal and karyotypically stable hUCB-derived neural stem cell line (hUCB-NSC) (Buzanska *et al.*, 2006). The hUCB-NSC model has been used to investigate neurodevelopmental processes. These cells were also applied in toxicological studies to assess their vulnerability to toxic compounds at various stages of development, i.e. from non-differentiated pluripotent stem cells, to committed neural progenitors and differentiated neuronal, astrocytic and oligodendroglial cells (Buzanska *et al.*, 2005; Ruiz *et al.*, 2008; Buzanska *et al.*, 2009, 2010).

In recent years, we have initiated studies of diverse biomaterials in an attempt to direct differentiation of hUCBs towards different neural cells. HUCB-NSCs readily adhere to patterned biomolecules, micro-contact printed on a PEO antifouling substrate (Ruiz *et al.*, 2007). We sought to determine what are the mechanisms that modulate stem cell fate decisions when they are attached to the patterned surface. Two factors have been found to be important: the configuration of the printed pattern and the nature of stem cell–surface interactions determined by the composition of the adhesive substrate. While cell receptor-mediated adhesion to ECM fibronectin promoted neuronal differentiation, a non-specific adhesion to polycationic molecules maintained the cells in a non-differentiated state. Additional factors, such as the presence of serum (used also as a negative control) and the presence of interconnecting lines between patterned active domains (the positive control) have been shown to influence proliferation and neuronal differentiation of hUCB-NSCs (Figure 9.5) (Buzanska *et al.* 2009; 2010; Zychowicz *et al.* 2011). Microspotting of bioactive domains enabled the study of different microenvironments on the same microarray. Interactions of hUCB-NSCs with different ECM proteins in domains that were further engineered to create a "smart" microenvironment with small signalling molecules immobilized to the surface gave additional insight into the molecular mechanisms that control stem cell fate. Activation of different intracellular pathways by small signaling molecules either committed the cells to a specific neuronal lineage or maintained them in an undifferentiated stage (Figure 9.6) (Zychowicz, unpublished; Buzanska *et al.*, 2010).

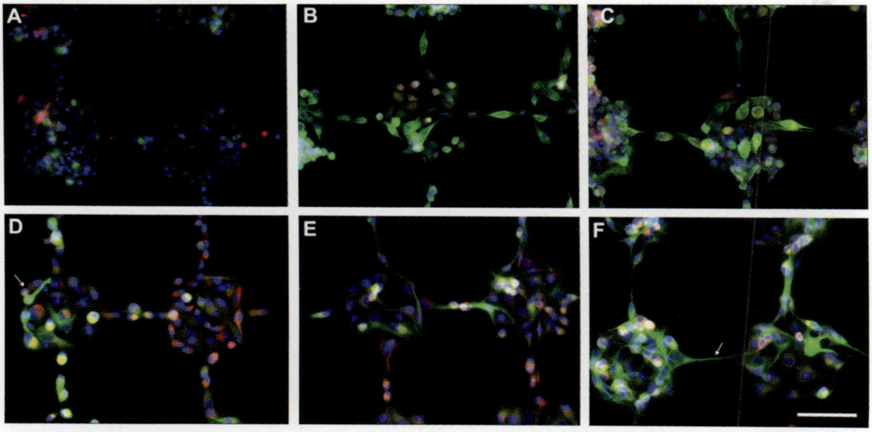

Figure 9.5. hUCB-NSC cultured for 7 days on (a–c) PLL and (c–e) fibronectin in standard control conditions — 2% FBS, 5% CO_2, 37°C (a and c), with the addition of neuronal differentiating factor — 100μM dBcAMP (b and e), in the presence of 100-μM dBcAMP and in serum-free conditions (c and f). Immunostaining for neuronal and astrocytic markers: β-tubulin III (green)/GFAP (red), respectively, cell nuclei contra-stained with Hoechst. Scale bar: 100 m. Adapted from Buzanska *et al.*, Toxicology 2010.

The protein microarray approach allowed for further study of the molecular mechanisms that promote hUCB-NSC fate decisions, but did not establish the microenvironmental conditions that might be needed for functional hUCB-NSC neuronal differentiation. Further insights into these conditions came from studies in which the cells were cultured with biomaterials organized into a 3D structure. The scaffold of biodegradable human keratin supported the maintenance and differentiation of hUCB-NSCs starting from the cells expressing genes characteristic of pluripotency (Oct4 and Sox2), to the genes associated with proliferating neuroblasts (Nestin, GFAP, and Ki67). When the scaffolds were placed on multi-electrode array (MEA) chips covered with poly-l-lysine and laminin, and incubated in differentiating medium (dibutyryl cyclic AMP + retinoic acid + brain-derived neurotrophic factor), neuroblasts migrated out of the scaffolds to the area of inserted electrodes, forming functional circuits and generating spontaneous field action potentials (Jurga *et al.*, 2009). Thus, a 3D environment is required for the functional maturation of neural stem cells derived from human cord blood.

The study on interactions between umbilical cord blood or umbilical cord-derived stem cells and biomaterials reveals the importance of scaffolding in cell development and demonstrates its potential for biomedical applications.

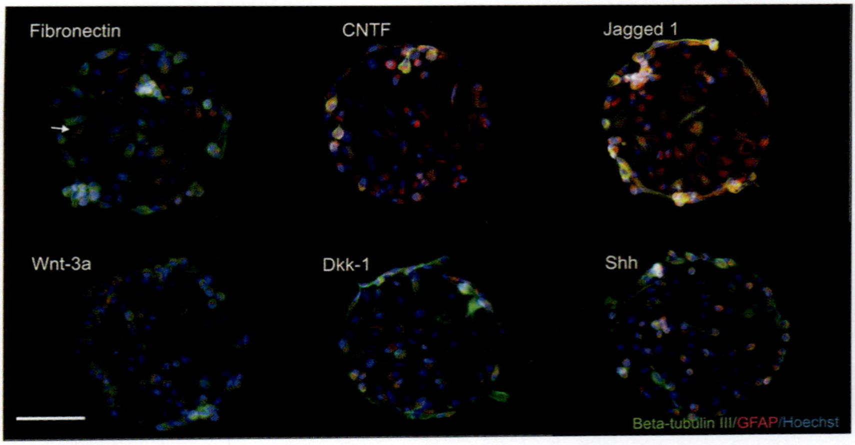

Figure 9.6. HUCB-NSC cultured for 3 days on protein array microspotted with fibronectin and fibronectin together with signaling molecules: CNTF, Jagged-1, Wnt-3a, Dkk-1 and Shh. Cells were immunostained for expression of -tubulin III (green) and GFAP (red). Cell nuclei were contra-stained with Hoechst (blue). Scale bar = 100 m. Adapted from Buzanska *et al.*, Toxicology 2010.

Mesenchymal stem cells, typically of the bone marrow origin (hUC-MSCs), can also be harvested from umbilical cord tissue (Wharton's Jelly) (Troyer and Weiss, 2008; Schneider *et al.*, 2010), or from the adherent, mononuclear fraction of cord blood (hUCB-MSCs) (Lee *et al.*, 2004). HUC-MSCs have potential to differentiate into adipogenic, osteogenic, chondrogenic neuronal and endothelial lineages *in vitro*. Recently, successful multi-lineage differentiation and delivery of hUC-MSCs via bioactive scaffolds for tissue engineering have been reported. When loaded into biomimetic artificial bone scaffold material and transplanted to immunodeficient mice, the hUCB-MSCs undergo neosteogenic differentiation leading to bone formation (Diao *et al.*, 2009). Zhao *et al.* (2010) reported the implementation of a new generation scaffold material biofunctionalized with calcium phosphate (CaP) minerals. This material bound to the native bone, provided the substrate for cell attachment and supported cell proliferation and expression of the osteoblast phenotype. Calcium phosphate cement (CPC) was proposed as the bioactive scaffold that can be molded *in situ* into the desired size. The resistance of CPC to fatigue and fracture can be improved with chitosan and polyglactin fibers. A CPC–chitosan–fiber scaffold supported the viability and osteogenic differentiation of the encapsulated hUC-MSCs, and displayed bone-like mechanical strength (Zhao *et al.*, 2010).

Interactions of the hUCB-MSCs with biomaterials and other cell types can promote other types of hUCB-MSCs differentiation. When co-seeded in Matrigel scaffold, together with hUCB-derived endothelial progenitors and transplanted into immunodeficient mice, hUCB-MSCs displayed the capacity to form vascular networks. Furthermore, co-transplanted endothelial and mesenchymal cells attained restricted positions in the vascular network. The endothelial progenitor cells formed the luminal aspect of the vessels. In contrast, mesenchymal progenitors remained adjacent to the lumen, confirming their role as perivascular cells (Melero-Martin *et al.*, 2008).

9.4. Applications of Biofunctionalized Domains — Stem Cells Systems

9.4.1. *Biotesting*

The dramatic increase in the number of toxicants and drug candidates available to the pharmaceutical and cosmetic industry has created a demand for more advanced testing procedures. Cell-based assays were developed for that purpose, as they offer the opportunity to perform screens of the vast chemical libraries across different cellular models.

Such tests based on the screening of the cellular processes (i.e. proliferation, self-renewal, differentiation, apoptosis) or a range of the cell molecular endpoints (i.e. gene and protein expression) have been proposed to detect acute and developmental toxicity (Coecke *et al.*, 2007; Bal-Price *et al.*, 2009) and to screen for drugs (Trosko and Chang, 2009). Conventional cell culture platforms based on 96-well plates and antibody-binding for tracking of cell fate decisions may not be readily applicable to high-throughput screening because their production is too costly, time consuming and they provide only limited functional information. Miniaturization technologies such as robotic spotting, soft-lithography and photo-assisted approaches are the preferred tools for high-throughput cell-based screening methods (Fernandes *et al.*, 2009). These approaches require smaller samples, enable the fast and efficient analysis of cellular responses and provide more information. Microfabricated screening devices (as described in Section 3) include: cellular (Buzanska *et al.*, 2010; Lee *et al.*, 2008), biomaterial (Anderson *et al.*, 2005) or transfection (Bailey *et al.*, 2006; Yoshikawa *et al.*, 2004) microarrays, as well as microwell (Chin *et al.*, 2004; Ochsner *et al.*, 2007; Moeller *et al.*, 2008) and microfluidic cell culture arrays (Hung *et al.*, 2005; Wada *et al.*, 2008). In addition to drug discovery and toxicology screening, microarrays, microwell and microfluidic systems can be used to further address the mechanisms that underlie the stem cells phenotype and drive differentiation. High-throughput optimization of stem cell environment is reviewed by Yang *et al.* (2009).

The drug discovery field further awaits the broader application of microarray or microfluidic cell culture systems. The pharmaceutical industry widely utilizes high-throughput cell-based screening assays, based on 96- or 384-well microtiter plates, in conditions that consume large amounts of chemicals and which are far from perfect in regard to stem cell–matrix interactions. Challenges include the miniaturization of biochips, the targeting of cells at the molecular level and the construction of sensor cell lines. Recently, miniaturized multiwell culture systems, with microscale tissue subunits and optimized architectures to maintain their functionality, were proposed for drug testing (Khetani and Bhatia, 2008). Another interesting solution implies cellular microarrays for small molecule screening, utilizing biodegradable poly-(D),(L)-lactide/glycolide disks impregnated with, and capable of gradually releasing, the compounds of interest (Gopalakrishnan *et al.*, 2003). The in-cell, on-chip immunofluorescence technique that combines the use of 3D cellular microarrays with sensitive immunofluorescence detection provides quantitative information of cell function (Fernandes *et al.*, 2008) and represents another interesting approach proposed for drug screening. In this particular study, a cellular microarray was designed to detect changes in the expression of hypoxia-inducible factor (HIF)- 1a, a transcription factor related to several types of cancer. However, different combinations of cell types, protein targets and small molecule modifiers could also be applied. The advantage of targeting cells at the molecular level for drug discovery comes from cell-based microarrays created for high-throughput screening of gene function. This type of approach exploits the transfection of stem cells with short interfering RNA (siRNA) or DNA immobilized to the active domains that the cells adhere to (Bailey *et al.*, 2006; Yoshikawa *et al.*, 2004). Also, the creation of a living array of variably stable transduced cell clusters allows one to determine cell responses to specific agents with a loss or gain of function of a particular gene.

What makes the high-throughput toxicology studies important in drug discovery and therapeutic application? The answer lies in the unacceptably high rate of toxicity among drug candidates in the late clinical stages, which should have been eliminated at the early, preclinical stage. One of the proposed high-throughput methods combines enzyme catalysis with cell-based screening on a microscale biochip platform (Lee *et al.*, 2008). In more detail, two complementary microarrays are prepared. One is a miniaturized 3D cell culture array with human cells encapsulated either in collagen or in alginate gels yielding 1080 individual cultures, as small as 20 nL (Data Chip). The second human P450 microarray, contains test compounds with combinations of human cytochrom P450 isoforms (Meta Chip) spotted in a compatible array to the Data Chip. Direct testing of compound toxicity, measured as the rate of P450-generated metabolites, requires the initial coupling of the Data Chip with the Meta Chip, followed by the removal of the Meta Chip and further incubation with the tested compound. The IC50 values obtained with this method were not different from the ones detected in a 96-well conventional plate, despite the nearly 2000-fold miniaturization!

Automated and high-throughput cell-based toxicity screening was also achieved using a microfluidic cell culture assay (Wada *et al.*, 2008). In this study, the microfluidic cell culture system was combined with sensor cells. Sensor cells were constructed by the incorporation of green fluorescent protein (GFP) under the promoter of HSP70B' (heat shock protein 70B') which is induced by a cytotoxic reagent. The implementation of microfluidic technology in biotesting allows for the analysis and isolation of compounds on a single chip, as well as the miniaturization, integration and automation of biochemical assays. This may be the method of choice for the next generation of biotools for drug discovery, such as the micro total analysis system (μTAS) that integrates all steps of the chemical analysis — sampling, sample pretreatment, transport, chemical reactions, product isolation and detection of the final result. Combined with automated fluidic manipulation, μTAS enables the regulation of feedback mechanisms and allows for high sample throughput (Dittrich and Manz, 2006).

Microfluidic techniques were a step towards the neuron-based functional chip (Morin *et al.*, 2006). For this purpose, 3D microfluidic systems were created in conjunction with planar microelectrode arrays (pMEAs). Microfluidic corridors guided axonal outgrowth, while microelectrodes recorded extracellular electrical signals. Such complex neuron-based biosensors may serve as testing platforms for toxicological and pharmacological screening (Ravula *et al.*, 2006). Recently, a new analysis platform called the network formation assay (NFA) addresses cell patterning for neurotoxicity testing (Frimat *et al.*, 2010). Uniformly spaced, hexagonal array of PDMS adhesive modules serve to pattern neuronal cells and support network formation by directing their axonal outgrowth between adhesive nodes. Network formation was inhibited in a dose-dependent manner by acrylamide, a neurotoxic reference compound. It was concluded that NFA met the demands of high-throughput applications in neurotoxicology and, if combined with stem cells developing into a neuronal lineage, may also have applications in developmental neurotoxicity (DNT).

What is the advantage of using stem cells, especially of human origin, for *in vitro* toxicology and drug development applications? Stem cells provide an unlimited source of highly uniform cell types for pharmaceutical applications, especially when compared with genetically abnormal transformed or tumor cell lines (Pouton and Haynes, 2005). Also from a developmental standpoint, many chronic diseases such as cancer, birth defects, and possibly adult diseases, may originate from stem cells that accumulate mutations. This is in agreement with the "*Baker hypothesis*" that the main causes leading to birth defects, neonatal death, or later chronic diseases are misbalances in the amount of stem cells in specific organs during development. Similarly, cancer stem cell theory is based on cumulative defects in developing organ-specific stem cell populations. Predictive studies related to human health demand toxicity testing with human stem cell models, most likely in a 3D microenvironment, mimicking the *in vivo* stem cell niche. Stem cells of human origin can be cultured and harvested for testing at different developmental stages. Thus, the availability of human stem cell lines (derived from embryonic, but also non-transformed somatic cells) (Conti *et al.*, 2005; Buzanska *et al.*, 2006) enables a comparative testing strategy based on studies of lineage-related populations of uniform cells at different developmental stages.

9.4.2. *Tissue engineering and regenerative medicine*

Tissue engineering is a rapidly expanding subfield of biotechnology and serves as the prominent vehicle for regenerative medicine. It combines the principles of stem cell and biomaterial technology to reconstruct, *in vitro*, tissue or part of the organ for transplantation. These attempts benefit largely from recent advances in stem cell and biomaterials technology. Modern biodegradable scaffolds are designed to drive stem cell fate decisions and promote the desired function in a spatially and temporally controlled manner. This can be achieved by incorporating microenvironmental signals directly into the structure of biomaterials and precisely positioning stem cells within the supporting material (Grayson *et al.*, 2009). Modern technologies for building stem cells/scaffolding constructs, such as spraying, "positioning layer by layer", or inkjet printing, make such an assembly possible (Facca *et al.*, 2008; Yang *et al.*, 2009). A future goal is the real-time conditional modulation of the scaffold structure and composition in order to induce the desired stem cell developmental responses (Grayson *et al.*, 2009; Kloxin *et al.*, 2009; Fisher *et al.*, 2010; Zhao *et al.*, 2011).

What are the specific requirements and quality criteria for the successful introduction of the stem cells into regenerative medicine? Donor cells should be of human origin and most likely from the patient, on demand, in order to diminish or eliminate problems of immunogenicity of the transplanted cell/biomaterial construct. Few alternative methods of obtaining autologous stem cells for cell replacement therapies have been proposed. These include therapeutic cloning, isolation of somatic, tissue-specific stem cells and recently, generation of induced pluripotent stem cells (iPSCs). The use of some of these methods remains restricted due to safety or ethical issues, or because they are invasive to the patient. However, stem cells obtained from somatic tissues are believed to be the safest and promising in terms of immunocompatibility and tumor formation risk (Hodgkinson *et al.*, 2009). Further quality criteria require Good Laboratory Practice (GLP) procedures and most likely, xeno-free media and supplements for the stem cell culture.

The biomaterials used to support transplanted cells are also suspected to be cytotoxic and cause immuno-adverse reactions. The classification of materials for biomedical applications is based on their source and physico-chemical properties, whether they are naturally derived or synthetic, hydrogels or minerals, and whether they are biodegradable. Further modifications or functional adaptation, such as surface charge, coating with ECM proteins or drug delivery, can be introduced to the materials to make them biocompatible with the stem cells they harbor and with the host tissue place of transplantation. Biomaterials are designed to act as carriers for the local delivery of stem cells but also, due to their ability to carry bioactive niche components, may actively participate in the formation of functional tissue that provides niche signals. The properties of transplantable materials may be important for preconditioning the damaged tissue area before stem cell transplantation and/or activation of the endogenous reservoirs of stem cells. The latter is of particular interest since the beneficial effect of transplanted stem cells is usually not only due to the differentiation and replacement of the lost cells, but also due to the modulation of the microenvironment of the site of transplantation. This may include production of neurotrophic growth factors, stimulation of angiogenesis and modulation of the inflammatory response. Thus, the implantation of biomaterial scaffolds, seeded with stem/progenitor cells, could potentially act as an artificial stem cell niche providing both structural and nutritive support (Walker *et al.*, 2009; Hill, *et al.*, 2006; Karp and Langer, 2007). Currently, there are advanced preclinical studies with different scaffolding constructs and human stem cells that involve connective tissue, skin, spinal cord and brain applications (Walker *et al.*, 2009; Hejcl *et al.*, 2010).

Stem cell/biomaterial constructs are specifically designed for different therapeutic applications, depending on the tissue target and disease (Vunjak-Novakovic and Scadden, 2011). Recent review papers on current tissue engineering, and novel therapeutic approaches using stem cells/biomaterial scaffolds directed to specific tissues, and organs, are listed in Table 9.1.

Ongoing clinical trials and applications of stem cell/scaffolding constructs are recorded for the reconstruction of connective tissues like cartilage and bone (Gersh *et al.*, 2009), or adipose tissue (Gautier *et al.*, 2008). Furthermore, due to advanced and successful preclinical studies (as in Table 9.1), stem cells/scaffolding constructs have also been applied in clinical trials for the treatment of skin (Mansbridge, 2009)

Table 9.1. Review papers on current tissue engineering and novel therapeutic approaches using stem cells/biomaterial scaffolds directed to specific tissues and organs.

Targeted tissue/organ	Review papers and protocols
Brain and spinal cord	Refs. Walker *et al.*, 2009; Bible *et al.*, 2009; Madigan *et al.*, 2009.
Skin	Refs. Boyce *et al.*, 2006; Gautier *et al.*, 2008; Mansbridge 2009; Huang and Fu 2008.
Cornea	Refs. Rama *et al.*, 2010; Marchini *et al.*, 2011
Bone and Cartilage	Refs. Granero-Molto *et al.*, 2008; El Tamer and Reis 2009; Richter 2009; Undale *et al.*, 2009; Iwasa 2009.
Blood vessels	Ref. Markowicz *et al.*, 2005
Myocardium	Refs. Gersh *et al.*, 2009; Reffelmann *et al.*, 2009; Yi *et al.*, 2010; Gonzales and Pedrazzini 2010; Shimizu *et al.*, 2009; Burba *et al.*, 2010.
Skeletal muscle	Refs. Deans and Elisseeff 2009, Liao and Zhou 2009, Koning *et al.*, 2009; Tedesco *et al.*,2009
Liver	Refs. Fiegel *et al.*, 2008; Kobayashi 2009; Zheng *et al.*, 2010.
Pancreas	Refs. Kobayashi 2008; Kroon *et al.*, 2008; Furth and Atala 2009.
Bladder	Refs. Fossum and Nordenskjöld 2010; Roth 2010; Tian *et al.*, 2010; Drewa *et al.*, 2010.
Trachea	Refs. Omori *et al.*, 2008; Macchiarini *et al.*, 2008; Alberti 2009; Lin *et al.*, 2009.

and brain tissues, or even to replace whole or parts of organs such as the bladder (Roth, 2008) and trachea (Omori *et al.*, 2008; Macchiarini *et al.*, 2008; Alberti, 2009).

9.5. Conclusions

The ability to probe biomaterial–stem cell interactions in either 2D or 3D arrangements provides a potent tool in identifying novel biochemical and biophysical regulators of stem cell fate. However, cell fate control by biofunctionalized surfaces is only the first step in understanding how to design and implement the tools necessary for modern biotesting and regenerative medicine. Biotesting requires high-throughput optimization of stem cell microenvironments by implementing combinatorial screening to assess cell material, cell–extracellular matrix proteins and cell–cell interactions. More advanced screenings will employ the identification of cellular responses to libraries of small signaling molecules and genetic material. Such optimization of stem cell microenvironments involves miniaturized cell growth platforms with patterned 2D or 3D bioactive domains and the ability to discover signals that influence stem cell fate in a high-throughput manner. The implementation of cellular microarrays for high-throughput toxicology and drug discovery screening has led to further technical advances both in terms of microscaling and biosensing. However, the power of these methods can only be fully realized if biological assays are developed and effectively adapted to the dimensions of microdevices. This has been achieved in modern lab-on-a-chip and micro total analysis systems (μTAS). Microscale technologies are emerging as powerful tools, not only for high-throughput screening, but also for tissue engineering. Scaffolding structures serve as an artificial microenvironment, controlling stem cells at different levels of physiological response. The first level of control concerns the direct interactions of the stem cell–material interface, which depend on the surface chemistry, topography and the attachment of biofunctional molecules in both 2D and 3D arrangements. The second level of control implements the microenvironmental architecture and geometry, while the third level relies on the properties of the biomaterials, including their ability to harbor and secrete active compounds that provide trophic support to the cellular components of the niche. More recently, such control has acquired a fourth (temporal) dimension, which enables the modulation of the biomaterial structural and functional properties, on demand, in a time-related manner.

To meet the requirements of safety and functionality in regenerative medicine, engineered tissues based on biomaterials and stem cells are expected to reach "biomimetic" levels, that is, conditions in the cell environment that authentically represent those of living organisms. Such systems should enable real-time control over the microenvironment with the possibility of conditional regulation of cellular and morphogenetic events. A new generation of scaffolding/stem cell constructs has already been applied and shows promising results in a number of ongoing preclinical and clinical studies. Progress in the fields of biomaterial and stem cell science, and their translation to medicine, is rapid and based on the multidisciplinary collaboration of cell biologists, biotechnologists and medical doctors. However, many basic questions remain unresolved, especially those regarding the safety of stem cells and biomaterials in human medical applications.

Acknowledgments

The authors would like to dedicate this paper to Dr. Francois Rossi for his support to this work. We also thank Marzena Zychowicz for her technical assistance and the other members of the laboratories in NanoBioScience Unit of IHCP, JRC, European Commission, Ispra, Italy and NeuroRepair Department of Mossakowski Medical Centre, Polish Academy of Sciences, Warsaw, Poland.

Bibliography

Alberti, C. 2009. Tissue engineering technologies: Just a quick note about transplantation of bioengineered donor trachea and augmentation cystoplasty by *de novo* engineered bladder tissue, *G Chir* 30:514–519.

Anderson, D. G., Levenberg, S. and Langer, R. 2004. Nanoliter-scale syntesis of arrayed biomaterials and application to human embryonic stem cells, *Nature Biotechnol* 22:863–866.

Anderson, D. G., Putnam, D., Lavik, E. B., Mahmood, T. A. and Langer, R. 2005. Biomaterial microarrays:rapidmicroscale screening of polymer-cell interaction, *Biomaterials* 26:4892–4897.

Bailey, S. N., Ali, S. M., Carpenter, A. E., *et al.* 2006. Microarrays of lentiviruses for gene function screens in immortalized and primary cells, *Nat Methods* 3:117–122.

Bal-Price, A. K., Hogberg, H. T., Buzanska, L., *et al.* 2009. *In vitro* developmental neurotoxicity (DNT) testing: Relevant models and endpoints, *Neurotoxicology* 31(5):545–554.

Bible, E., Chau, D. Y., Alexander, M. R., *et al.* 2009. Attachment of stem cells to scaffold particles for intra-cerebral transplantation, *Nat Protoc* 4:1440–1453.

Boyce, S. T., Kagan, R. J., Greenhalgh, D. G., *et al.* 2006. Cultured skin substitutes reduce requirements for harvesting of skin autograft for closure of excised, fullthickness burns, *J Trauma* 60:821–829.

Bretagnol, F., Kylian, O., Hasiwa, M., *et al.* 2007. Micro-patterned surfaces based on plasma modification of PEO-like coating for biological applications, *Sensors and Actuators B* 123:283–292.

Burba, I., Devanna, P. and Pesce, M. 2010. When cells become a drug. Endothelial progenitor cells for cardiovascular therapy: Aims and reality, *Recent Pat Cardiovasc Drug Discov* 1:1–10.

Buzanska, L., Machaj, E. K., Zabłocka, B., *et al.* 2002. Human cord blood-derived cells attain neuronal and glial features *in vitro*, *J Cell Sci* 15:2131–2138.

Buzanska, L., Habich, A., Jurga, M., *et al.* 2005. Human cord blood — derived neural stem cell line — possible implementation in studying neurotoxicity, *Toxicol In Vitro* 19:991–999.

Buzanska, L., Jurga, M., Stachowiak, E. K., *et al.* 2006. Neural stem-like cell line derived from a nonhematopoietic population of human umbilical cord blood, *Stem Cells Dev* 15(3):391–406.

Buzanska, L., Sypecka, J., Nerini-Molteni S., *et al.* 2009. A human stem cell-based model for identifying adverse effects of organic and inorganic chemicals on the developing nervous system, *Stem Cells* 27(10):2591–2601.

Buzanska, L., Ruiz, A., Zychowicz, M., *et al.* 2009. Patterned growth and differentiation of human cord blood-derived neural stem cells on bio-functionalised surfaces, *ActaNeurobiol Exp* 69:1–14.

Buzanska, L., Zychowicz, M., Ruiz, A., *et al.* 2010. Neural stem cells from human cord blood on bioengineered surfaces — novel approach to multiparameter bio-tests, *Toxicology* 270:35–42.

Buzanska, L., Jurga, M., Stachowiak, E. K., *et al.* 2006. Neural stem-like cell line derived from a nonhematopoietic population of human umbilical cord blood, *Stem Cells Dev* 15(3):391–406.

Cajal, S. 1928. *Degeneration and Regeneration of the Nervous System.* Haffner Publishing Co, New York.

Ceriotti, L., Buzanska, L., Rauscher, H., *et al.* 2009. Fabrication and characterization of protein arrays for stem cell patterning, *Soft Matter* 5:1406–1416.

Chen, C. S., Mrksich, M., Huan, S., *et al.* 1997. Geometric control of cell life and death, *Science* 276:1425–1428.

Chin, V. I., Taupin, P., Sanga, S., *et al.* 2004. Microfabricated platform for studying stem cell fates, *Biotechnol Bioeng* 88:399–415.

Coecke, S., Goldberg, A. M., Allen, S., *et al.* 2007. Workgroup report: Incorporating *in vitro* alternative methods for developmental neurotoxicity into international hazard and risk assessment strategies, *Environ Health Perspect* 115:924–931.

Conti, L., Pollard, S. M., Gorba, T., *et al.* 2005. Niche-independent symmetrical self-renewal of mammalian tissue stem cell, *PLOS Biology* 3:e283.

Dalby, M. J., Andar, A., Nag, A., *et al.* 2008. Genomic expression of mesenchymal stem cells to altered nanoscale topographies, *J R Soc Interface* 5:1055–1065.

Deans, T. L. and Elisseeff, J. H. 2009. Stem cells in musculoskeletal engineered tissue, *Curr Opin Biotechnol* 20:537–544.

Diao, Y., Ma, Q., Cui, F., *et al.* 2009. Human umbilical cord mesenchymal stem cells: Osteogenesis in vivo as seed cells for bone tissue engineering, *J Biomed Mater Res A* 91:123–131.

Dittmar, T., Kassmer, S. H., Kasenda, B., *et al.* 2010. Stem cell migration in health and disease, in Entschladen, F., Zänker, K. S., (eds.), *Transl Res Biomed.* Basel, Karger. pp. 7–27.

Dittrich, P. S. and Manz, A. 2006. Lab-on-a-chip: Microfluidics in drug discovery, *Nat Rev Drug Discov* 5(3):210–218.

Drewa, T., Chlosta, P. and Czajkowski, R. 2010. Will tissue-engineered urinary bladders change indications for a laparoscopic cystectomy? *Surg Innov* 17(4):295–299.

El, Tamer., M. K. and Reis, R. L. 2009. Progenitor and stem cells for bone and cartilage regeneration, *J Tissue EngRegen Med* 3:327–337.

El-Ali, J., Sorger, P. K. and Jensen, K. F. 2006. Cells on chips, *Nature* 442:403–411.

Eriksson, P. S., Perfilieva, E., Björk-Eriksson, T., *et al.* 1998. Neurogenesis in the adult human hippocampus, *Nat Med* 4(11):1313–1317.

Facca, S., Gillet, P., Stoltz, J. F., *et al.* 2008. Three-dimensional sprayed active biological gels and cells for tissue engineering application, *Biomed Mater Eng* 18(4–5):231–235, Review.

Falconnet, D., Csucs, G., Grandin, H. M. and Textor M. 2006. Surface engineering approaches to micropattern surfaces for cell-based assays, *Biomaterials* 27:3044–3063.

Fernandes, T. G., Kwon, S. J., Lee, M. Y., *et al.* 2008. On-chip, cell-based microarray immunofluorescence assay for high-throughput analysis of target proteins, *Anal Chem* 80:6633–6639.

Fernandes, T. G., Diogo, M. M., Clark, D. S., *et al.* 2009. High-throughput cellular microarray platforms: Applications in drug discovery, toxicology and stem cell research, *Trends Biotechnol* 27:342–349.

Fernandes, T. G., Kwon, S. J., Bale, S. S., *et al.* 2010. Three-dimensional cell culture microarray for high-throughput studies of stem cell fate, *Biotechnol Bioeng* 106(1):106–118.

Ffrench-Constant, D. and Colognato, H. 2004. Integrins: Versatile integrators of extracellular signals, *Trends Cell Biol* 14:678–686.

Fiegel, H. C., Kaufmann, P. M., Bruns, H. *et al.* 2008. Hepatic tissue engineering: From transplantation to customized cell-based liver directed therapies from the laboratory, *J Cell Mol Med* 12:56–66.

Fisher, O. Z., Khademhosseini, A., Langer, R. and Peppas, N. A. 2010. Bioinspired materials for controlling stem cell fate, *Acc Chem Res* 43(3):419–428.

Flaim, C. J., Chien, S. and Bhatia, S. N. 2005. An extracellular matrix for probing cellular differentiation, *Nature Methods* 2:119–125.

Flaim, C. J., Teng, D., Chien, S. and Bhatia, S. N. 2008. Combinatorial signalling microenvironments for studying stem cell fate, *Stem Cells Dev* 17:29–30.

Fossum, M. and Nordenskjöld, A. 2010. Tissue-engineered transplants for the treatment of severe hypospadias, *Horm Res Paediatr* 73:148–152.

Frimat, J. P., Sisnaiske, J., Subbiah, S., *et al.* 2010. The network formation assay: A spatially standardized neurite outgrowth analytical display for neurotoxicity screening, *Lab Chip* 10(6):701–709.

Furth, M. E. and Atala, A. 2009. Stem cell sources to treat diabetes. *J Cell Biochem* 1:507–511.

Gadegaard, N., Mosler, S. and Larsen, N. B. 2003. Biomimetic polymer nanostructures by injection moulding, *Macromol Mater Eng* 288:76–83.

Gautier, S., Xhauflaire-Uhoda, E., Gonry, P. *et al.* 2008. Chitin-glucan, a natural cell scaffold for skin moisturization and rejuvenation. *Int J CosmetSci* 30(6):459–469.

Gerecht, S., Brudick, J. A., Ferreira, L. S., *et al.* 2007. Hyaluronic acid hydrogel for controlled self-renewal and differentiation of human embryonic stem cells, *PNAS* 104:11298–11303.

Gersh, B. J., Simari, R. D., Behfar, A., *et al.* 2009. Cardiac cell repair therapy: A clinical perspective, *Mayo Clin Proc* 84:876–892.

Goldman, S. 2005. Stem and progenitor cell-based therapy of the human central nervous system, *Nat Biotechnol* 23:862–871.

Gonzales, C. and Pedrazzini, T. 2009. Progenitor cell therapy for heart disease, *Exp Cell Res* 1:3077–3085.

Gopalakrishnan, S. M., Moreland, R. B., Kofron, J. L., *et al.* 2003. A cell-based microarrayed compound screening format for identifying agonists of G-protein coupled receptors, *Anal Biochem* 321:192–201.

Granero-Molto, F., Weis, J. A., Longobardi, L., *et al.* 2008. Role of mesenchymal stem cells in regenerative medicine: Application to bone and cartilage repair, *Expert Opin Biol Ther* 8:255–268.

Grayson, W. L., Martens, T. P., Eng, G. M., *et al.* 2009. Biomimetic approach to tissue engineering, *Semin Cell Dev Biol* 20:665–673.

Habich, A., Jurga, M., Markiewicz, I., *et al.* 2006. Early appearance of stem/progenitor cells with neural-like characteristics in human cord blood mononuclear fraction cultured *in vitro*, *Exp Hematol* 34(7):914–925.

Hejcl, A., Sedý, J., Kapcalová, M., *et al.* 2010. HPMA-RGD hydrogels seeded with mesenchymal stem cells improve functional outcome in chronic spinal cord injury, *Stem Cells Dev* 19(10):1535-46. Madigan, N. N., McMahon, S., O'Brien, T., *et al.* 2009. Current tissue engineering and novel therapeutic approaches to axonal regeneration following spinal cord injury using polymer scaffolds, *Respir Physiol Neurobiol* 30:183–199.

Hill, E., Boontheekul, T. and Mooney, D. J. 2006. Regulating activation of transplanted cells controls tissue regeneration, *Proc Natl Acad Sci U S A* 103:2494–2499.

Hodgkinson, T., Yuan, X. F. and Bayat A. 2009. Adult stem cells in tissue engineering, *Expert Rev Med Devices* 6:621-640.

Huang, S. and Fu, X. 2010. Naturally derived materials-based cell and drug delivery systems in skin, regeneration, *J Control Release* 3:149–159.

Hung, P. J., Lee, P. J., Sabounchi, P., *et al.* 2005. Continuous perfusion microfluidic cell culture array for high-throughput cell-based assays, *Biotechnol Bioeng* 89:1–8.

Ingber, D. 2006. Mechanical control of tissue morphogenesis during embryological development, *Int J Dev Bio* 50:255–266.

Iwasa, J., Engebretsen, L., Shima, Y., *et al.* 2009. Clinical application of scaffolds for cartilage tissue engineering knee, *Surg Sports Traumatol Arthrosc* 17(6):561–577.

Jurga, M., Lipkowski, A. W., Lukomska, B., *et al.* 2009. Generation of functional neural artificial tissue from human umbilical cord blood stem cells, *Tissue Eng Part C Methods* 15:365–372.

Karp, J. M. and Langer, R. 2007. Development and therapeutic applications of advanced biomaterials, *Curr Opin Biotechnol* 18(5):454–459.

Karp, J. M., Yeh, J., Eng, G., *et al.* 2007. Controlling size, shape and homogeneity of embryoid bodies using poly(ethylene glycol) microwells, *Lab Chip* 7:786–794.

Keller, G. 2005. Embryonic stem cell differentiation: Emergence of a new era in biology and medicine, *Genes Dev* 19:1129–1155.

Keselowsky, B. G., Collard, D. M. and Garcia, A. J. 2003. Surface chemistry modulates fibronectin conformation and directs integrin binding and specificity to control cell adhesion, *J Biomed Mater Res A* 66:247–259.

Keselowsky, B. G., Collard, D. M. and Garcia, A. J. 2004. Surface chemistry modulates focal adhesion composition and signaling through changes in integrin binding, *Biomaterials* 25:5947–5954.

Keselowsky, B. G., Collard, D. M. and Garcia, A. J. 2005. Integrin binding specificity regulates biomaterial surface chemistry effects on cell differentiation, *Proc Natl Acad Sci USA* 102:5953–5957.

Khademhossein, A. 2005. Chips to hits: Microarray and microfluidic technologies for high-throughput analysis and drug discovery, *Expert Rev Mol Diagn* 5:843–846.

Khetani, S. R. and Bhatia, S. N. 2008. Microscale culture of human liver cells for drug development, *Nat Biotechnol* 26:120–126.

Kloxin, A. M., Kasko, A. M., Salinas, C. N., *et al.* 2009. Photodegradable hydrogels for dynamic tuning of physical and chemical properties, *Science* 3:59–63.

Kobayashi, N. 2008. Bioartificial pancreas for the treatment of diabetes, *Cell Transplant* 17:11–17.

Kobayashi, N. 2009. Life support of artificial liver: Development of a bioartificial liver to treat liver failure, *J Hepatobiliary Pancreat Surg* 16:113–117.

Koepsel, J. T. and Murphy, W. L. 2009. Patterning discrete stem cell culture environments via localized self-assembled monolayer replacement, *Langmuir* 25:12825–12834.

Kollar, K., Cook, M. M., Atkinson, K., *et al.* 2009. Molecular mechanisms involved in mesenchymal stem cell migration to the site of acute myocardial infarction, *Int J Cell Biol* pp. 1–8.

Koning, M., Harmsen, M. C., van Luyn, M. J., *et al.* 2009. Current opportunities and challenges in skeletal muscle tissue engineering, *J Tissue EngRegen Med* 3:407–415.

Kroon, E., Martinson, L. A., Kadoya, K., *et al.* 2008. Pancreatic endoderm derived from human embryonic stem cells generates glucose-responsive insulin-secreting cells *in vivo*, *Nat Biotechnol* 26:443–452.

Kucia, M., Halasa, M., Wysoczynski, M., *et al.* 2007. Morphological and molecular characterization of novel population of CXCR4+ SSEA-4+ Oct-4+ very small embryonic-like cells purified from human cord blood: Preliminary report. *Leukemia* 21(2):297–303.

Kuschel, C., Steuer, H., Maurer, A. N., *et al.* 2006. Cell adhesion profiling using extracellular matrix protein microarrays, *BioTechniques* 40:523–530.

Lee, O. K., Kuo, T. K., Chen, W. M., *et al.* 2004. Isolation of multipotent mesenchymal stem cells from umbilical cord blood, *Blood* 103:1669–1675.

Lee, M. Y., Kumar, R. A., Sukumaran, S. M., *et al.* 2008. Three-dimensional cellular microarray for high-throughput toxicology assays, *Proc Natl Acad Sci U S A* 8:59–63.

Lees, J. G., Lim, S. A., Croll, T., *et al.* 2007. Transplantation of 3D scaffolds seeded with human embryonic stem cells: Biological features of surrogate tissue and teratoma-forming potential, *Regenerative Med* 2:289–300.

Li, Y. J., Chung, E. H., Rodriguez, R. T., *et al.* 2006. Hydrogels as artificial matrices for human embryonic stem cell self-renewal, *J Biomed Mater Res* 79:1–5.

Li, X., Hou, S., Feng, X., *et al.* 2009. Patterning of neural stem cells on poly(lactic-co-glycolic acid) film modified by hydrophobin, *Colloids and Surfaces B: Biointerfaces* 74:370–374.

Liao, H. and Zhou, G. Q. 2009. Development and progress of engineering of skeletal muscle tissue, *Tissue Eng Part B Rev* 15:319–331.

Lin, C. M., Jiang, T. X., Baker, R. E., *et al.* 2009. Spots and stripes: Pleomorphic patterning of stem cells via p-ERK-dependent cell chemotaxis shown by feather morphogenesis and mathematical simulation, *Dev Biol* 334:369–382.

Lin, C. H., Hsu, S. H., Huang, C. E., *et al.* 2009. A scaffold-bioreactor system for a tissue-engineered trachea, *Biomaterials* 30:4117–4126.

Liu, W. F. and Chen, C. S. 2007. Cellular and multicellular form and function, *Advanced Drug Delivery Reviews* 59:1319–1328.

Luo, W., Jones, S. R. and Yousaf, M. 2008. Geometric control of stem cell differentiation rate on surfaces, *Langmuir* 24:12129–12133.

Lutolf, M. P., Gilbert, P. M. and Blau, H. M. 2009. Designing materials to direct stem-cell fate, *Nature* 462(7272):433–441.

Macchiarini, P., Jungebluth, P., Go, T., *et al.* 2008. Clinical transplantation of a tissue-engineered airway, *Lancet* 13:2023–2030.

Machado, C. B., Ventura, J. M. G., Lemos, A. F., *et al.* 2007. Three chitosan-gelatin-chondroitin porous scaffold improves osteogenic differentiation of mesenchymal stem cells, *Biomed Mater* 2:124–131.

Magnusson, A. K., Linderholm, P., Vieider, C., *et al.* 2008. Surface protein patterns govern morphology, proliferation, and expression of cellular markers but have no effect on physiological properties of cortical precursor cells, *J Neurosci Res* 86:2363–2375.

Mansbridge, J. N. 2009. Tissue-engineered skin substitutes in regenerative medicine. *Curr Opin Biotechnol* 20:563–567.

Markowicz, M., Heitland, A., Steffens, G. C., *et al.* 2005. Effects of modified collagen matrices on human umbilical vein endothelial cells *Int J Artif Organs* 28(12):1251–1258.

Marchini, G., Pedrotti, E., Pedrotti, M. *et al.* 2011. Long-term effectiveness of autologous cultured limbal stem cell grafts in patients with limbal stem cell deficiency due to chemical burns, *Clin Experiment Ophthalmol* doi:10.111/j.1442-9071.2011.02609.

McGuckin, C. P., Forraz, N., Baradez, M. O., *et al.* 2005. Production of stem cells with embryonic characteristics from human umbilical cord blood, *Cell Prolif* 38:245.

Melero-Martin, J. M., De Obaldia, M. E., Kang, S. Y., *et al.* 2008. Engineering robust and functional vascular networks *in vivo* with human adult and cord blood-derived progenitor cells, *Circ Res* 103:194–202.

Metallo, C. M., Azarin, S. M., Ji, L., *et al.* 2008. Engineering tissue from human embryonic stem cells, *J Cell Mol Med* 12:709–729.

Moeller, H. C., Mian, M. K., Shrivastava, S., *et al.* 2008. A microwell array system for stem cell culture, *Biomaterials* 29:752–763.

Moors, M., Rockel, T. D., Abel, J. *et al.* 2009. Human neurospheres as three-dimensional cellular systems for developmental neurotoxicity testing, *Environ Health Perspect* 117(7):1131–1138.

Morin, F., Nishimura, N., Griscom, L., *et al.* 2006. Constraining the connectivity of neuronal networks cultured on microelectrode arrays with microfluidic techniques: A step towards neuron-based functional chips, *Biosens Bioelectron* 21:1093–1100.

Nakajima, M., Ishimuro, T., Kato, K., *et al.* 2007. Combinatorial protein display for the cell-based screening of biomaterials that direct neural stem cell differentiation, *Biomaterials* 28:1048–1060.

Nelson, C. M., Raghavan, S., Tan, J. L., *et al.* 2003. Degradation of micropatterned surfaces by cell-dependent and -indipendent processes, *Langmuir* 19:1493–1499.

Neuss, S., Apel, C., Buttler, P., *et al.* 2008. Assessment of stem cell/biomaterial combinations for stem cell-based tissue engineering, *Biomaterials* 29:302–313.

Ochsner, M., Dusseiller, M. R., Grandin, H. M., *et al.* 2007. Micro-well arrays for 3D shape control and high resolution analysis of single cells, *Lab Chip* 7:1074–1077.

Oh, J., Recknor, J. B., Recknor, J. C., *et al.* 2009. Soluble factors from neocortical astrocytes enhance neuronal differentiation of neural progenitor cells from adult rat hippocampus on micropatterned polymer substrates, *J Biomed Mater Res A* 91(2):575–585.

Omori, K., Tada, Y., Suzuki, T., *et al.* 2008. Clinical application of in situ tissue engineering using a scaffolding technique for reconstruction of the larynx and trachea, *Ann Otol Rhinol Laryngol* 117:673–678.

Ostuni, E., Chen, C. S., Ingber, D. E., *et al.* 2001. Selective deposition of proteins and cells in arrays of microwells, *Langmuir* 17:2828–2834.

Park, K. I., Teng, Y. D. and Snyder, E. Y. 2002. The injured brain interacts reciprocally with neural stem cells supported by scaffolds to reconstitute lost tissue, *Nat Biotechnol* 20:1111–1117.

Phillips, J. E., Petrie, T. A., Creighton, F. P., and García, A. J. 2010. Human mesenchymal stem cell differentiation on self-assembled monolayers presenting different surface chemistries, *Acta Biomaterialia* 6:12–20.

Pluchino, S., Zanotti, L., Deleidi, M., *et al.* 2005. Neural stem cells and their use as therapeutic tool in neurological disorders, *Brain Res Brain Res Rev* 48:211–219.

Pompe, T., Keller, K., Mothes, G., *et al.* 2007. Surface modification of poly(hydroxybutyrate) films to control cell-matrix adhesion *Biomaterials* 28(1):28–37.
Pouton, C. W. and Haynes, J. M. 2005. Pharmaceutical applications of embryonic stem cells, *Adv Drug Deliv Rev* 57(13):1918–1934.
Qian, L. and Saltzman, W. M. 2004. Improving the expansion and neuronal differentiation of mesenchymal stem cells through culture surface modification, *Biomaterials* 25:1331–1337.
Rama, P., Matsuka, S., Puganoni, G. *et al.* 2010. Limbal stem-cell therapy and long-term corneal regeneration, *N Engl J Med* 363(2):147–155.
Ravula, S. K., McClain, M. A., Wang, M. S., *et al.* 2006. A multi electro demicrocompartment culture platform for studying signal transduction in the nervous system, *Lab Chip* 6:1530–1536.
Recknor, J. B., Sakaguchi, D. S. and Mallapragada, S. K. 2006. Directed growth and selective differentiation of neural progenitor cells on micropatterned polymer substrates, *Biomaterials* 27:4098–4108.
Reffelmann, T. and Kloner, R. A. 2009. Intracoronary blood- or bone marrow-derived cell transplantation in patients with ischemic heart disease, *Regen Med* 4:709–719.
Richter, W. 2009. Mesenchymal stem cells and cartilage in situ regeneration, *J Intern Med* 266:390–405.
Rosenthal, A., Macdonald, A. and Voldman, J. 2007. Cell patterning chip for controlling the stem cell microenvironment, *Biomaterials* 28:3208–3216.
Roth, C. C. 2010. Urologic tissue engineering in pediatrics: From nanostructures to bladders, *Pediatr Res* 67(5):509–513
Ruiz, S. A. and Chen, C. S. 2007. Microcontact printing: A tool to pattern, *Soft Matter* 3:1–11.
Ruiz, S. A. and Chen, C. S. 2008. Emergence of patterned stem cell differentiation within multicellular structures, *Stem Cells* 26:2921–2927.
Ruiz, A., Ceriotti, L., Buzanska, L., *et al.* 2007. Controlled micropatterning of biomolecules for cell culturing, *Microelectronic Engineering* 84:1733–1736.
Ruiz, A., Buzanska, L., Gilliland, D., *et al.* 2008. Micro-stamped surfaces for the patterned growth of neural stem cells, *Biomaterials* 29:4766–4774.
Ruiz, A., Zychowicz, M., Buzanska, L., *et al.* 2009. Single stem cell positioning on polylysine and fibronectin microarrays, *Micro Nanosyst* 1:50–56.
Saltzman, W. M. 2000. Principles of tissue engineering. Cell interactions with polymers, in Lanza, R. P., Langer, R. S., Vacanti, J., (eds.), *Principles of Tissue Engeneering*. 2nd Edition: Academic Press, San Diego. pp. 7–27.
Sasaki, J. E., Shimizu, T., Masuda, S., *et al.* 2009. Mass preparation of size-controlled mouse embryonic stem cell aggregates and induction of cardiac differentiation by cell patterning method, *Biomaterials* 30:4384–4389.
Scadden, D. T. 2006. The stem-cell niche as an entity of action, *Nature* 441:1075–1079.
Schneider, R. K., Puellen, A., Kramann, R., *et al.* 2010. The osteogenic differentiation of adult bone marrow and perinatal umbilical mesenchymal stem cells and matrix remodelling in three-dimensional collagen scaffolds, *Biomaterials* 31:467–480.
Serfozo, P., Schlarman, M. S., Pierret, C., *et al.* 2006. Selective migration of neuralized embryonic stem cells to stem cell factor and media conditioned by glioma cell lines, *Cancer Cell Int* 25:6–10.

Shimizu, T., Sekine, H., Yamato, M., *et al.* 2009. Cell sheet-based myocardial tissue engineering: New hope for damaged heart rescue, *Curr Pharm Des* 15:2807–2814.

Silva, G. A., Czeisler, C., Niece, K. L., *et al.* 2004. Selective differentiation of neural progenitor cells by high-epitope density nanofibers, *Science* 303:1352–1355.

Singhvi, R., Kumar, A., Lopez, G. P., *et al.* 1994. Engineering cell shape and function, *Science* 264:696–698.

Soen, Y., Mori, A., Palmer, T. D. and O'Brown, P. 2006. Exploring the regulation of human neural precursor cell differentiation using arrays of signaling microenvironments, *Molecular Systems Biology* 2:E1–E14.

Stillaert, F., Di Bartolo, C., Hunt, J., *et al.* 2007. Adipose precursor cells seeded on hyaluronic scaffolds: A pilot clinical trial. Ibrahim, N. A., Abu Osman, J., Usman and Kadri, N. A. (Eds.): *Biomed 06, IFMBE Proceedings* 15:688–691.

Tedesco, F. S., Dellavalle, A., Diaz-Manera, J., *et al.* 2010. Repairing skeletal muscle: Regenerative potential of skeletal muscle stem cells, *J Clin Invest* 120:11–19.

Thonhoff, J. R., Lou, D. I., Jordan, P. M., *et al.* 2008. Compatibility of human fetal neural stem cells with hydrogel biomaterials *in vitro*, *Brain Res* 2:42–51.

Tian, H., Bharadwaj, S., Liu, Y., *et al.* 2010. Myogenic differentiation of human bone marrow mesenchymal stem cells on a 3D nano fibrous scaffold for bladder tissue engineering, *Biomaterials* 31(5):870–877.

Tolosa, J. N., Park, D. H., Eve, D. J., *et al.* 2010. Mankind's first natural stem cell transplant, *J Cell Mol Med* 14(3):488–495

Trosko, J. E. and Chang, C. C. 2009. Factors to consider in the use of stem cells for pharmaceutic drug development and for chemical safety assessment, *Toxicology* 270(1):18–34

Troyer, D. L. and Weiss, M. L. 2008. Wharton's jelly-derived cells are a primitive stromal cell population, *Stem Cells* 26:591–599.

Undale, A. H., Westendorf, J. J., Yaszemski, M. J., *et al.* 2009. Mesenchymal stem cells for bone repair and metabolic bone diseases, *Mayo Clin Proc* 84:893–902.

Vunjak-Novakovic, G. and Scadden, D. T. 2011. Biomimetic platforms for human stem cell research, *Cell Stem Cell* 8(3):252–261.

Wada, K., Taniguchi, A., Kobayashi, J., *et al.* 2008. Live cells-based cytotoxic sensorchip fabricated in a microfluidic system, *Biotechnol Bioeng* 99:1513–1517.

Walker, P. A., Aroom, K. R., Jimenez, F., *et al.* 2009. Advances in progenitor cell therapy using scaffolding constructs for central nervous system injury, *Stem Cell Rev Rep* 5:283–300.

Wang, M., Yang, Y., Yang, D., *et al.* 2009. The immunomodulatory activity of human umbilical cord blood-derived mesenchymal stem cells *in vitro*, *Immunology* 126:220–232.

Whitesides, G. M., Ostumi, E., Takayama, S., Jiang, X. Y. and Ingber, D. E. 2001. Soft lithography in biology and biochemistry, *Annu Rev Biomed Eng* 3:335–373.

Xie, J., Willerth, S. M., Li, X., *et al.* 2009. The differentiation of embryonic stem cells seeded on electrospunnanofibers into neural lineages, *Biomaterials* 30:354–362.

Xu, R. H., Chen, X., Li, D. S., *et al.* 2002. BMP4 initiates human embryonic stem cell differentiation to trophoblast, *Nat Biotechnol* 20:1261–1264.

Xu, R. H., Peck, R. M., Li, D. S., *et al.* 2005. Basic FGF and suppression of BMP signaling sustain undifferentiated proliferation of human ES cells, *Nat Methods* 2:185–190.

Yang, F., Mei, Y., Langer, R., *et al.* 2009. High throughput optimization of stem cell microenvironments, *Comb Chem High Throughput Screen* 12(6):554–561.

Yi, B. A., Wernet, O. and Chien, K. R. 2010. Pregenerative medicine: Developmental paradigms in the biology of cardiovascular regeneration, *J Clin Invest* 120:20–28.

Yoshikawa, T., Uchimura, E., Kishi, M., *et al.* 2004. Transfection microarray of human mesenchymal stem cells and on-chip siRNA gene knockdown, *J. Control. Release* 96:227–232.

Zhao, L., Burguera, E. F., Xu, H. H., *et al.* 2010. Fatigue and human umbilical cord stem cell seeding characteristics of calcium phosphate-chitosan-biodegradable fiber scaffolds, *Biomaterials* 31:840–847.

Zhao, X., Kim, J., Cezar, C. A., *et al.* 2011. Active scaffolds for on-demand drug and cell delivery, *Proc Natl Acad Sci U S A* 108(1):67–72.

Zhao, L., Weir, M. D. and Xu H. H. K. 2010. Human umbilical cord stem cell encapsulation in calcium phosphate scaffolds for bone engineering, *Biomaterials* 31(14):3848–3857

Zheng, M. H., Ye, C., Braddock, M., *et al.* 2010. Liver tissue engineering: Promises and prospects of new technology, *Cytotherapy* Jan 7. [Epub ahead of print].

Ziauddin, J. and Sabatini, D. M. 2001. Microarrays of cells expression defined cDNAs, *Nature* 411:107–110.

Zychowicz, M., Mehn, D., Ruiz, A., *et al.* 2011. Proliferation capacity of cord blood derived neural stem cell line on different micro-scale biofunctional domains, *Acta Neurobiol Exp (Wars)* 71(1):12–23.

Human Umbilical Cord Blood-Derived Cells: Implications in Experimental and Clinical Therapy of CNS Disorders

10

Barbara Lukomska, Leonora Buzanska, and Krystyna Domanska-Janik

Cell replacement therapy is a widely discussed and relatively novel form of medical treatment in regard to various diseases that are characterized by the loss of functional cells. Such a therapy would be especially important in the treatment of central nervous system deficits with focal or disperse cell loss as the main pathological mechanism. Although brain-derived NSCs have received the most attention as a source of cells for neural cell replacement, the other sources of non-CNS–derived somatic stem cells capable of terminal neural differentiation are also the subjects of active investigation. Such alternative sources of NSCs can include bone marrow, peripheral blood, skin, adipose tissue or umbilical cord blood and tissues. In contrast to the majority of cells isolated from adults, stem cells in cord blood are situated at the border between fetal and newborn tissues, thus they are less mature, and have greater potential for proliferation and differentiation. In addition to this biological "superiority", cord blood is abundant and routinely harvested with little risk to the donor, as viral infection is a rare exception. In the past few years, many research papers have reported on how subsets of cord blood stem cells, under defined conditions, differentiate into neurons, astrocytes, and oligodendroglia. Equally impressive and just as frequent, are reports of the myriad of trophic factors secreted by cord blood cells for initiating and maintaining the process of CNS repair. Several issues in basic and experimental research must be resolved before stem cell-based therapies of neurological diseases can be tested in patients. It is preferred that the cells considered for transplantation into patients are of human origin. This presents a predicament for testing the human cells in animals. Xenotransplantation of human cells into rodents is a widely used research paradigm that often produces encouraging results in terms of reduction of behavioral deficits, despite of the possibility to accelerate immuno-dependent graft degeneration and rejection. This chapter provides an overview of the current knowledge about the origin and fate of cord blood progenitor/stem cells *in vitro*. Special attention is paid to the use of different types of cells derived from human cord blood in the preclinical and experimental medical treatment of neurological disorders. It also discusses the current theories of the possible origin of stem cells in human cord blood, as well as the proposed molecular mechanisms underlying their neuronal differentiation.

10.1. Introduction

Cell replacement therapy is a widely discussed and relatively novel form of medical treatment in various diseases characterized by the loss of functional cells. Traumatic and ischemic injuries to the central nervous system are attributed to the progressive degeneration of neurons and glial cells. Some interventions during the acute phase of the insult, such as thrombolytic agents, have been recognized to improve the outcome of patients. However, since there is no treatment to halt neurodegeneration, stem cell-based approaches could be used therapeutically to restore CNS function. This might be achieved by the replacement of lost cells and/or by inducing endogenous neurogenesis in the adult brain. Aside from these mechanisms, grafted stem cells could encourage functional improvement by releasing different molecules that are either neuroprotective, or modulate inflammation.

Based on several studies, neural stem cells (NSCs) are regarded as the ideal cell type for the repair of damaged neural tissue because they most closely resemble the cell type that is lost and may better respond to the differentiating and guidance cues from the local microenvironment. Although brain-derived NSCs have received the most attention as a source of cells for neural cell replacement, other sources of non-CNS–derived somatic stem cells capable of terminal neural differentiation are also subject to active investigation. The alternative sources of NSCs include bone marrow (Jiang *et al.*, 2003; Hermann *et al.*, 2004), peripheral blood (Zhao *et al.*, 2003), umbilical cord blood (Sanchez-Ramos *et al.*, 2001; Buzanska *et al.*, 2002; Kögler *et al.*, 2004; McGuckin *et al.*, 2004; Ali *et al.*, 2009), amniotic fluid (Tsai *et al.*, 2006; De Coppi *et al.*, 2007), Wharton's jelly (Mitchell *et al.*, 2007; Wang *et al.*, 2004), skin (Joannides *et al.*, 2004; Metcalfe *et al.*, 2008) or adipose tissue (Safford *et al.*, 2002; Kim *et al.*, 2007). For tissue regeneration, these stem cells are either pre-differentiated to the required neural tissue type before application or they are applied directly without prior differentiation, following the concept that signals in the injured tissue will induce stem cell differentiation *in situ*. In addition to the differentiation potential of the stem/progenitor cells, other parameters also have an impact on their suitability for therapy. These include cell numbers, safety and the low immunogenicity of transplanted cells.

During the past decade, the human umbilical cord has been of great interest as a novel valuable source for stem cells with several unique characteristics (Sanberg *et al.*, 2005; Zandonella, 2005; McGuckin *et al.*, 2008; Copeland *et al.*, 2009). Reports of cord blood stem cells turning into brain cells, both *in vitro* and *in vivo*, have been consistent and incessant. Many studies have shown how subsets of cord blood stem cells differentiate, under defined conditions, into neurons, astrocytes, and oligodendrocytes. Equally impressive and just as common, are reports of the myriad of trophic factors secreted by cord blood cells that can initiate and maintain the process of CNS repair (Borlongan *et al.*, 2004; Fan *et al.*, 2005). In contrast to bone marrow and other adult stem cells, those derived from cord blood are less mature and have a higher potential to proliferate. This has been associated with an extended lifespan and longer telomeres (Kögler *et al.*, 2004; Lindvall *et al.*, 2004). In addition to this biological superiority, cord blood is abundantly available and routinely harvested with little risk to the donor, with viral infection as a rare exception.

Taking into consideration the above facts, in addition to freshly isolated neural progenitors, this laboratory has successfully derived a non-immortalized cell line with properties of neural stem cells (NSCs) from human umbilical cord blood (Buzanska *et al.*, 2006). This cell line serves as a useful prototype for the ideal transplantable material and, based on our *in vitro* studies, it has a strong potential for clinical use (Jurga *et al.*, 2006; Domanska-Janik *et al.*, 2008; Jurga *et al.*, 2009; Sypecka *et al.*, 2009). This chapter provides an overview of the current knowledge relating to the characterization, and experimental and clinical uses, of different types of stem and precursor cells derived from human cord blood. The theories regarding the possible origin of stem cells in human cord blood, and the molecular fingerprints underlying their neuronal differentiation, are also discussed.

10.2. Different Types of Stem and Precursor Cells in Human Cord Blood

For over 3 decades, human umbilical cord blood (HUCB) has been known to harbor hematopoietic stem cells that are restricted to differentiate only into the cells that are typical for blood tissue, and are used in hematopoietic reconstruction (Ende and Ende, 1972). However, the specific postnatal ontogenic position of HUCB in fetal tissue encouraged scientists to search for subpopulations of stem and progenitor cells with a flexible phenotype, even with embryonic-like properties, that are able to differentiate into other lineages. Indeed, many reports starting from early 2000 have shown that a mononuclear cell fraction from HUCB, in addition to hematopoietic stem cells, contains different non-hematopoietic (CD45−, CD34−) adherent stem cell populations. These include cord blood mesenchymal stromal cells (CB MSCs) with similar characteristic to MSCs from the bone marrow (Yang *et al.*, 2004) and pluripotent, embryonic-like stem cells that are able to differentiate into cell types derived from all three germ layers and express human embryonic stem cell markers. The latter was reported by several investigators and named unrestricted somatic stem cells (USSC) (Kögler *et al.*, 2004), hematopoietic lineage-negative (LinNeg) stem cells (McGuckin *et al.*, 2005), early aggregating non-hematopoietic mononuclear cells (Habich *et al.*, 2006) and very small embryonic-like stem cells (VSELs) (Kucia *et al.*, 2007).

Various procedures, mainly based on the immunorecognition of surface markers, are employed to isolate diverse stem cell populations from human cord blood. Hematopoietic stem cells are characterized by the expression of surface proteoglycan CD34 antigen (only about 5% of the whole population of mononuclear cells) and their isolation is based on immunomagnetic or immunofluorescent cell sorting. Clonally expanded hematopoietic $CD34^+CD45^+CD133^+CD38^+$ give rise to T- and B-lymphoid, myeloid, and endothelial cells (Ramos *et al.*, 2010). However endothelial progenitors have also been derived from primary cord blood $AC133^+$ cells (Janic *et al.*, 2010), mesodermal cord blood progenitor cells (Trombi *et al.*, 2009; Xu *et al.*, 2010), and lineage-negative (LinNeg) fractions of cord blood mononuclear cells (Chua *et al.*, 2009). Umbilical cord blood (UCB)-derived very small embryonic/epiblast-like stem cells (VSELs) are a very interesting subpopulation, although they are especially difficult to isolate due to their small size and scarcity. To preserve these cells in final UCB preparations, a three-step isolation protocol has been proposed (Zuba-Surma *et al.*, 2010), that allows the recovery of approximately 60% of the initial number of $Lin^-/CD45^-/CD133^+$ UCB-VSELs.

Recently two different molecular methods were described that distinguish between non-hematopoietic mononuclear stem cell populations in human cord blood. The first procedure is based on the expression profile of DLK-1, δ-like 1/preadipocyte factor 1 (DLK-1/PREF1) that is present in multiple embryonic tissues. Pluripotent, embryonic-like stem cells express DLK-1 to a high or moderate level, while DLK-1 is silenced in the already committed MSC fraction (Kluth *et al.*, 2010). The second method was used to distinguish functionally different pluripotent and MSC populations derived from the non-hematopoietic fraction of mononuclear cells, via the use of the so-called "biological fingerprint" based on HOX code (Liedtke *et al.*, 2010). Cord blood-derived MSCs, similar to bone marrow MSCs, are HOX-positive whereas pluripotent USSCs, that resemble the H9 embryonic stem cell line, appear HOX-negative. In parallel, a new method for the simultaneous expansion and harvest of hematopoietic stem cells, and mesenchymal stem cells derived from umbilical cord blood, was established (Kedong *et al.*, 2010). This is important for future clinical applications since, as shown in both animal and human studies, umbilical cord blood (UCB)-derived mesenchymal stem cells (MSCs) facilitate the engraftment of human hematopoietic stem cells when transplanted simultaneously (Yim *et al.*, 2010). Recently, Kogler and coworkers generated USSCs from fresh cord blood according to good manufacturing practices (GMP), fulfilling the safety requirements for clinical grade applications and allows cell expansion to the clinical scale (Aktas *et al.*, 2010). It was shown, that these USSC-derived multipotent mesenchymal stem cell lines significantly support proliferation of HSCs in an *in vitro* feeder layer assay (Jeltsch *et al.*, 2010).

Even though the existence and isolation of stem cells with pluripotent or neural characteristics have been well documented, human umbilical cord blood was shown to be suitable for the genetic induction of pluripotent stem cells (iPS). For that purpose, different mononuclear fractions were used. Cord blood-derived cells positive for endothelial markers (CD31, CD146) and negative for monocyte markers (CD45, CD14) were induced to pluripotency with lentiviral vectors expressing Oct4, Sox2, Nanog, and LIN28 (Haase *et al.*, 2009). Enhancing pluripotency in human cord blood, unrestricted somatic stem cells was done via the retroviral delivery of transcription factors Oct4, Sox2, Klf4, and c-Myc (Zaehres *et al.*, 2010). A reduced set of transforming factors, containing only Oct4 and Sox2 and applied to immunoselected CD133 positive cells, have been shown to be sufficient for obtaining iPS clones (Giorgetti *et al.*, 2010). The induction of pluripotency was also successful when applied to the hematopoietic CD34 positive cells of healthy donors or patients with acquired blood disorders (Ye *et al.*, 2009). For the $CD34^+$ fraction, it was also shown that constitutive inhibition of p53 expression results in an increased number of iPS cell colonies (Takenaka *et al.*, 2010). Due to possible future therapeutic applications, our group pays special attention to the generation of iPS cells from cord blood in viral-free, xeno-free and physiological oxygen tension conditions (Szablowska-Gadomska *et al.*, 2011, and unpublished). The induction of pluripotency in cord blood cells with three reprogramming proteins, followed by epigenetic modifications, resulted in a partial reprogramming event (Wong *et al.*, 2010). This could indicate that the transgene-free generation of iPS from cord blood is possible.

10.3. *In Vitro* Neurogenic Potential of Human Umbilical Cord Blood Stem Cells

In early 2000, the challenge facing this and other research groups was to isolate an ethically uncontroversial human somatic stem cell population suitable for long-lasting *in vitro* expansion and attaining neural features in both *in vitro* and *in vivo* experimental studies. Human umbilical cord blood, because of its cellular immaturity and fetal origin, was considered the best candidate. For the first time, EGF-based expansion of a $CD34^-$ mononuclear fraction allowed for the isolation of neural progenitors from human umbilical cord blood (Buzanska *et al.*, 2002, 2001). Furthermore, it was shown that neural progenitors can be propagated from an unselected whole mononuclear cell fraction (Sanchez-Ramos *et al.*, 2001) or different subpopulations of cord blood. These include hematopoietic lineage negative (LinNeg) fractions (McGuckin *et al.*, 2004) as well as the hematopoietic (CD 34^+, CD 133^+) cells (Jang *et al.*, 2004) and cord blood-derived mesenchymal stem cells — adherent cells expressing MSCs-related antigens such as SH2, CD13, CD29, and ASMA (Jeong *et al.*, 2004). Neurally directed cells were also reported from so-called FSFl (grown in the presence of FGF4, SCF, and Flt-3 ligand) cell populations, despite of their origin from CD45 positive fractions of cord blood (Rogers *et al.*, 2007). The exact origin of all of these CB-derived stem/progenitor subsets is still unclear and controversial. However, from the practical viewpoint for both experimental and therapeutic applications, the challenge is to obtain a neural progenitor population that is suitable for adequate expansion.

The first expanding, indefinitely growing non-transformed HUCB-NSC line was derived by this group from a mononuclear $CD34^-$ population (Buzanska *et al.*, 2006) through repeated passages of floating, round-shaped, nestin-positive and clonogenic cells. The phenotypic and functional identity of this HUCB-NSC line, their stability, clonogenity, ability to form neurospheres and differentiate into all neural cell types (Buzanska *et al.*, 2006; Jurga *et al.*, 2006; Buzanska *et al.*, 2006b) were confirmed in further studies. Critical *in vitro* analysis involved contrast-phase microscopic observations, and gene expression studies using cDNA microarray and quantitative RT-PCR followed by the immunocytochemical evaluation of protein expression. The non-transformed character of the HUCB-NSC cell line was proved by karyotypic analysis and transplantations into immunodeficient non-obese diabetic/severe combined immunodeficient (NOD/SCID) mice. After transplantation into NOD/SCID, mice tumor formation was not observed. In addition, enhanced expression of the classical proto-oncogenes in HUCB-NSCs was not found when compared to the starting population of HUCB mononuclear cells. The stability of the HUCB-NSC cell line was confirmed by the lack of significant differences in the growth rates, clonogenicity, and differentiation potency between the early and late passages. Single cells isolated from the long-term expanded HUCB-NSC produced clones (cloning efficiency approximately 10%) of multipotent neural progenitors differentiating into neuronal, astrocytic, and oligodendroglial lineages. Standardized conditions for the lineage-directed differentiation of HUCB-NSCs were established by the subsequent screening of

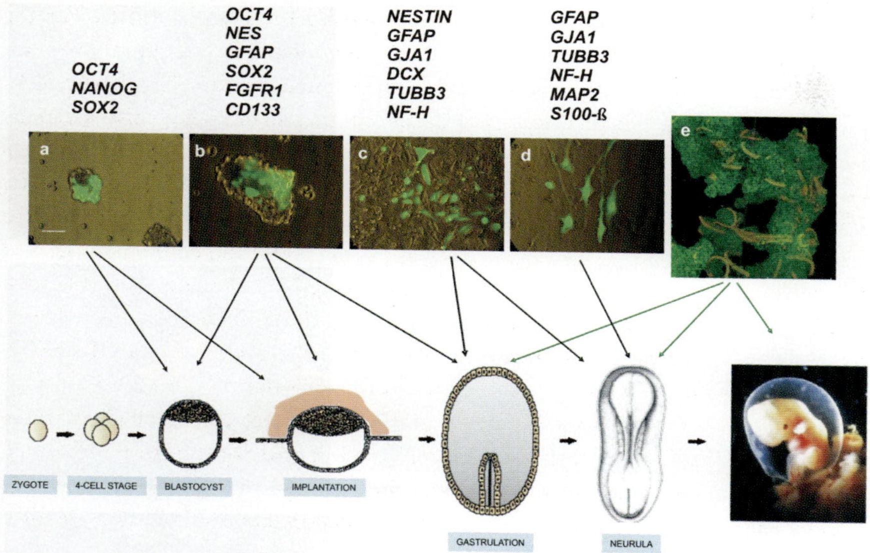

Figure 10.1. Extrapolation of the stages of development of human umbilical cord blood neural stem cell (HUCB-NSC) line (upper panel) with human neural development (lower panel). Molecular signature of the specific stages of HUCB-NSC development is indicated as the expression of typical genes. (a–d) Phase-contrast and fluorescence images (green — green fluorescent protein) of HUCB-NSCs at different developmental stages: (a–b) undifferentiated cells, (c) committed progenitors, (d) growth factor-differentiated cells, (e) HUCB-NSC grown on 3D keratin scaffold. Scale bar = 50 μm.

a panel of neuromorphogenes and neurotrophic factors (Figure 10.1). The ability of HUCB-NSCs to form neurospheres, a hallmark of genuine neural stem cells, further confirmed stem character and neural commitment of cord blood-derived line. Non-adherent HUCB-NSC spheres (Jurga and Domanska-Janik, 2004; Buzanska *et al.*, 2005), are lineage negative for advanced neuronal, astroglial or oligodendroglial markers, but contain cells expressing markers of uncommitted human neural stem cells, such as Nestin and GFAP.

The maintenance of HUCB-NSC at different developmental stages, starting from a pluripotent-like, non-adherent cell population, through neurally committed adherent non-differentiated cells, to differentiated (into neuronal, astrocytic and oligodendroglial lineages) was shown by the stage-dependent gene expression profile (Figure 10.2) (Buzanska *et al.*, 2006, 2009a), and by changes in the epigenetic make up of genes typical for pluripotent stem cells (Habich, unpublished). The neural character of the HUCB-NSC line was evaluated by comparative analysis of the transcriptional profile (Affymetrix DNA microarray probes HG-U133, covering

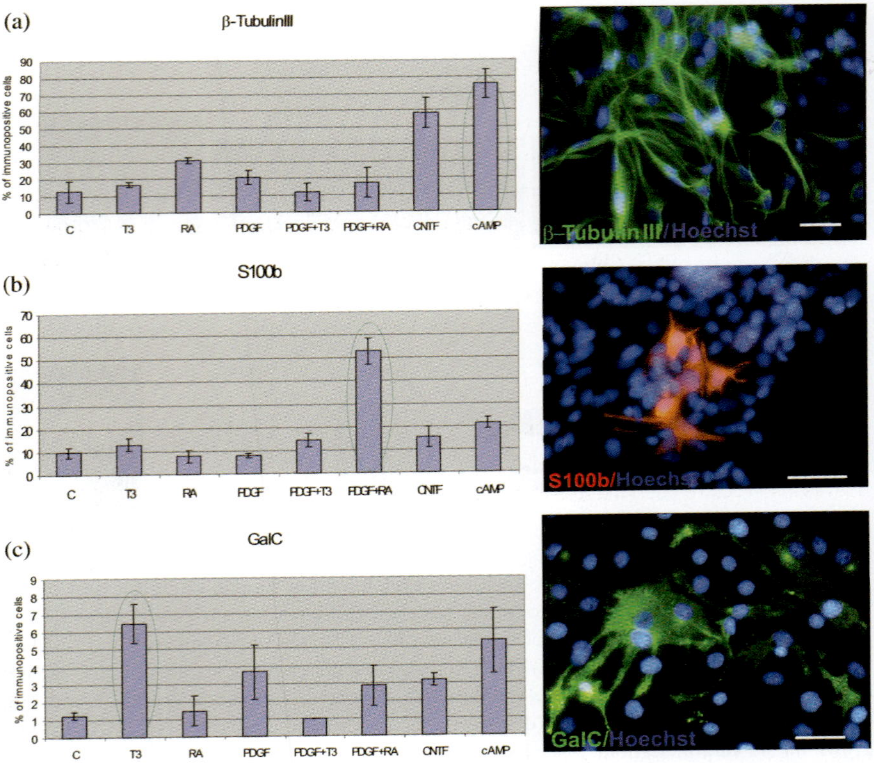

Figure 10.2. Right panel: Immunostaining of neuronal (β-tubulinIII), astroglial (S100β) and oligodendroglial (GalC) markers in HUCB-NSC differentiated by exposure to the indicated factors for 2 weeks. Left panel: (a) Percentage of β-tubulinIII, (b) S100β, and (c) GalC-expressing cells. The best differentiating conditions are encircled. Scale bars = 50 μm.
*C: control (no factors), CNTF: ciliary neurotrophic factor.

the expression of about 40,000 genes), between the free-floating, non-differentiated HUCB-NSC and its starting, non-selected cell population — mononuclear cell (MNC) fraction of human cord blood. It was shown that more than 90% of stem and neural-related genes are upregulated in HUCB-NSC as compared to its reference population of MNC. "Stemness" upregulated genes in HUCB-NSCs encode proteins involved in Wnt, Lif and Notch signaling pathways, which are known to regulate stem cell self-renewal and proliferation. They are represented mainly by frizzled, wingless, Lif and notch gene families (Fzd8, Fzd7, Wnt5a, Ctnmb1, Lif, LifR, Erbb2, Jak3, Notch2, Notch3, Notch4, Jag2, Dll3, Hey1, Pnn1 and 2, Adam15, 17, 23). In addition to activation of pro-neural genes (e.g. NeuroD1, Otx1, and Msi), studies revealed the expression of Oct4, Sox2, Mdr1, Rex1 — the genes characteristic of pluripotent embryonic stem cells. However, at the same time, there was a lack of expression of genes typical for mesodermal (T gene) and endodermal differentiation (Foxa1) (Buzanska *et al.*, 2006; Habich *et al.*, 2006). Thus, HUCB-NSCs can retain their potential for pluripotency in their molecular profile, while executing exclusively neural-restricted gene expression and differentiation. The preliminary pluripotent stage of the "stemness" genes in the HUCB-NSC line (Oct3/4, Nanog and Sox2) and their subsequent silencing during dBcAMP differentiation was accompanied by changes in the methylation state of their promoters. HUCB-NSCs, after neural differentiation, revealed significantly higher methylation status in promoter regions of the "stemness" genes than in undifferentiated cells. Those changes negatively correlated with the expression of Oct3/4, Nanog and Sox2 gene and protein in cultured cells (Habich, unpublished). It was also recently shown to this group that the expression of pluripotent-like genes at early developmental stages of HUCB-NSCs is significantly enhanced in the low oxygen level conditions, mimicking the *in vivo* situation (Szablowska-Gadomska, unpublished).

The gene expression profile of HUCB-NSCs changed substantially upon neuronal differentiation in the presence of dBcAMP. Activated genes (as compared to non-differentiated HUCB-NSCs) were mainly related to the signaling pathways, neurotransmitters, receptors, or channels involved in neuronal functions (Buzanska *et al.* 2006; 2006b; Sun *et al.* 2005). The upregulation of genes was followed by the expression of corresponding proteins, which included: GluR2, dopamine, serotonin and acetylcholine receptors, Gad 65, GABARAPL1, and 3, Tau, SV2, TH and many more involved in synaptic transmission and function of GABAergic, glutamatergic and serotoninergic neurons. Mechanisms underlying the neuronal differentiation of HUCB-NSCs are thought to resemble the mechanisms operating in the developing mammalian brain, and thus involve the bHLH cascade of transcriptional regulation. In gain-of-function experiments, it was shown that ID1 (an endogenous inhibitor of bHLH transcription factors) inhibits neurogenesis. Negative regulation of HUCB-NSC neuronal differentiation is coupled with translocation of ID1 protein from the cytoplasm to the nucleus, suggesting its possible usage as a stage-specific marker of neurogenesis (Jurga *et al.*, 2006a).

Of special interest is the high activity of the FGF signaling pathway in HUCB-NSCs. Upregulation of FgfR1 was shown to be a hallmark of neural commitment of clonally related stem cells (Hsieh and Gage, 2004). In non-differentiated HUCB-NSC, expression of the FgfR1 gene was increased up to 600-fold. Since the FGF signaling pathway was also shown to be involved in the epigenetic control of self-renewal activity of stem cells (Hsieh and Gage, 2004; Rao, 2004), significantly increased expression of FgfR1 may be important not only for neural commitment, but also for the self-renewal capacity of HUCB-NSCs. Targeting of the novel integrative nuclear FGF receptor 1 signaling (INFS) pathway by nanoparticle-mediated FGFR1 gene transfer was recently shown to influence early neurogenesis *in vivo* (Stachowiak *et al.*, 2009) This is in line with the previous observations by Stachowiak's group of the nuclear accumulation of FGFR1 during the cAMP-induced neuronal differentiation of the HUCBC-NSCs. Also, an ectopic expression of FGFR1 in the nuclear compartment inhibited the cell proliferation while committing HUCB-NSCs to differentiate along the neuronal pathway (Fang *et al.*, 2005) Using the novel "biomimetic" approach of stem cell fate control by biofunctional surfaces, this group recently showed that tailoring the architecture, content of microenvironment (Buzanska *et al.*, 2009b) and stimulation of selected intracellular pathways by specific signaling molecules (Zychowicz, unpublished; Buzanska *et al.*, 2010), resulted in the differentiation of hUCB-NSCs to either a neuronal or astroglial lineage. To attain functionality, HUCB-NSCs must not only develop morphological characteristics of neurons, but also voltage- and ligand- gated ion channels that will allow them to function within a neural network and respond to neurotransmitters released from neighboring neurons. The whole-cell patch clamp of the 2D neuronally differentiating HUCB-NSCs in the presence of dBcAMP, revealed two types of voltage sensitive currents: (1) inward rectifying current (Kir) and (2) outward-rectifying potassium current (IK+) together with several ligand-gated currents typical for neuronal cells. However, action potentials typical of terminally differentiating neurons were not recorded (Sun *et al.*, 2005). The functional neuronal network of HUCB-NSC-derived cells with action potential-related field activity sensitive to TTX was obtained only in 3D scaffold-based culture conditions (Jurga *et al.*, 2009). The important point for possible therapeutic applications appeared from the work of Chua *et al.* (2009), who established serum-free and feeder-free cell culture conditions to derive neural progenitors, neurons and oligodendrocytes from human umbilical cord blood.

10.4. Expansion and Epigenetics of Cord Blood Stem Cells

As described above, many research papers have shown how different subsets of cord blood stem cells differentiate under defined conditions into hematopoietic, endothelial, mesodermal or neuroepithelial directions. Thus, it is plausible to suggest that the most primitive stem cells residing in CB would express pluripotency similar to embryonic SCs. After exposure to different agents or *in vitro* conditions, these cells would be able to express antigens of diverse lineages. But even if cord blood contains a subset of cells resembling pluripotent, embryonic-like precursors with respect to their multilineage differentiation potential, there are still obstacles to expand them *in vitro* for its production in the quantities needed in therapeutic applications. This is because CB progenitors, like any other stem cells of the adult type, lose their "stemness" character including the ability to indefinitely self-renew and grow *in vitro*. On the other hand, there are increasing data pointing to the almost unlimited plasticity of different progenitors present in adult stem cell niches which can be activated only under still poorly defined conditions *in vitro* or even *in vivo*. It is still unclear whether exposure to such specific factors *in vitro* would make HUCB cells disregard their hematopoietic identity due to epigenetic reprogramming in the mechanism called trans- or dedifferentiation (Abkowitz, 2002; Poulsom *et al.*, 2002; Terskikh *et al.*, 2001) or whether they acquire new features by directed expansion of a preexisting, embryonic-like, pluripotent stem cell subpopulation positively selected during employed experimental procedures (McGuckin *et al.*, 2004; Habich *et al.*, 2006), Such phenotypic transitions involving both genetic and epigenetic mechanisms, although mechanically unclear, can be correlated with defined molecular switches from the patterns typical of pluripotent-like stages to that characteristic of more committed progenitors and (at least) mature cells. Transcription factors initiating and regulating such events are better described in embryonic SCs and include Nanog, Sox2 and Oct3/4. More recently, this genetic set has been proposed to play a similar role in adult (including CB) stem cells (Leeb *et al.*, 2010). However, many experiments have come up with conflicting results (Liedtke *et al.*, 2007; Redshaw and Strain, 2010; Pacini *et al.*, 2010). In this respect, among the different types of most frequently tested somatic cells, freshly isolated, as well as cultured *in vitro* cord blood preparations, have given the most promising and reliable results. On the other hand, and probably because of this, there are several reports showing that CB cells may be particularly valuable for the generation of induced pluripotency in adult cells (Haase *et al.*, 2009; Zaehres *et al.*, 2010; Giorgetti *et al.*, 2010; Ye *et al.*, 2009) These cells — which are easily accessible, well tolerated after transplantation and with more primitive molecular characteristics resulting in their great flexibility — might be highly appreciated for further preclinical studies and eventual therapeutic applications.

10.5. *In Vivo* Potential of Human Umbilical Cord Blood Cells

The first experimental transplantation studies performed with human umbilical cord blood (HUCB) cells have reported promising results after ischemic insults

(Chen *et al.*, 2001; Willing *et al.*, 2003) Infusion of cord blood cells into rats in the commonly used middle carotid artery occlusion (MCAO) model of stroke reversed many of the physical and behavioural deficits associated with the disease. In addition to stroke, cord blood cells have been used in other nervous system injury models. Cord blood stem cell grafts could also alleviate symptoms of newborn cerebral palsy in a rat model, with improved neurological effects (Meier *et al.*, 2006). Spinal cord injured (SCI) rats infused with cord blood stem cells have shown significant improvements 5 days post-treatment compared to untreated animals. The donor cells were observed at the site of injury but not at uninjured regions of the spinal cord (Saporta *et al.*, 2003). This finding was supported by another study demonstrating that cord blood cells transplanted into spinal cord injured animals differentiated into various neural cells, improving axonal regeneration and motor function (Kuh *et al.*, 2005). Veeravalli *et al.* (2009) observed that treatment with HUCB after SCI altered the expression of several MMPs in rats. Induced upregulation of MMP-2 by HUCB transplantation reduced formation of the glial scar at the site of spinal cord injury along with reduced immunoreactivity to chondroitin sulfate proteoglycans, thereby creating an environment suitable for endogenous repair mechanisms.

Furthermore, therapeutic benefits of cord blood infusion have been reported in chronic neurodegenerative diseases after transplantation. In the Cu/Zn superoxide dismutase (SOD1) G93A mutant mouse model of amyotrophic lateral sclerosis (ALS), it has been shown to delay disease progression and increase the lifespan of the mice (Ende *et al.*, 2000; Garbuzova-Davis *et al.*, 2003). The use of human cord blood in ameliorating symptoms of Parkinsonism in animal models has also been investigated. Among the studies with stem cells, Newman *et al.* reported on the isolation and culture of CD133-positive stem cells from HUCB, followed by their transplantation into a 6-OHDA rat model. Intrastriatal transplantation of CD133-positive stem cells significantly improved behavioral motor functions similar to that seen with a mononuclear fraction of HUCB (Newman *et al.*, 2009). With the extension of this study, Kelly *et al.* (2009) also showed that intrastriatal implantation of a HUCB-derived mononuclear fraction and neural-induced CD133-positive stem cells produced significant improvements in neurological behavioral tests compared to transplantation of human embryonic stem cells (ESC) in a 6-OHDA rodent model. Furthermore, the HUCB-derived stem cell grafts appeared to be more widely distributed and migrated further into the contralateral side than human ESC.

Human cells transplanted into the CNS appeared to migrate into the brain parenchyma or spinal cord, and expressed neural and glial markers. In the experiments in this laboratory, human umbilical cord blood-derived neural stem cells (HUCB-NSC), after transplantation into the corpus callosum of adult rats, could be mainly encountered along the white matter tracts, whereas the density of donor cells in the rest of the brain was low and apparently random (Jablonska *et al.*, 2010). A similar pattern of distribution has been observed after grafting of human neural stem cells (Burnstein *et al.*, 2004; Wennersten *et al.*, 2004; Olstorn *et al.*, 2007). Implantation of HUCB cells in rats with brain ischemia induces apparent migration of donor cells into the lesion probably due to an increase of chemokines in the ischemic brain area (Kozłowska *et al.*, 2007). Jiang *et al.* (2008) have found that monocyte chemoattractant protein-1 (MCP-1) and macrophage inflammatory protein-1 (MIP-1α) expression was significantly increased in the ischemic hemisphere of rat brain, and significantly promoted HUCB cell migration compared to the contralateral side. It was also found that chemokine receptors were constitutively expressed on the surface of HUCB cells. The HUCB cell migration was neutralized with polyclonal antibodies against MCP-1or MIP-1α. These results suggested that the increased chemokines in the ischemic areas can bind surface receptors on exogenously delivered HUCB cells and attract these cells to localized areas of insult (Park *et al.*, 2009).

HUCB-NSCs infused into the adult rat brain continued to proliferate in the host tissue (Kozłowska *et al.*, 2007; Jablonska *et al.*, 2010). The rate of proliferation was comparable to what has been reported when xenografting embryonic or adult neural human neural precursors into the rodent brain (Olstorn *et al.*, 2007; Le Belle *et al.*, 2004). The results were in agreement with studies from other laboratories where human cord blood cells were noticed after their direct injection into the brain of xenogeneic recipients (Coenen *et al.*, 2005; Nystedt *et al.*, 2006; De Paula *et al.*, 2009). HUCB-NSCs, when kept under *in vitro* conditions favoring differentiation, were shown to differentiate into cells expressing the antigens of mature neurons, astrocytes and oligodendrocytes (Jurga *et al.*, 2006). Compared with this, differentiation *in vivo* was protracted because mature neurons were never found among the descendants of the transplanted HUCB-NSCs into the rat brain. The fact that rather slow progression was observed in neuronal differentiation 7 days after HUCB-NSC transplantation was visualized by NF200 and β-tubulin III marker expression in only few donor cells indicates that, given the short time period of possible observation due to the elimination of human cells by the discordant recipient, the end-stage may not have been reached (Jablonska *et al.*, 2010). Incomplete neuronal differentiation was observed in other studies where a relatively low number of xenogeneic cells transplanted into the brain of adult rats expressed antigens typical for mature neurons (Kuh *et al.*, 2005; Liu *et al.*, 2006; Hicks *et al.*, 2007).

The limited survival of human cells in rats, possibly due to immune rejection, is a major setback. While several reports ascribe a low immunogenic profile to HUCB cells *in vitro*, the immune-privileged status of these cells *in vivo* remains controversial (Park *et al.*, 2009; Walczak *et al.*, 2004). The local inflammatory response in the rat brain associated with surgical injury and/or xenogeneic cell transplantation seems to work against the donor cells as well as endogenous progenitors (Ekdahl *et al.*, 2003; Kelly *et al.*, 2004; Hicks *et al.*, 2009a). Indeed, the majority of the reported HUCB cell transplantation into adult or aged rat brains was associated with vigorous rejection (Walczak *et al.*, 2004). The involvement of immune cells in human transplant paradigms is well illustrated in the evaluation of the graft fate in the rodent tissues. The innate immune response serves as the first-line defence in any transplantation scenario, whether it involves xenogeneic or allogeneic, or even autologous transplantation (Bergwerf *et al.*, 2009; Glover *et al.*, 2010). The central nervous system strongly relies on microglia for innate protection. The results of our studies indicated that there is a massive activation of ED1-positive cells in the adult rat brain 24 hours after implantation of HUCB-NSCs (Kozłowska *et al.*, 2007). Furthermore, additional immunostaining, including CD15 for neutrophils and CD5 for T-lymphocytes revealed the involvement of these inflammatory cell types in the immune-mediated response in the rat brain. The increased activation of macrophage/microglia was observed for a few days after HUCB transplantation whereas, the number of lymphocytes diminished over time (Jablonska *et al.*, 2010). Immunoreactivity, as shown by the activation of microglia and lymphocyte markers, has been demonstrated after the transplantation of human cells into the adult rat brain (Olstorn *et al.*, 2007; Walczak *et al.*, 2004; Chen and Palmer, 2008). However, the results suggest that microglia, rather than T cells, are the main contributors to xenograft rejection in CNS. It was also shown by the xenografting of human cells into NOD/SCID mouse, which is completely depleted of competent lymphocytes (Walczak *et al.*, 2007). These severe immunodeficient mice possessed the ability to reject the graft since HUCB cells were no longer detected in the brain of the recipients 4 weeks post grafting. This suggests that the disappearance of the implanted cells was not due to a T- and B cell-mediated immune response. Although there is no evidence for the involvement of T cells in clearing human cells in the rat brain, one cannot exclude their subsequent effectors functioning in the CNS.

Humoral mechanisms may also contribute to the rapid rejection of xenogeneic transplants in the CNS. Previous studies have shown the presence of IgM immunoglobulins in the rat brain after the transplantation of porcine neural precursor cells, especially in the area displaying features of graft rejection (Armstrong *et al.*, 2001). In our studies, the intense IgM immunoreactivity was observed around the graft of human cells into the rat brain and this suggests that antibodies directed against specific epitopes on HUCB-NSC–derived cells constitute a component of the response (Jablonska *et al.*, 2010). However, on the basis of these data, the possibility that some IgM immunoreactivity may result from plasma leakage of a range of non-specific IgM cannot be excluded. IgM is produced in the first phase of the primary immune response and also constitutes the principal class of the natural, preformed xenoreactive antibodies that exist in all species (Larsson and Widner, 2000). Thus specific IgM binding may be generated *de novo* as part of a primary response against HUCB-NSC grafts, or may reflect the binding of pre-existing antibodies. Additionally, IgM may be produced by locally infiltrating cells of the B cell lineage. The identity of these cells has been noted previously in neural xenografts (Larsson *et al.*, 1999). Thus far, experiments have not been able to demonstrate any B cells in the rat brain tissue within 2 weeks after HUCB-NSC transplantation. The results of the studies suggest that antibodies play a role in the initial phase of graft rejection. This is in concordance with the Larsson's report where porcine neural tissue was grafted to immunoglobulin-deficient mice and the rejection was delayed for around 4 weeks compared to normal mice. However, the transplanted cells were eventually rejected by 6 weeks (Larsson *et al.*, 1999). The deteriorate effect of IgM in discordant xenografts seems to be multifunctional. Antibodies may not only attack the grafted cells by themselves but may also prime cells for lysis by host phagocytes. The significant correlation of IgM staining intensity with ED1 grading in rat brain tissue, after HUCB-NSC transplantation, could be related to indirect cytotoxic effect of antibodies.

Several reports describe the need for immunosuppressive therapy in order to avoid rejection of xenogenic implants in the CNS (Hayashi *et al.*, 2005; Ronsyn *et al.*, 2007). However, multiple studies, including work of our group, described low human cell survival in the rat brain despite the use of classical cyclosporine A (CyA) (Kögler *et al.*, 2004; Walczak *et al.*, 2004; Pan *et al.*, 2005; Janowski *et al.*, 2006; Niranjan *et al.*, 2007; Hicks *et al.*, 2009b). This suggests that CyA immunosuppression is not an effective strategy for enhancing survival. The other strategy to avoid the deleterious reaction of the host to the transplant included triple-immunosuppression with CyA, azathioprin (Aza) and methylprednisolone (MP) (Finsen personal communication); however according to this observation and experience, $CyA^{+}Aza^{+}MP$-treated rats became ill at 2 weeks and had to be killed in accordance with the rules of the animal license. Graft survival could, therefore, not be assessed beyond this time (unpublished). In the context of cell therapies, even if immunosuppression could prevent the rejection of non-autologous grafts, in parallel the restorative immune response is eliminated together with the immune-mediated guidance cues that are required for donor cells to migrate and proliferate. Recent studies have shown that the infiltration of CNS by immune cells following injury promotes neuroregeneration (Ziv *et al.*, 2006; Glezer *et al.*, 2007; Kulbatski, 2010). The contradictory function of immune cells results in their detrimental or beneficial roles for the repair of CNS damage — this has been debated over the past decade, arguing in favor or against the use of immunosuppressive drugs.

Previous studies showed that xenogenic transplants into the developing rat brain were more successful than the same transplant in aged animals and immunosuppression was not decisive for the survival and differentiation of human donor cells (Coenen *et al.*, 2005; Walczak *et al.*, 2004; Englund *et al.*, 2002; Zigova *et al.*, 2002). In trying to ameliorate the negative consequences of cell graft rejection by suppressing the host immune response, HUCB-NSC was transplanted into neonates with incompletely developed immunological systems.

The results were compatible with these studies which showed that HUCB-NSCs transplanted into newborn rats persisted in the host brain for a longer period of time with no histopathological signs of graft rejection. In the study, depending on survival time, transplanted HUCB-NSCs into the newborn rat brain were found around the injection site in the anterior part of the SVZ and along the initial part of the RMS. These findings confirm the ability of HUCB-NSC to follow the migratory route of endogenous neural progenitors. Within 1 week, Ki67-positive cells were observed among the graft, suggesting that the donor cells are still actively proliferating. At the same time, many of the HUCB-NSCs acquired a neuron-specific NF200 and TUJ-1 immunoreactivity (Jablonska *et al.*, 2010). This suggests that transplantation of HUCB-NSCs into the neurogenically active region of the brain (SVZ) may support neural differentiation of these cells. In the experiments, human cells transplanted into neonatal rats survived in the brain up to 5 weeks, being visible in the most remote locations from the injection site. Morphologically, a few single donor cells displayed long processes accompanied by the presence of neuronal MAP-2 protein. These data are consistent with previous studies showing that human cells transplanted into the SVZ of neonatal rats become immunopositive for neuronal antigens (Englund *et al.*, 2002; Zigova *et al.*, 2002). Also, HUCB injected into newborn rats after the induction of neonatal hypoxia-ischemia (HI) brain damage have shown to have therapeutic potential (Pimentel-Coelho *et al.*, 2010). Transplantation of HUCB cells resulted in a better performance in two of three developmental sensorimotor reflexes after the HI insult. Moreover, the authors observed that in the treated group, there was a reduction in cell death in the striatum and a decrease in the number of activated microglial cells in the cerebral cortex. This suggests that HUCB cells have a neuroprotective role in the hypoxia-ischemia model, and as a consequence, the treated animals have better functional recovery than the untreated group.

Several studies have demonstrated that direct transplantation of stem cells into the brain was not required and in fact, beneficial effects could be observed after systemic injection of cord blood cells into recipients subjected to CNS injury. Functional improvements from cord blood administration by intravenous or intraperitoneal infusion have been widely reported (Willing *et al.*, 2003; Meier *et al.*, 2006; Vendrame *et al.*, 2004; Xiao *et al.*, 2005; Newcomb *et al.*, 2006). These data were consistent with the observations in our laboratory. The results from the study demonstrated that intracarotid artery infusion of human umbilical cord blood cells, depleted of hematopoietic precursors and directed *in vitro* toward a neural phenotype (HUCB-NP) (Habich *et al.*, 2006), ameliorates neurological deficits associated with stroke-like, ouabain-induced brain injury. Rats transplanted with HUCB-NPs showed significant behavioral improvements in walking beam and rotarod performance after 4 week, with further progress in recovery of locomotor deficits observed after 4 weeks (Janowski *et al.*, 2004; Gornicka-Pawlak *et al.*, 2011). The functional therapeutic benefit was correlated with obvious reduction in lesion volume in transplanted rats in comparison to control, non-transplanted counterparts. Furthermore, therapeutic benefits of cord blood infusion have been reported in chronic neurodegenerative diseases. It has been shown that infusion of HUCB cells delayed the progression of amyotrophic lateral sclerosis (ALS) and increased the lifespan of the diseased, transgenic mice (Ende *et al.*, 2000; Garbuzova-Davis *et al.*, 2003). Systemic delivery of HUCB cells has the ability to prevent circulatory shock and the resulting brain damage in the heatstroke animal model (Chen *et al.*, 2001).

The mechanisms behind the observed beneficial effects afforded by cord blood therapies are unclear. Interestingly, some human cord blood cells had entered the site of brain injury and expressed neuronal and astroglial markers (Pimentel-Coelho *et al.*, 2010; Vendrame *et al.*, 2004; Nan *et al.*, 2005). However, several investigators reported behavioral improvement of neurological disorders with no detection of HUCB cells infused systemically in the host brain (Borlongan *et al.*, 2004; Nystedt *et al.*, 2006; Xiao *et al.*, 2005; Gornicka-Pawlak *et al.*, 2011). A "bystander" function of systemically infused cells may be largely derived from the secretion of neurotrophic survival factors or the suppression of cell apoptosis in CNS lesion sites. Human umbilical cord blood cells as well as their neural derivatives produce different neurotrophic factors [i.e. brain-derived neurotrophic factor (BDNF), glial-derived neurotrophic factor (GDNF), nerve growth factor (NGF), neurotrophin 3 (NT3) and neurotrophin 4-5 (NT4-5)] (Fan *et al.*, 2005). The results of Borlongan *et al.* demonstrated that intravenous HUCB infusion significantly increase GDNF, BDNF and NGF brain levels in stroke animals as compared with non-transplanted counterparts (Borlongan *et al.*, 2004). It is also possible that, in addition to the role of already known or unknown factors secreted by the transplanted cord, the blood cells themselves are responsible for functional improvement following CNS lesion. Recently, it has been shown that mononuclear cord blood-derived neuronal progenitor cells decreased free radicals and induced antioxidants in an oxygen and glucose deprivation insult model (Arien-Zakay *et al.*, 2009).

Another potential mode of action promoted by cord blood cell transplantation is neovascularization around ischemic tissue. Angiogenesis is an important healing process of wounds including tissue ischemia, and it can attenuate local tissue ischemia and improve clinical outcomes. Furthermore, angiogenesis appears to promote neurogenesis in neurological disease (Rizvanov *et al.*, 2008; Teng *et al.*, 2008). There is accumulating data that transplantation of HUCB cells induce neovascularization, followed by functional improvement in the animal ischemic disease model (Hu *et al.*, 2006). VEGF was increased in the experimental group even though it was unclear whether transplanted HUCB cells directly released it or stimulated other cells for indirect secretion. HUCB contains high numbers of CD34-positive endothelial progenitor cells, which can give rise to mature endothelial cells and induce angiogenesis in ischemic tissues (Pesce *et al.*, 2003). Taguchi *et al.* demonstrated that CD34-positive cells derived from HUCB-induced angiogenesis, followed by endogenous neurogenesis in a mouse stroke model (Taguchi *et al.*, 2004). Ding *et al.* revealed that HUCB-MSCs differentiated into glial, neuronal, and vascular endothelial cells as well as promoted the formation of new vessel angiogenesis in the rat stroke model (Ding *et al.*, 2007).

Recently, there has been increasing evidence of an anti-inflammatory effect of HUCB after transplantation, beyond the ability of neurogenic and angiogenic actions. Apart from ambitious goals of functional neural, glial or endothelial cell replacement via transplantation engraftment, optimizing the ability of transplanted (or infused) cord blood stem cells to contribute to the survival and regenerative capacity of CNS lesion sites by suppressing immune response is perhaps a more achievable short-term goal. Mesenchymal cells derived from the human umbilical cord blood (HUCB-MSC) could serve this purpose.

The cord blood MSC were found to have powerful migratory activity, moving from the primary implantation site to the ischemic damaged area in adult rats (Koh *et al.*, 2008) or hippocampus in rat neonates (Xia *et al.*, 2010). *In vitro* culture of HUC-MSC showed immunoreactivity for various extracellular matrix proteins, such as intercellular adhesion molecule (ICAM)-1, vascular cell adhesion molecule (VCAM)-1, homing cell adhesion molecule (HCAM), and vimentin, which affect homing and engraftment of transplanted cells (Koh *et al.*, 2008).

Several lines of evidence support the hypothesis that transplanted MSC modulate both the host immune response to injury and to the graft itself. Yoo *et al.* demonstrated that HUCB-MSCs were capable of suppressing mitogen-induced T cell proliferation *in vitro* (Yoo *et al.*, 2009). Cord blood cell administration in stroke models reduced inflammation by lowering parenchymal levels of astrocytic/microglial activation and granulocyte/macrophage infiltration (Newcomb *et al.*, 2006; Vendrame *et al.*, 2006). Also, HUCB-MSC treated APP/PS1 (a double transgenic model of Alzheimer's disease) mice exhibited decreased levels of neurotoxic pro-inflammatory cytokines (TNFα, IL-1β) associated with classically activated microglia, and increased levels of neuroprotective cytokines (IL-4, IL-13) characteristic of alternatively activated microglia (Lee *et al.*, 2010).

Apart from the immunosuppressive effect on the host immune system and alteration of the relative level of pro- and anti-inflammatory cytokine expression by T cells, MSCs are capable of enhancing the endogenous repair potential of injured tissue. HUC-MSC created hUC-MSCs–produced neurotrophic G-CSF, VEGF, GDNF, and BDNF, which are well known to have neuroprotective effects on stroke outcome before, during, and after neuronal differentiation. Moreover, after HUC-MSC implantation, Nestin expression in the hippocampus was significantly increased in the donor cells-implanted group which suggests that HUC-MSC might enhance endogenous neurogenesis and reduce the final infarct volume. Indeed, the final infarct volume in ischemic stroke model rats was significantly reduced after HUC-MSC transplantation (Koh *et al.*, 2008). This was in concordance with the results of other authors. The alleviation of brain tissue injury was found after HUCB-MSC treatment of fetal hypoxic-ischemic brain injured (HIBI) rats (Xia *et al.*, 2010). Also, intra-arterially delivered HUCB-MSC in canine cerebral ischemia demonstrated their efficiency by reducing the infarction lesion volume and through earlier recovery from the neurological deficits (Chung *et al.*, 2009).

Most HUCB-MSCs transplanted into the brain of rodents remained undifferentiated. The data of cord blood MSC differentiation to neural cells are controversial. Koh *et al.* observed that some HUC-MSCs transplanted into the brain of ischemic stroke model rats were reactive for the neuronal markers, but in spite of the morphological differentiation, these neurons were not functionally active (Koh *et al.*, 2008). In contrast, intracerebral infusion of HUCB-MSCs into newborn rats with HIBI revealed differentiation of MSC to astrocytes but not neurons (Xia *et al.*, 2010). However, the neurobehavioral function of the brain injured rats improved. This suggests that the improvement of neurological deficit following cord blood MSC transplantation in CNS disorders might be related to neurotrophic factors secreted from the injected donor cells, rather than from novel network formation between host neurons and implanted human cord blood MSC.

10.6. Current Status of Cord Blood Clinical Transplantation

Transplantation of umbilical cord blood in babies with infantile Krabbe's disease that favourably altered their natural history could be the first, promising example from the human clinic. Infants who underwent HUCB cell graft before the development of symptoms showed progressively increasing central myelination and continued gains in developmental skills and cognitive function (Escolar *et al.*, 2005, 2007) These therapeutic benefits of cord blood infusion may hold hope for slowing the progression of other chronic neural diseases.

The observation that cord blood stem cells can become different types of nervous system cells extends their utility to other areas of neurological damage, including spinal cord injury. In a recently reported clinical use of cord blood cells to treat a patient with a spinal cord injury (Kang *et al.*, 2005), it was stated that transplantation of these cells improved her sensory perception and mobility in the hip and thigh regions. Both computed tomography and magnetic resonance imaging studies revealed regeneration of the spinal cord at the injury site. Since the cord blood cells were allogenic in origin, it will be significant to determine if immune rejection or other immune-mediated problems occured that might jeopardize the early improvement. Neither additional patients nor additional studies in this area have been reported. However, the use of cord blood cells for spinal cord injury seems to be the next logical clinical trial. Large numbers of children are unfortunate enough to suffer a spinal cord injury at an early age (e.g. diving into a pool, car accidents, falls etc.) and it would be expected that a significant number would have autologous cord blood banked and available for treatment.

A pilot non-randomized clinical study at Duke University is currently assessing the safety of autologous umbilical cord blood MNC cells in newborn infants with hypoxic-ischemic encephalopathy. To date, 188 patients were infused with autologous (and a few allogeneic) cord blood units (aged 1 week to 9 years old) with a minimum cell dose of 1×10^7 cells/kg. Infusions were well tolerated and no clinical adverse effects were reported. Children are followed to determine neurodevelopmental outcome and for functional MRI imaging at Duke University (clinicaltrials.gov ref#NCT00593242). Another study, a randomized observer-blinded crossover clinical trial has just started at the Medical College of Georgia for children with diagnosed cerebral palsy whose parents have saved their infants cord blood in a stem cell bank. Patients to be included are those between the ages of 2 and 12 without evidence of a non-progressive motor disability (clinicaltrials.gov ref# NCT01072370). The McGuckin and Simeoni groups recently designed a multi-centric clinical study to assess the safety and efficacy of umbilical cord blood MNC infusion for neonates with hypoxic encephalopathy in >10 University Hospitals in France (McGuckin, personal communication). This covers cell infusions and provides the necessary phases of safety to efficacy evaluation. It is strongly considered that subpopulations of the MNC, including the endothelial progenitors and the mesenchymal compartment are most beneficial (Leeb *et al.*, 2010). The first intra-cerebroventricular transplantation of autologous, neurally committed cord blood cells was performed in a 16 month-old child with global ischemic brain injury (Jozwiak *et al.*). The neuroimaging data demonstrated that committed cord blood-derived cells transplanted intra-cerebroventricularly remained in the host brain for at least 4 months. Moreover, this autologous neurally committed cord blood cell transplantation into the lateral ventricle of the brain seems to be safe when done under neuronavigation and may represent a novel and promising tool for the management of brain ischemic injury.

10.7. Conclusions

Human umbilical cord blood was shown to be a source of pluripotent, as well as lineage-specific hematopoietic, mesenchymal and neural stem cell populations. Neural progenitors can arise *in vitro* from all of these different subsets of stem cells and are able to differentiate further into neuronal, astrocyti,c and oligodendroglial directions. Discovery of this phenomenon widens the applicability of cord blood cells in therapies of neurological diseases. The mechanisms underlying phenotypic conversion of cord blood-derived stem/progenitor cells remain to be elucidated. The molecular switch driving the phenotypic change of cord blood stem cells towards neuronal fate involves epigenetic silencing of "stemness"-like genes and activation of intracellular pathways including novel integrative nuclear FGF-R1 signaling (INFS) and bHLH cascade. Targeting those mechanisms may provide for an effective control on cell expansion, differentiation, prevent teratomas and thus advance possible clinical applications. Studies by this research group performed over the past decade have demonstrated the potential of cord blood cells for differentiation into fully functional neurons *in vitro*. Beyond this, cord blood-derived cells were shown to be responsible for strong neurotrophic and immunomodulatory properties that promote brain repair by suppressing inflammation together with inhibiting neuronal death and glial scar formation. Based on these data, cord blood cell therapy is currently a matter of several trials performed in children with cerebral palsy or other cerebral impairments.

Bibliography

Abkowitz, J. L. 2002. Can human hematopoietic stem cells become skin, gut, or liver cells? *N Engl J Med* 346(10):770–772.

Aktas, M., Buchheiser, A., Houben, A., *et al.* 2010. Good manufacturing practice-grade production of unrestricted somatic stem cell from fresh cord blood, *Cytotherapy* 12(3):338–348.

Ali, H., Jurga, M., Kurgonaite, K., *et al.* 2009. Defined *in vitro* serum-free culturing conditions for neural tissue engineering from umbilical cord blood stem cells, *Acta Neurobiol Exp* (Wars) 69:11–23.

Arien-Zakay, H., Lecht, S., Bercu, M. M., *et al.* 2009. Neuroprotection by cord blood neural progenitors involves antioxidants, neurotrophic and angiogenic factors, *Exp Neurol* 216(1):83–94.

Armstrong, R. J., Harrower, T. P., Hurelbrink, C. B., *et al.* 2001. Porcine neural xenografts in the immunocompetent rat: Immune response following grafting of expanded neural precursor cells, *Neuroscience* 106(1):201–216.

Bergwerf, I., De Vocht, N., Tambuyzer, B., *et al.* 2009. Reporter gene-expressing bone marrow-derived stromal cells are immune-tolerated following implantation in the central nervous system of syngeneic immunocompetent mice, *BMC Biotechnol* 9:1.

Borlongan, C. V., Hadman, M., Sanberg, C. D., *et al.* 2004. Central nervous system entry of peripherally injected umbilical cord blood cells is not required for neuroprotection in stroke, *Stroke* 35:2385–2389.

Burnstein, R. M., Foltynie, T., He, X., *et al.* 2004. Differentiation and migration of long term expanded human neural progenitors in a partial lesion model of Parkinson's disease, *Int J Biochem Cell Bio* 36:702–713.

Buzanska, L., Machaj, E. K., Zablocka, B., *et al.* 2001. Human cord blood derived neurons, astrocytes and oligodendrocytes, *J Neurochem* 78 Suppl. 1:58.

Buzanska, L., Machaj, E. K., Zablocka, B., *et al.* 2002. Human cord blood-derived cells attain neuronal and glial features *in vitro*, *J Cell Sci* 115, 2131–2138.

Buzanska, L., Habich, A., Jurga, M., *et al.* 2005. Human cord blood — derived neural stem cell line — possible implementation in studying neurotoxicity. *Toxicol In Vitro* 19:991–999.

Buzanska, L., Jurga, M., Stachowiak, E. K., *et al.* 2006. Neural stem-like cell line derived from a nonhematopoietic population of human umbilical cord blood, *Stem Cells Dev* 15(3):391–406.

Buzanska, L., Jurga, M. and Domanska-Janik, K. 2006b. Neuronal differentiation of human umbilical cord blood neural stem-like cell line, *Neurodegener Dis* 3:19–26.

Buzanska, L., Sypecka, J., Nerini-Molteni, S., *et al.* 2009a. A human stem cell-based model for identifying adverse effects of organic and inorganic chemicals on the developing nervous system, *Stem Cells* 27(10):2591–2601.

Buzanska, L., Ruiz, A., Zychowicz, M., *et al.* 2009b. Patterned growth and differentiation of Human Cord Blood-derived Neural Stem Cells on bio-functionalised surfaces, *Acta Neurobiol Exp* 69:1–14.

Buzanska, L., Zychowicz, M., Ruiz, A., *et al.* 2010. Neural stem cells from human cord blood on bioengineered surfaces — novel approach to multiparameter bio-tests, *Toxicology* 270:35–42.

Chen, J., Sanberg, P. R., Li, Y., *et al.* 2001. Intravenous administration of human umbilical cord blood reduces behavioral deficits after stroke in rats, *Stroke* 32:2682–2688.

Chen, Z. and Palmer, T. D. 2008. Cellular repair of CNS disorders: An immunological perspective, *Hum Mol Genet* 17(R1):R84–R92.

Chua, S. J., Bielecki, R., Wong, C. J., *et al.* 2009. Neural progenitors, neurons and oligodendrocytes from human umbilical cord blood cells in a serum-free, feeder-free cell culture, *Biochem Biophys Res Commun* 6;379(2):217–221.

Chung, D. J., Choi, C. B., Lee, S. H., *et al.* 2009. Intraarterially delivered human umbilical cord blood-derived mesenchymal stem cells in canine cerebral ischemia, *J Neurosci Res* 87(16):3554–3567.

Coenen, M., Kögler, G., Wernet, P., *et al.* 2005. Transplantation of human umbilical cord blood-derived adherent progenitors into the developing rodent brain, *J Neuropathol Exp Neurol* 64:681–688.

Copeland, N., Harris, D., Gaballa, M. A. 2009. Human umbilical cord blood stem cells, myocardial infarction and stroke, *Clin Med* 9:342–345.

De Coppi, P., Bartsch, G. Jr., Siddiqui, M. M., *et al.* 2007. Isolation of amniotic stem cell lines with potential for therapy, *Nat Biotechnol* 25:100–106.

De Paula, S., Vitola, A. S., Greggio, S., *et al.* 2009. Hemispheric brain injury and behavioral deficits induced by severe neonatal hypoxia-ischemia in rats are not attenuated by intravenous administration of human umbilical cord blood cells, *Pediatr Res* 65:631–635.

Ding, D. C., Shyu, W. C., Chiang, M. F., *et al.* 2007. Enhancement of neuroplasticity through upregulation of beta1-integrin in human umbilical cord-derived stromal cell implanted stroke model, *Neurobiol Dis* 27(3):339–353.

Domanska-Janik, K., Buzanska L. and Lukomska, B. 2008. A novel, neural potential of non-hematopoietic human cord blood stem cells, *Int J Dev Biol* 52:237–248.

Ekdahl, C. T., Claasen, J. H., Bonde, S., *et al.* 2003. Inflammation is detrimental for neurogenesis in adult brain, *Proc Natl Acad Sci U S A* 100:13632–13637.

Ende, M. and Ende, N. 1972. Hematopoietic transplantation by means of fetal (cord) blood. A new method. Virginia Medical Monthly (1918), 99(3):276–280.

Ende, N., Weinstein, F., Chen, R., *et al.* 2000. Human umbilical cord blood effect on sod mice (amyotrophic lateral sclerosis), *Life Sci* 67(1):53–59.

Englund, U., Björklund, A. and Wictorin, K. 2002. Migration patterns and phenotypic differentiation of long-term expanded human neural progenitor cells after transplantation into the adult rat brain, *Brain Res Dev Brain Res* 134:123–141.

Escolar, M. L., Poe, M. D., Provenzale, J. M., *et al.* 2005. Transplantation of umbilical-cord blood in babies with infantile Krabbe's disease, *N Engl J Med* 352(20):2069–2081.

Escolar, M. L., Poe, M. D., Martin, H. R., *et al.* 2007. A staging system for infantile Krabbe disease to predict outcome after unrelated umbilical cord blood transplantation, *Pediatrics* 118(3):e879–e889.

Fan, C. G., Zhang, Q. J., Tang, F. W., *et al.* 2005. Human umbilical cord blood cells express neurotrophic factors, *Neurosci Lett* 380:322–325.

Fang X., Stachowiak, E. K., Dunham-Ems, S. M., *et al.* 2005. Control of CREB-binding protein signaling by nuclear fibroblast growth factor receptor-1, a novel mechanism of gene regulation, *J Biol Chem* 280:28451–28462.

Garbuzova-Davis, S., Willing, A. E., Zigova, T., *et al.* 2003. Intravenous administration of human umbilical cord blood cells in a mouse model of amyotrophic lateral sclerosis: Distribution, migration, and differentiation, *J Hematother Stem Cell Res* 12:255–270.

Giorgetti, A., Montserrat, N., Rodriguez-Piza, I., *et al.* 2010. Generation of induced pluripotent stem cells from human cord blood cells with only two factors: Oct4 and Sox2, *Nat Protoc* 5(4):811–820.

Glezer, I., Simard, A. R. and Rivest, S. 2007. Neuroprotective role of the innate immune system by microglia, *Neuroscience* 147:867–883.

Glover, J. C., Boulland, J. L., Halasi, G., *et al.* 2010. Chimeric animal models in human stem cell biology, *ILAR J* 51:62–73.

Gornicka-Pawlak, E. B., Janowski, M., Habich, A. *et al.*, 2011. Systemic treatment of focal brain injury in the rat by human umbilical cord blood cells being at different level of neural commitment, *Acta Neurobiol Exp* 71(1):46–64.

Haase, A., Olmer, R., Schwanke, K., *et al.* 2009. Generation of induced pluripotent stem cells from human cord blood, *Cell Stem Cell* 5(4):434–441.

Habich, A., Jurga, M., Markiewicz, I., *et al.* 2006. Early apperance of stem/progenitor cells with neural-like characteristics in human cord blood mononuclear fraction cultured *in vitro*, *Exp Hematol* 34:914–925.

Hayashi, Y., Shumsky, J. S., Connors, T., *et al.* 2005. Immunosuppression with either cyclosporine A or FK506 supports survival of transplanted fibroblasts and promotes growth of host axons into the transplant after spinal cord injury, *J Neurotrauma* 22:1267–1281.

Hermann, A., Gastl, R. and Liebau, S., *et al.* 2004. Efficient generation of neural stem cell-like cells from adult human bone marrow stromal cells, *J Cell Sci* 117:4411–4422.

Hicks, A. U., Hewlett, K., Windle, V., *et al.* 2007. Enriched environment enhances transplanted subventricular zone stem cell migration and functional recovery after stroke, *Neuroscience* 146(1):31–40

Hicks, A. U., MacLellan, C. L., Chernenko, G. A., *et al.* 2009a. Long-term assessment of enriched housing and subventricular zone derived cell transplantation after focal ischemia in rats, *Brain Res* 1231:103–112.

Hicks, A. U., Lappalainen, R. S., Narkilahti, S., *et al.* 2009b. Transplantation of human embryonic stem cell-derived neural precursor cells and enriched environment after cortical stroke in rats: Cell survival and functional recovery, *Eur J Neurosci* 29(3):562–574.

Hu, C. H., Wu, G. F., Wang, X. Q., *et al.* 2006. Transplanted human umbilical cord blood mononuclear cells improve left ventricular function through angiogenesis in myocardial infarction, *Chin Med J (Engl)* 119(18):1499–1506.

Hsieh, J. and Gage, F. H. 2004. Epigenetic control of neural stem cell fate, *Curr Opin Genet Dev* 14:461–469.

Jablonska, A., Kozlowska, H. and Markiewicz, I., *et al.* 2010. Transplantation of neural stem cells derived from human cord blood to the brain of adult and neonatal rats, *Acta Neurobiol Ex* 70:337–350.

Jang, Y., Park, J., Lee, M., Yoon, B., Yang, Y., Yang, S. and Kim, S. 2004. Retinoic acid-mediated induction of neurons and glial cells from human umbilical cord-derived hematopoietic stem cells, *J Neurosci Res* 75:573–584.

Janic, B., Guo, A. M., Iskander, A. S., *et al.* 2010. Human cord blood-derived AC133+ progenitor cells preserve endothelial progenitor characteristics after long term *in vitro* expansion, *PLoS One* 11;5(2):e9173.

Janowski, M., Lukomska, B., Kozlowska, H., *et al.* 2004 Intra-carotic administration of neural progenitors derived from human umbilical cord blood mononuclear fraction gives robust behavioural recovery in ouabain model of stroke, *Exp Neurology* 187:210.

Janowski, M., Kozlowska, H., Jurga, M., *et al.* 2006. Xenotransplantation of human umbilical cord blood neural stem cells (HUCB-NSC) into immunospressed rat brain, *Acta Neurobiol Exp (Wars)* 66:402–403.

Jeltsch, K. S., Radke, T. F., Laufs, S., *et al.* 2010. Unrestricted somatic stem cells: Interaction with CD34(+) cells *in vitro* and *in vivo*, expression of homing genes and exclusion of tumorigenic potential, *Cytotherapy* 18.

Jeong, J., Gang E., Hong, S., *et al.* 2004. Rapid neural differentiation of human cord blood derived mesenchymal stem cells, *Neuroreport* 15:1731–1734.

Jiang, Y., Henderson, D., Blackstad, M., *et al.* 2003. Neuroectodermal differentiation from mouse multipotent adult progenitor cells, *Pro Natl Acad Sci U S A* 100:11854–11860.

Jiang, L., Newman, M., Saporta, S., *et al.* 2008. MIP-1alpha and MCP-1 Induce Migration of Human Umbilical Cord Blood Cells in Models of Stroke, *Curr Neurovasc Res* 5(2):118–124.

Joannides, A., Gaughwin, P., Schwiening, C., *et al.* 2004. Efficient generation of neural precursors from adult human skin: Astrocytes promote neurogenesis from skin-derived stem cells, *Lancet* 364:172–178.

Jozwiak, S., Habich, A., Kotulska, K., *et al.* 2011. Intracerebroventricular transplantation of cord blood-derived neural progenitors in a child with severe global brain ischemic injury", *Cell Medicine: Cell Transplantation Part B* 1:71–89.

Jurga, M. and Domanska-Janik, K. 2004. Neurospheres of human umbilical cord blood derived neural stem cells — *in vitro* model of stem cell niche, *FENS Lisbon* 2; A040:16.

Jurga, M., Markiewicz, I., Sarnowska, A., *et al.* 2006. Neurogenic potential of human umbilical cord blood: Neural-like stem cells depend on previous long-term culture conditions, *J Neurosci Res* 83:627–637.

Jurga, M., Buzanska, L., Malecki, M., *et al.* 2006a. Function of ID1 protein in human cord blood-derived neural stemlike cells, *J Neurosci Res* 84(5):993–1002.

Jurga, M., Lipkowski, A. W., Lukomska, B., *et al.* 2009. Generation of functional neural artificial tissue from human umbilical cord blood stem cells, *Tissue Eng Part C Methods* 15:365–372.

Kang, K. S., Kim, S. W., Oh, Y. H., *et al.* 2005. A 37-year-old spinal cord-injured female patient, transplanted of multipotent stem cells from human UC blood, with improved sensory perception and mobility, both functionally and morphologically: A case study, *Cytotherapy* 7(4):368–373.

Kedong, S., Xiubo, F., Tianqing, L., *et al.* 2010. Simultaneous expansion and harvest of hematopoietic stem cells and mesenchymal stem cells derived from umbilical cord blood, *J Mater Sci Mater Med* PMID: 20924776.

Kelly, S., Bliss, T. M., Shah, A. K., *et al.* 2004. Transplanted human fetal neural stem cells survive, migrate, and differentiate in ischemic rat cerebral cortex, *Proc Natl Acad Sci U S A* 100:11839–11844.

Kelly, L., Newman, M., Smith, A., *et al.* 2009. Transplantation of human umbilical cord blood and embryonic stem cells into a 6-hydroxydopamine rodent model of Parkinson's disease: To repair, replace, or regenerate dopaminergic tissue in the nigral-striatal pathway, *Cell Transplant* 18:220.

Kim, J. M., Lee, S. T., Chu, K., *et al.* 2007. Systemic transplantation of human adipose stem cells attenuated cerebral inflammation and degeneration in a hemorrhagic stroke model, *Brain Res* 1183:43–45.

Kluth, S. M., Buchheiser, A., Houben, A. P., *et al.* 2010. DLK-1 as a marker to distinguish unrestricted somatic stem cells and mesenchymal stromal cells in cord blood, *Stem Cells Dev* 19(10):1471–1483.

Kögler, G., Sensken, S., Airey, J. A., *et al.* 2004. A new human somatic stem cell from placental cord blood with intrinsic pluripotent differentiation potential, *J Exp Med* 200:123–135.

Koh, S. H., Kim, K. S., Choi, M. R., *et al.* 2008. Implantation of human umbilical cord-derived mesenchymal stem cells as a neuroprotective therapy for ischemic stroke in rats, *Brain Res* 1229:233–248.

Kozłowska, H., Jabłonka, J., Janowski, M., *et al.* 2007. Transplantation of a novel human cord blood-derived neural-like stem cell line in a rat model of cortical infarct, *Stem Cells Dev* 16:481–488.

Kucia, M., Halasa, M., Wysoczynski, M., *et al.* 2007. Morphological and molecular characterization of novel population of CXCR4+ SSEA-4+ Oct-4+ very small embryonic-like cells purified from human cord blood: Preliminary report, *Leukemia* 21(2):297–303.

Kulbatski, I. 2010. Stem/precursor cell-based CNS therapy: The importance of circumventing immune suppression by transplanting autologous cells, *Stem Cell Rev* 6:405–410.

Kuh, S. U., Cho, Y. E., Yoon, D. H., *et al.* 2005. Functional recovery after human umbilical cord blood cells transplantation with brain-derived neutrophic factor into the spinal cord injured rat, *Acta Neurochir (Wien)* 147:985–992.

Larsson, L. C., Czech, K. A., Widner, H., *et al.* 1999. Discordant neural tissue xenografts survive longer in immunoglobulin deficient mice, *Transplantation* 68(8):1153–1160.

Larsson, L. C. and Widner, H. 2000. Neural tissue xenografting, *Scand J Immunol* 52:249–256.

Le Belle, J. E., Caldwell, M. A. and Svendsen, C. N. 2004. Improving the survival of human CNS precursor-derived neurons after transplantation, *J Neurosci Res* 76(2):174–183.

Lee, H. J., Lee, J. K., Lee, H., *et al.* 2010. Human umbilical cord blood-derived mesenchymal stem cells improve neuropathology and cognitive impairment in an Alzheimer's disease mouse model through modulation of neuroinflammation, *Neurobiol Aging* [Epub ahead of print].

Leeb, C., Jurga, M., McGuckin, C., *et al.* 2010. Promising new sources for pluripotent stem cells, *Stem Cell Rev* 6(1):15–26.

Liedtke, S., Enczmann, J., Waclawczyk, S., *et al.* 2007. Oct4 and its pseudogenes confuse stem cell research, *Cell Stem Cell* 1:364–366.

Liedtke, S., Buchheiser, A., Bosch, J., *et.al* 2010. The HOX Code as a "biological fingerprint" to distinguish functionally distinct stem cell populations derived from cord blood, *Stem Cell Res* 5(1):40–50.

Lindvall, O., Kokaia, Z., Martinez-Serrano, A. 2004. Stem cell therapy for human neurodegenerative disorders-how to make it work, *Nat Med* 10 (Suppl):42–50.

Liu, H. Y., Zhang, Q. J., Li, H. J., *et al.* 2006. Effect of intracranial transplantation of CD34+ cells derived from human umbilical cord blood in rats with cerebral ischemia, *Chin Med J* (Engl) 119:1744–1748.

McGuckin, C. P., Forraz, N., Allouard, Q., *et al.* 2004. Umbilical cord blood stem cells can expand hematopoietic and neuroglial progenitors *in vitro*, *Exp Cell Res* 295:350–359.

McGuckin, C. P., Forraz, N., Baradez, M. O., *et al.* 2005. Production of stem cells with embryonic characteristics from human umbilical cord blood, *Cell Proliferation*, 38(4):245–255.

McGuckin, C., Jurga, M., Ali, H., *et al.* 2008. Culture of embryonic-like stem cells from human umbilical cord blood and onward differentiation to neural cells *in vitro*, *Nat Protoc* 3:1046–1055.

Meier, C., Middelanis, J., Wasielewski, B., *et al.* 2006. Spastic paresis after perinatal brain damage in rats is reduced by human cord blood mononuclear cells, *Pediatr Res* 59(2):244–249.

Metcalfe, A. D., Ferguson, M. W., *et al.* 2008. Molecular and cellular basis of regeneration and tissue repair: skin stem and progenitor cells: Using regeneration as a tissue-engineering strategy, *Cell Mol Life Sci* 65:24–32.

Mitchell, K. E., Weiss, M. L., Mitchell, B. M., *et al.* 2007. Matrix cells from Wharton's jelly form neurons and glia, *Stem Cells* 21:50–60.

Nan, Z., Grande, A., Sanberg, C. D., *et al.* 2005. Infusion of human umbilical cord blood ameliorates neurologic deficits in rats with hemorrhagic brain injury, *Ann NY Acad Sci* 1049:84–96.

Newcomb, J. D., Ajmo, C. T. Jr., Sanberg, C. D., *et al.* 2006. Timing of cord blood treatment after experimental stroke determines therapeutic efficacy, *Cell Transplant* 15(3):213–223.

Newman, M. B., Smith, A. P., Kelly, L. P., *et al.* 2009. Nearly complete motor function recovery after the transplantation of neural-induced CD133 stem cells from human umbilical cord blood, in a rodent model of Parkinson's disease, *Cell Transplant* 18:228.

Niranjan, A., Fellows, W., Stauffer, W., *et al.* 2007. Survival of transplanted neural progenitor cells enhanced by brain irradiation, *J Neurosurg* 107:383–391.

Nystedt, J., Mäkinen, S., Laine, J., *et al.* 2006. Human cord blood CD34+ cells and behavioral recovery following focal cerebral ischemia in rats, *Acta Neurobiol Ex* 66:293–300.

Olstorn, H., Moe, M. C., Røste, G. K., *et al.* 2007. Transplantation of stem cells from the adult human brain to the adult rat brain, *Neurosurgery* 60:1089–1098.

Pacini, S., Carnicelli, V., Trombi, L., *et al.* 2010. Constitutive expression of pluripotency-associated genes in mesodermal progenitor cells (MPCs), *PLoS One* 25;5(3):e9861.

Pan, Y., Nastav, J. B., Zhang, H., *et al.* 2005. Engraftment of freshly isolated or cultured human umbilical cord blood cells and the effect of cyclosporin A on the outcome, *Exp Neurol* 192:365–372.

Park, D. H., Borlongan, C. V., Willing, A. E., *et al.* 2009. Human umbilical cord blood cell grafts for brain ischemia, *Cell Transplant* 18:985–998.

Pesce, M., Orlandi, A., Iachininoto, M. G., *et al.* 2003. Myoendothelial differentiation of human umbilical cord blood-derived stem cells in ischemic limb tissues, *Circ Res* 93(5):e51–e62.

Pimentel-Coelho, P. M., Magalhães, E. S., Lopes, L. M., *et al.* 2010. Human cord blood transplantation in a neonatal rat model of hypoxic-ischemic brain damage: Functional outcome related to neuroprotection in the striatum, *Stem Cells Dev* 19(3):351–358.

Poulsom, R., Alison, M. R., Forbes, S. J. *et al.* 2002. Adult stem cell plasticity, *J Pathol* 197(4):441–456.

Ramos, A. L., Darabi, R., Akbarloo, N., *et al.* 2010. Clonal analysis reveals a common progenitor for endothelial, myeloid, and lymphoid precursors in umbilical cord blood, *Circ Res* 107:1460–1469.

Rao, M. 2004. Conserved and divergent paths that regulate self-renewal in mouse and human embryonic stemcells, *Dev Biol* 275:269–286.

Redshaw, Z. and Strain, A. J. 2010. Human haematopoietic stem cells express Oct4 pseudogenes and lack the ability to initiate Oct4 promoter-driven gene expression, *J Negat Results Biomed* 31;9(1):2.

Rizvanov, A. A., Kiyasov, A. P., Gaziziov, I. M., *et al.* 2008. Human umbilical cord blood cells transfected with VEGF and L(1)CAM do not differentiate into neurons but transform into vascular endothelial cells and secrete neurotrophic factors to support neuro-genesis-a novel approach in stem cell therapy, *Neurochem Int* 53(6–8):389–394.

Rogers, I., Yamanaka N., Bielecki R., *et al.* 2007. Identification and analysis of *in vitro* cultured CD45- positive cells capable of multi-lineage differentiation, *Exp Cell Res* 313:1839–1852.

Ronsyn, M. W., Daans, J., Spaepen, G., *et al.* 2007. Plasmid-based genetic modification of human bone marrow-derived stromal cells: Analysis of cell survival and transgene expression after transplantation in rat spinal cord, *BMC Biotechnol* 7:90.

Safford, K. M., Hicok, K. C., Safford, S. D., *et al.* 2002. Neurogenic differentiation of murine and human adipose-derived stromal cells, *Biochem Biophys Res Commun* 294:371–379.

Sanberg, P. R., Willing, A. E., Garbuzova-Davis, S., *et al.* 2005. Umbilical cord blood-derived stem cells and brain repair, *Ann NY Acad Sci* 1049:67–83.

Sanchez-Ramos, J. R., Song, S., Kamath, S. G., *et al.* 2001. Expression of neural markers in human umbilical cord blood, *Exp Neurol* 171:109–115.

Saporta, S., Kim, J. J., Willing, A. E., *et al.* 2003. Human umbilical cord blood stem cells infusion in spinal cord injury: Engraftment and beneficial influence on behavior, *J Hematother Stem Cell Res* 12(3):271–278.

Stachowiak, E. K., Roy, I., Lee, Y. W., Capacchietti, M., Aletta, J. M., Prasad, P. N. and Stachowiak, M. K. 2009. Targeting novel integrative nuclear FGFR1 signaling by nanoparticle-mediated gene transfer stimulates neurogenesis in the adult brain, *Integr Biol (Camb)* 1(5–6):394–403.

Sun, W., Buzanska, L., Domanska-Janik, K., *et al.* 2005. Voltage-sensitive and ligand-gated channels in differentiating fetal neural stem cells isolated from the non-hematopoietic fraction of human umbilical cord blood, *Stem Cells* 23:931–945.

Sypecka, J., Dragun-Szymczak, P., Zalewska, T., *et al.* 2009. Laminin promotes oligogliogenesis and increases MMPS activity in human neural stem cells of HUCB-NSC line, *Acta Neurobiologiae Exp* 69(1):37–45.

Szablowska-Gadomska, I. Zayat, V., and Buzanska, L. 2011. Influence of low oxygen tensions on expression of pluripotency genes in stem cells, *Acta Neurobiol Exp* 71(1):86–93.

Taguchi, A., Soma, T., Tanaka, H., *et al.* 2004. Administration of CD34+ cells after stroke enhances neurogenesis via angiogenesis in a mouse model, *J Clin Invest* 114(3):330–338.

Takenaka, C., Nishishita, N., Takada, N., *et al.* (2010) Effective generation of iPS cells from CD34+ cord blood cells by inhibition of p53, *Exp Hematol* 38(2):154–162.

Teng, H., Zhang, Z. G., Wang, L., *et al.* 2008. Coupling of angiogenesis and neurogenesis in cultured endothelial cells and neural progenitor cells after stroke, *J Cereb Blood Flow Metab* 28(4):764–771.

Terskikh, A. V., Easterday, M. C., Li, L., *et al.* 2001. From hematopoiesis to neuropoiesis: evidence of overlapping genetic programs, *Proc Natl Acad Sci USA* 98(14):7934–7939.

Trombi, L., Pacini, S., Montali, M., *et al.* 2009. Selective culture of mesodermal progenitor cells, *Stem Cells Dev* 18(8):1227–1234.

Tsai, M. S., Hwang, S. M., Tsai, Y. L., *et al.* 2006. Clonal amniotic fluid-derived stem cells express characteristics of both mesenchymal and neural stem cells, *Biol Reprod* 74:545–551.

Veeravalli, K. K., Dasari, V. R., Tsung, A. J., *et al.* (2009) Human umbilical cord blood stem cells upregulate matrix metalloproteinase-2 in rats after spinal cord injury, *Neurobiol Dis* 36(1):200–212.

Vendrame, M., Cassady, J., Newcomb, J., *et al.* 2004. Infusion of human umbilical cord blood cells in a rat model of stroke dose-dependently rescues behavioral deficits and reduces infarct volume, *Stroke* 35(10):2390–2395.

Vendrame, M., Gemma, C., Pennypacker, K. R., *et al.* 2006. Cord blood rescues stroke-induced changes in splenocyte phenotype and function, *Exp Neurol* 199(1):191–200.

Walczak, P., Chen, N., Hudson, J. E., *et al.* 2004. Do hematopoietic cells exposed to a neurogenic environment mimic properties of endogenous neural precursors? *J Neurosci Res* 76:244–254.

Walczak, P., Chen, N., Eve, D., *et al.* 2007. Long-term cultured human umbilical cord neural-like cells transplanted into the striatum of NOD SCID mice, *Brain Res Bull* 74:155–163.

Wang, H. S., Hung, S. C., Peng, S. T., *et al.* 2004. Mesenchymal stem cells in the Wharton's jelly of the human umbilical cord, *Stem Cells* 22:1330–1337.

Wennersten, A., Meier, X., Holmin, S., *et al.* 2004. Proliferation, migration, and differentiation of human neural stem/progenitor cells after transplantation into a rat model of traumatic brain injury, *J Neurosurg* 100:88–96.

Willing, A. E., Lixian, J., Milliken, M., *et al.* 2003. Intravenous versus intrastriatal cord blood administration in a rodent model of stroke, *J Neurosci Res* 73:296–307.

Wong, C. J., Casper, R. F. and Rogers, I. M. 2010. Epigenetic changes to human umbilical cord blood cells cultured with three proteins indicate partial reprogramming to a pluripotent state, *Exp Cell Res* 1;316(6):927–939.

Xia, G., Hong, X., Chen, X., *et al.* 2010. Intracerebral transplantation of mesenchymal stem cells derived from human umbilical cord blood alleviates hypoxic ischemic brain injury in rat neonates, *J Perinat Med* 38(2):215–221.

Xiao, J., Nan, Z., Motooka, Y., *et al.* 2005. Transplantation of a novel cell line population of umbilical cord blood stem cells ameliorates neurological deficits associated with ischemic brain injury, *Stem Cells Dev* 14(6):722–733.

Xu, Y., Meng, H., Li, C., *et al.* 2010. Umbilical cord-derived mesenchymal stem cells isolated by a novel explantation technique can differentiate into functional endothelial cells and promote revascularization, *Stem Cells Dev* 19(10):1511–1522.

Yang, S. E., Ha, C. W., Jung, M., *et al.* 2004. Mesenchymal stem/progenitor cells developed in cultures from UC blood, *Cytotherapy* 6:476–486.

Ye, Z., Zhan, H., Mali, P., Dowey, S., *et al.* 2009. Human-induced pluripotent stem cells from blood cells of healthy donors and patients with acquired blood disorders, *Blood* 2009 114(27):5473–5480.

Yim, Y. S., Noh, Y. H., Kim, D. H., *et al.* 2010. Correlation between the immature characteristics of umbilical cord blood–derived mesenchymal stem cells and engraftment of hematopoietic stem cells in NOD/SCID mice, *Transplant Proc* 42:2753–2758.

Yoo, K. H., Jang, I. K., Lee, M. W., *et al.* 2009. Comparison of immunomodulatory properties of mesenchymal stem cells derived from adult human tissues, *Cell Immunol* 259(2):150–156.

Zandonella, C. 2005. Stem cell-therapies: The first wave, *Nature* 435:877–878.

Zaehres, H., Kögler, G., Arauzo-Bravo, M. J., *et al.* 2010. Induction of pluripotency in human cord blood unrestricted somatic stem cells, *Exp Hematol* 38(9):809–818, 818.e1-2.

Zhao, Y., Glesne, D. and Huberman, E. 2003. A human peripheral blood monocyte-derived subset acts as pluripotent stem cells, *Proc Natl Acad Sci U S A* 100:2426–2431.

Zigova, T., Song, S., Willing, A. E., *et al.* 2002. Human umbilical cord blood cells express neural antigens after transplantation into the developing rat brain, *Cell Transplant* 11:265–274.

Ziv, Y., Avidan, H., Pluchino, S., *et al.* 2006. Synergy between immune cells and adult neural stem/progenitor cells promotes functional recovery from spinal cord injury, *Proc Natl Acad Sci USA* 103:13174–13179.

Zuba-Surma, E. K., Klich, I., Greco, N., *et al.* 2010. Optimization of isolation and further characterization of umbilical-cord-blood-derived very small embryonic/epiblast-like stem cells (VSELs), *Eur J Haematol* 1;84(1):34–46.

Triggering Neuronogenesis by Endogenous Brain Stem Cells with DNA Nanoplexes

11

Ewa K. Stachowiak, Indrajit Roy, and Michal K. Stachowiak

An understanding of the mechanisms that control the transition of neural stem/progenitor cells (NS/PC) into functional neurons could potentially be used to recruit endogenously produced NS/PCs for neuronal replacement in a variety of neurological diseases. The elucidation and therapeutic control of neuronogenic mechanisms requires the ability to manipulate gene expression in NS/PCs *in vivo*. Nanotechnology has provided new therapeutic tools such as non-viral vectors for DNA delivery. We have employed organically modified silica-DNA nanoplexes capable of forming complexes with plasmid DNA for effective *in vivo* transfection of NS/PC in the mouse subventricular zone (SVZ). Using DNA nanoplexes for transfection, neuronogenesis can be effectively stimulated by genes engineered to target the integrative nuclear FGF receptor-1 signaling (INFS) pathway, a novel developmental mechanism that enhances the latent potential of NS/PCs to undergo neuronal differentiation. This novel nanobiological technology may open a new horizon for the treatment of a broad range of neurological disorders.

11.1. Can Neuronal Loss in Neurodegenerative Disorders of the Brain be Reversed?

There are many debilitating neurological disorders associated with extensive neuronal loss in the adult brain including progressive neurodegenerative disorders such as Parkinson's and Alzheimer's, aging-associated dementias, acute brain injuries and stroke. Accompanying the damaged brain tissue is functional deterioration of complex neurological, emotional, and mental capacities. These disorders are a major healthcare issue for the world's aging population. Similarly, neurodevelopmental disorders such as autism and schizophrenia are characterized by the diminished production of neuronal cells and the disruption of normal neuronal structure. A potential therapeutic option for patients of these and other related conditions involve the targeting of endongenous stem-like cells already present in the central nervous system (CNS), which may potentially be promoted to regenerate missing neurons. The instruction of stem cells to differentiate into neurons capable of repopulating diseased brain areas and restoring behavior may be paramount for the treatment of a variety of neurodegenerative and neurodevelopmental disorders.

11.2. Stem Cells in the Mature Brain Have Diminished Neuronogenic Capability

The CNS develops as a tubular structure with a central canal before subdividing into the cerebrum, brain stem, and spinal cord. Throughout development, these structures retain their central ventricular spaces. Groups of germinal cells in the superficial layers of the ventricular walls produce neurons and glia that will eventually populate the entire developing CNS. In the developed, mature brain, new neurons are generated only in two germinal "niches" — the subventricular zone (SVZ) (Gates *et al.*, 1995; Kuhn *et al.*, 1997) of the cerebral lateral ventricles and the subgranual zone (SGZ) of the hippocampal dental gyrus. In this chapter, we will focus on the SVZ niche. The SVZ contains a self-renewing population of neural stem cells that divide asymmetrically, producing replicate stem cells and a population of rapidly proliferating, transit amplifying progenitor cells capable of differentiating into glial and neuronal lineage cells (Gates *et al.*, 1995; Kuhn *et al.*, 1997) (Figure 11.1).

In the developing brain, this population of neural stem/progenitor cells (NS/PC) develop into neuroblasts (progeny committed specifically to the neuronal lineage), which ascend to generate layers of cortical and subcortical neurons. This process stops shortly after birth, and in the mature brain, the SVZ-derived neuroblasts migrate almost exclusively towards the olfactory bulb along a well-defined pathway, the rostral migratory stream (RMS) (Lois and Alvarez-Buylla, 1994) (Figure 11.1b). In the olfactory bulb, newly generated cells ascend radially, differentiate into interneurons in the granule cell and glomerular layers, and functionally integrate into olfactory circuitries. In other brain areas, adult neuronogenesis is very limited and relatively few new neurons are added to the cortical and subcortical circuits. In these regions, the progeny of the NS/PC from the SVZ differentiate mainly into astroglia and oligodendroglia.

A great challenge for neuroscience and medicine has been to determine whether neuronogenesis on a global level may be effectively reactivated in the adult brain in order to replenish neuronal populations in diseased brain regions.

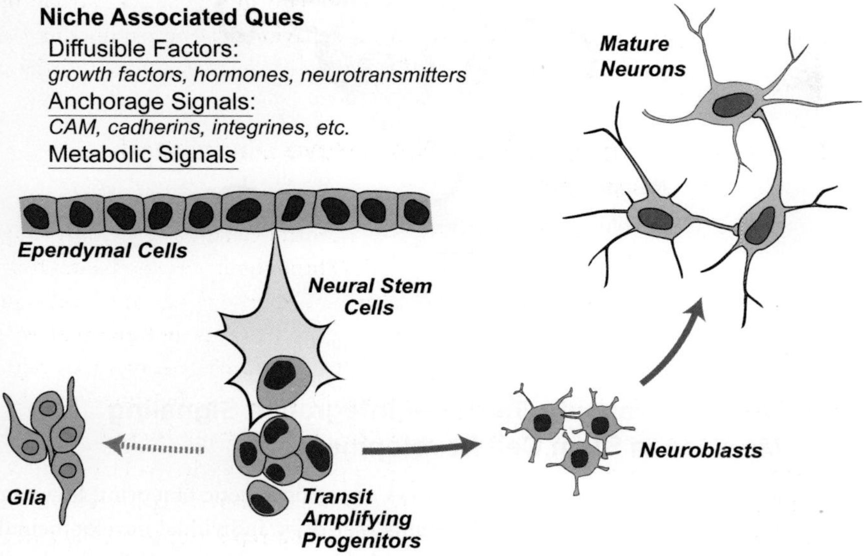

(a)

Figure 11.1. (a) The subventricular zone forms a structural and functional "nidus" in the brain lateral ventricle. The neural stem cells residing in the SVZ give rise to transit amplifying neural progenitor cells which may differentiate to a glial lineage or to migratory neuroblasts which develop into neurons. In the developing brain, these new neurons populate different regions, including the brain cortex and subcortical basal ganglia. However, in the mature brain, they are incorporated only into the neuronal networks of olfactory bulb. In the SVZ nidus, stem and progenitor cells are exposed to a vast plethora of factors including cell and extracellular matrix adhesion molecules, mechanical forces, neurotrophins, morphogens, hormones, vitamins and neurotransmitters. How are these stochastic signals integrated to elicit coordinate multigene developmental programs? Universal, integrative network modules responsible for self-renewal and mitotic expansion have been described. Our investigation has revealed an analogous signal-integrating module, integrative FGF-receptor 1 (FGFR1) signaling (INFS), which promotes postmitotic development of neural stem and progenitor cells into neurons (Chapter 4) (Fang *et al.*, 2005; Stachowiak *et al.*, 2003; Stachowiak, *et al.*, 2007). (b) *LV: lateral ventricles, OB: olfactory bulb, RMS: rostral migratory stream, STR: striatum, SVZ: subventricular zone. Red arrows indicate routes which supply neurons in developing brain and black arrows indicate neuronogenic route which remains active in developed brains.

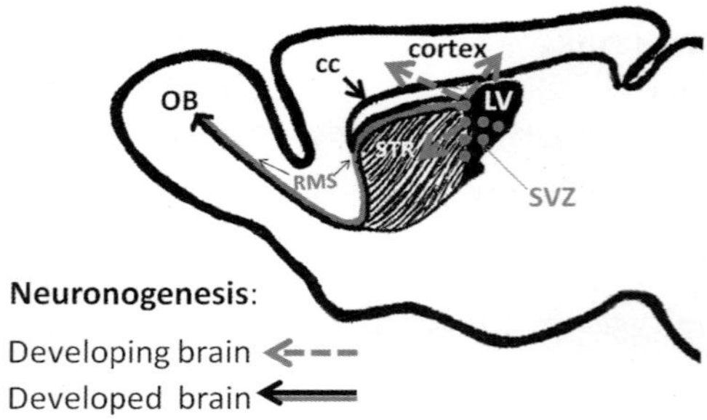

(b)

Figure 11.1. (*Continued*)

11.3. Novel Therapeutic Target — Integrative Signaling Modules in Stem Cell Development

It is believed that neural development follows a specific genetic blueprint, however little is known about how the gene programs that drive individual developmental phases such as stem cell self-renewal, mitotic expansion-amplification and differentiation are coordinately switched on and off. Within the neurogenic niche of the SVZ, there is a complex milieu of signals that navigate cells through the phases of neurogenesis (Figure 11.1a). These signals include mechanical forces exerted by vascular, neuronal, glial and ciliary microstructures, anchorage signals from cell surface and extracellular matrix-associated immobile molecules, diffusible growth factors, cytokines, neurotransmitters, and hormones. In experiments in which some of these individual factors were manipulated in isolation, changes in NS/PC proliferation and/or differentiation have been observed, thus supporting the roles of niche-supplied factors in neurogenesis.

In vivo, NS/PCs are exposed to a vast plethora of signals, some of which may promote different, mutually exclusive, developmental outcomes. It is of critical concern to determine how these stochastic niche-generated signals are integrated by recipient cells to elicit coordinate gene programs for cell self-renewal, proliferation and lineage development. Universal, integrative network modules responsible for self-renewal and mitotic expansion have been identified (see Chapters 1, 3 and 4), at least, in part. However, manipulation of these mechanisms in order to promote mitogenic expansion of NS/PC in the adult brain has had only a minor effect on the production of new mature neurons, perhaps due to inefficient NS/PC differentiation and survival. Our laboratory has identified an analogous universal regulatory module called integrative FGF-receptor 1 (FGFR1) signaling (INFS), which promotes postmitotic neuronal development and may offer an efficient means for increasing neuronogenesis (Chapter 4) (Fang *et al.*, 2005; Stachowiak *et al.*, 2003; Stachowiak, *et al.*, 2007). Central to this mechanism is the release of newly synthesized FGFR1 molecules from the endoplasmic reticulum into the cytosol, enabled by an atypical hydrophilic transmembrane domain. A portion of the released FGFR1 molecules are targeted for nuclear translocation by other means (Stachowiak *et al.*, 1996a,b; Dunham-Ems *et al.*, 2009; Myers *et al.*, 2003; Reilly and Maher, 2001; Bharali *et al.*, 2005). In cultured NS/PC and other stem-like cells, nuclear accumulation is promoted by diverse intra- and extracellular stimuli that induce neuronal development of those cells (Stachowiak, *et al.*, 2007). This neuronal differentiation is blocked by tyrosine kinase-deleted dominant negative FGFR1engineered to block the INFS, specifically in the cell nucleus. Thus, the INFS mechanism is essential for neuronal differentiation. Further investigation has determined that INFS acts through a "Feed-Forward-And-Gate" mechanism to execute stem cell transition from the proliferating multipotent or pluripotent state into the differentiated neuronal phenotype (Chapter 4 — see Figure 4.10) (Fang *et al.*, 2005). In this mechanism, diverse membrane receptors of hormones, growth factors, neurotransmitters and cell contacts transduce signals to sequence-specific transcription factors via classical cytoplasmic-nuclear signaling cascades.The same stimuli trigger nuclear translocation of FGFR1, a parallel signal that "feeds forward" the stimulation to the essential common transcriptional coactivator, CREB-binding protein (CBP). The CBP is present in limited quantities and is the functional equivalent of the "And Gate". By releasing CBP from inactive complexes, nuclear FGFR1 enables the activation of diverse genes and neuronal differentiation. Nuclear accumulation of FGFR1 and its interaction with the CBP occurs in brain NS/PCs, accompanying differentiation and providing an essential common gating mechanism through which diverse neuronogenic systems activate neuronal differentiation of human neural stem-like cells in culture (Fang *et al.*, 2005; Stachowiak *et al.*, 2003). Our collaborative studies showed that INFS also operates in embryonic stem cells (ESC) and controls neuronal commitment of these cells.

Technologies have been developed to directly access the INFS and control neuronal differentiation, thus bypassing diverse neuronogenic membrane receptors. One such technology involves transfection of the high molecular weight (23 kDa) form of fibroblast growth factor-2 (HMW-FGF2). This nucleus-targeted ligand stimulates nuclear accumulation by endogenous FGFR1, binds to FGFR1 in the nucleus and activates gene programs coordinately with FGFR1. Another effective technology involves transfection of an engineered recombinant [FGFR1(SP-/NLS)]. In FGFR1(SP-/NLS), the replacement of the membrane-targeting signal peptide (SP) with the nuclear localization signal (NLS) prevents receptor insertion into cytoplasmic membranes and directs it instead, exclusively to the nucleus to switch on neuronal differentiation (Fang *et al.*, 2005; Stachowiak *et al.*, 2007; Peng *et al.*, 2001, 2002). In contrast, "membrane-arrested" FGFR1/R4, in whichthe atypical transmembrane domain of FGFR1 is replaced with the typical hydrophobic transmembrane domain of FGFR4, does not enter the nucleus and fails to induce differentiation (Stachowiak *et al.*, 2003). The INFS has been shown to be both necessary and sufficient to direct *in vitro* neuronogenesis in human brain-derived NS/PC or umbilical cord blood-derived neural stem-like cells (Stachowiak *et al.*, 2003; Hu *et al.*, 2004; Fang *et al.*). Also, activation or reconstitution of INFS in different cancer cells was shown to induce their neuronal-like differentiation. In contrast, stimulation of cell surface FGFR1 with exogenous FGF2 (the 18 kDa secreted form) increases NS/PC or cancer cell proliferation (Stachowiak *et al.*, 2003, 2007).

11.4. INFS is Active in Developing Brain Cells

Factors which have been shown to activate INFS-mediated neuronal differentiation include: (i) neurotransmitter receptors like Ach (ii) secondary messengers (i.e. cAMP, Ca++), (iii) neurotrophins, (iv) morphogens like BMP7, (v) nuclear receptors of the retinoic acid (vitamin A), and (vi) changes in cell contact (see chapter 4). These factors are present in the SVZ niche and are thought to control neurogenic function *in vivo*. Hence, our goals have been to determine whether INFS is operating *in vivo* in SVZ neuronogenesis and whether these processes may be harnessed to augment neuronal supply in the adult brain. Our studies have aimed at providing a proof of concept for an endogenous stem cell-gene therapy approach to treat neuronal loss in the adult brain.

Immunocytochemistry combined with fluorescent or laser confocal microscopy allows for insight into the INFS mechanism in the developing and adult mouse brains. In the early postnatal (Figure 11.2i) and mature brain, FGFR1 in SVZ cells is concentrated predominantly in the cytoplasm and cell periphery. In contrast, in cells migrating out from the SVZ [i.e. in the rostral migratory stream (RMS)], FGFR1 is predominantly nuclear. Cells with nuclear FGFR1 do not incorporate bromodeoxyuridine (BrdU) into their DNA and thus are no longer dividing (Figure 11.2ii). This pattern is observed in the early postnatal and mature brain. However, in the developing brain (i.e. postnatal days 1–5), cells with FGFR1 localized in the nucleus are present in several different regions in addition to the RMS. For instance, in the midbrain substantia nigra compacta (SNc), where immature dopamine neurons grow their axons to the distant striatum, FGFR1 is nuclear and colocalizes with its partner protein CBP (Figure 11.2iiia).

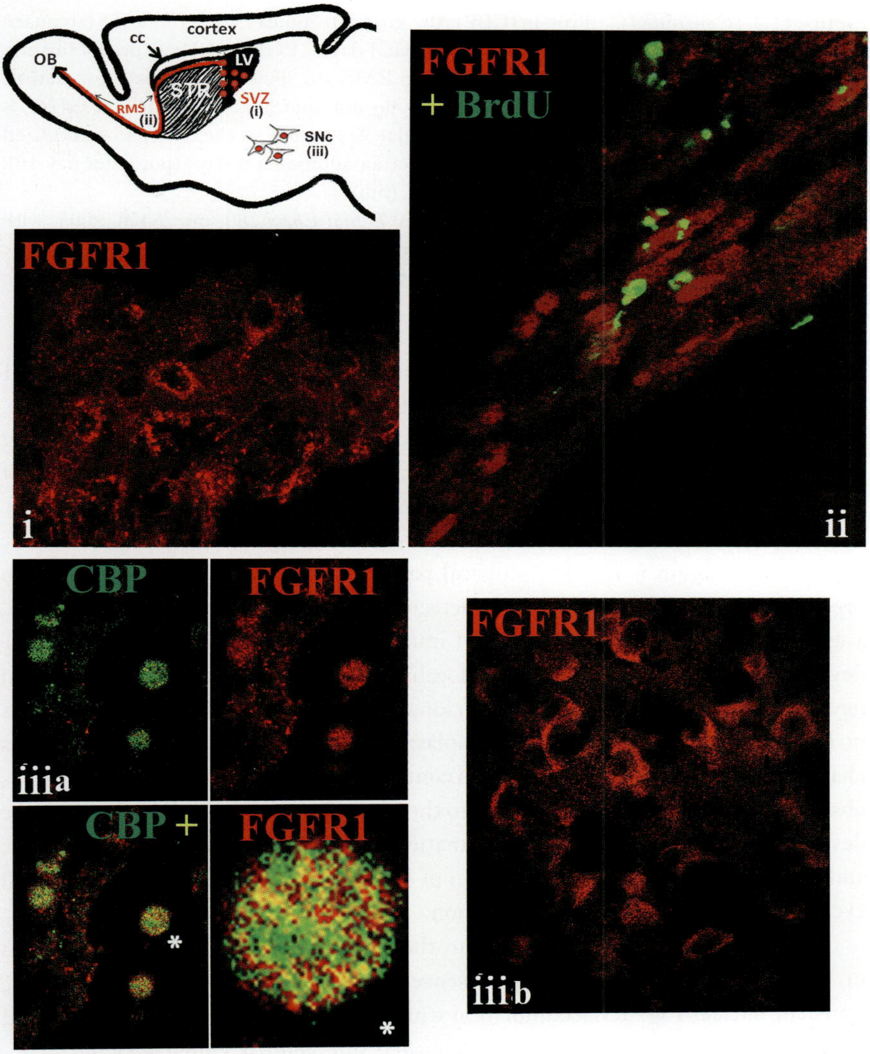

Figure 11.2. Nuclear FGFR1 accumulation correlates with postmitotic neuronal development *in vivo*. In the mouse brain, this diagram illustrates localization of SVZ with the neural stem/progenitor cells (red circles), RMS, SNc with developing nigro-striatal projection neurons and STR. Stained sections are derived from the indicated brain regions at postnatal day 4 (i–iiia) or postnatal day 10 (iiib). Mice were injected with BrdU (100 mg/kg i.p.) and were sacrificed 2 hours later. Brain sections were incubated with polyclonal (C-term) FGFR1 Ab/goat anti-rabbit-Cy3 Ab (red) and were co-stained with BrdU Ab or anti-CREB Binding

Figure 11.2. (*Continued*). Protein (CBP) Ab/ goat anti-mouse-Alexa Fluor488 (an image of 10-fold enlarged nucleus* is shown). At postnatal day 4, FGFR1-IR was predominantly SVZ (i) but is nuclear in cells that migrated to RMS (ii). Few RMS cells show mitotic activity (BrdU+). The mitotically active nuclei do not colocalize FGFR1(ii). FGFR1 is predominantly nuclear in developing (postnatal day 4) neurons of the SNc and colocalized with its nuclear partner protein, CBP (iiia). After an additional 6 days (postnatal day 10), FGFR1-IR in SNc is predominantly cytoplasmic (iiib).
Figures 11.2(i) and 11.2(iii) were reprinted from *J. Biol Chem* 280, pp. 28456, 2005, with permission of American Society for Biochemistry and Molecular Biology.
*RMS: rostral migratory stream, SNc: substantia nigra compacta, STR: striatum, SVZ: subventricular zone.

However as the development of dopamine neurons nears completion, FGFR1 again becomes cytoplasmic (Fang *et al.*, 2005; Stachowiak *et al.*, 2003; Hu *et al.*, 2004) (Figure 11.2iiib). Interference with the FGFR1 signaling in these developing dopamine neurons in a transgenic mouse model substantiates the central role of FGFR1 in neuronal development and behaviour (Klejbor *et al.*, 2006, 2009).

Another neuronogenic route active in the developing brain but largely inactive in the mature brain is the SVZ-cortical pathway. Developing cortical neurons in prenatal and early postnatal life can be detected with antibodies against β-III tubulin, a cytoskeletal protein characteristic of immature neurons (Figure 11.3b). These developing neurons express FGFR1 localized largely in the nuclei. At postnatal day 15 and later in life, cortical cells no longer express β-III tubulin and FGFR1 is no longer localized in the nucleus. Cytoplasmic (perinuclear) FGFR1 localization is characteristic of mature cortical cells. In contrast, accumulation of nuclear FGFR1 is observed in the new neurons supplied to the olfactory bulb via the RMS in both the developing brain (postnatal day 5) and mature brain (postnatal day 15, Figure 11.3c; and later in life). Nuclear accumulation of FGFR1 accompanies exit from the cell cycle, RMS migration and differentiation.

Together, these results conform to the earlier observations in cultured cells in which there is a cytoplasmic presence of FGFR1 in proliferating NS/PC, transient nuclear FGFR1 accumulation which drives neuronal differentiation, and depletion in mature neurons with established intercellular contacts (Fang *et al.*, 2005; Stachowiak *et al.*, 2003; Hu *et al.*, 2004; Fang *et al.*; Stachowiak *et al.*, 1997).

11.5. Reactivation of INFS in Mature Brain by Transfections of DNA Nanoplexes

The elucidation and therapeutic control of neuronogenic mechanisms *in vivo* requires the ability to manipulate gene expression. There are generally two types of transgene carriers: viral and non-viral. Although viral vectors such as *Herpes*

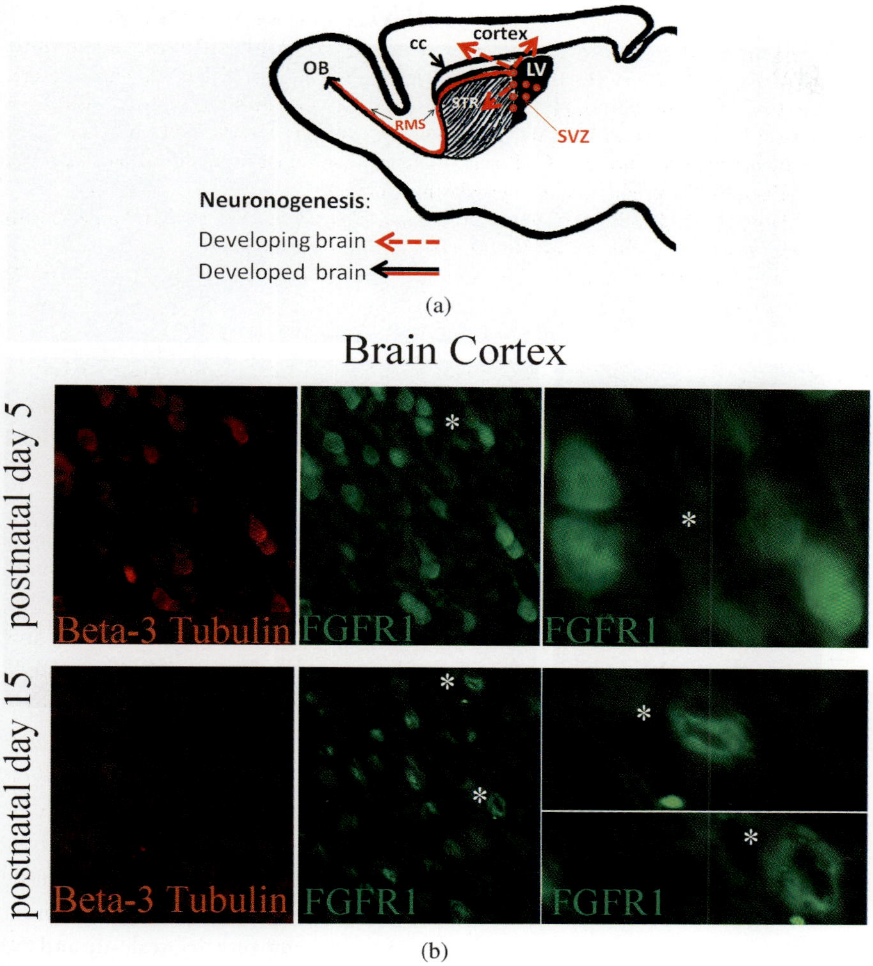

Figure 11.3. Nuclear FGFR1 accumulation correlates with postmitotic neuronal development *in vivo.* (b) INFS activity and cortical neurogenesis. At postnatal day 5, developing cortical neurons express β-III tubulin (red), a marker protein of immature neurons and FGFR1 (green) localized largely in the nuclei At Pd15, the cortical cells no longer express β-III tubulin and FGFR1 is cytoplasmic. (*fivefold enlarged cells). (c) In the SVZ-RMS-OB route which supplies new olfactory neurons in developing as well as mature brains, nuclear accumulation of FGFR1 is observed in cells exiting from the SVZ into RMS> OB at both postnatal days 5 and 15. More cells with nuclear FGFR1 appear in the OB on postnatal day 5 than 15.

*LV: lateral ventricles, OB: olfactory bulb, RMS: rostral migratory stream, STR: striatum, SVZ: subventricular zone. Red arrows indicate routes which supply neurons in developing brain and black arrows neuronogenic route which remains active in developed brain.

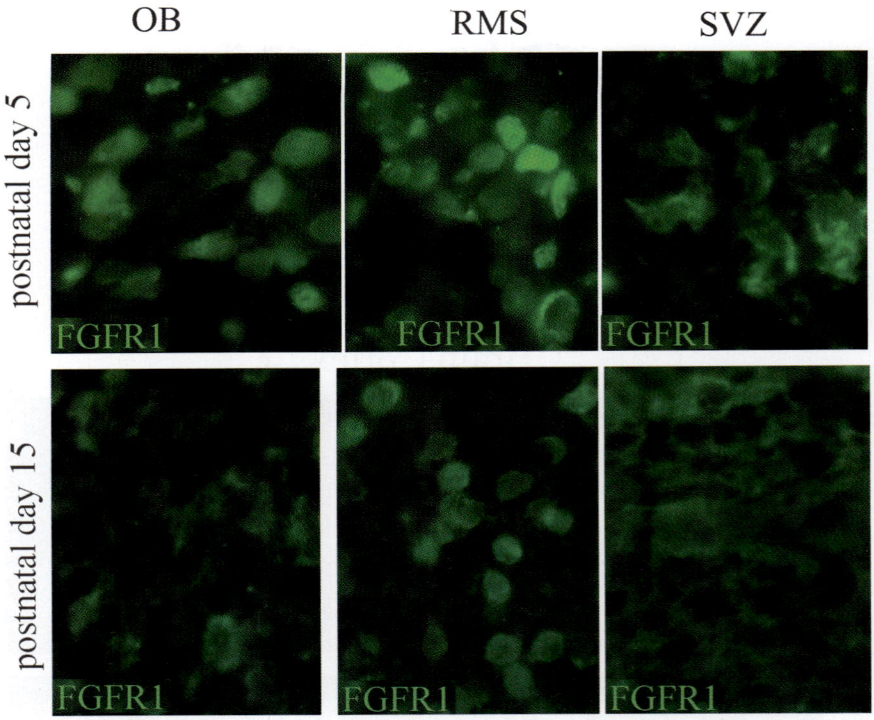

(c)

Figure 11.3. (*Continued*)

simplex-based viruses, Lentiviruses and retrovirus are attractive in terms of scientific strategy because of their natural mechanisms, such systems suffer from inherent difficulties of effective pharmaceutical processing, immunogenicity, scale-up and the possibility of reversion of an engineered virus to the wild type (Davis, 1997). Actual deaths have occurred in human trials, leading to a halt in further use of viral vectors for gene transfection (Check, 2003; Marwick, 2003). There is also the potential risk of excessive immune response (adenovirus) and insertional mutagenesis (retroviruses) when viruses are used as transfection vectors. Given the potential of gene transfer as a therapeutic tool, development of effective and safe transfection vectors for controlling brain stem cells *in vivo* is urgently needed. A number of non-viral chemical carriers like cationic or anionic liposome vectors have been explored including polylysine (Zauner *et al.*, 1998), polyornithine (Ramsay *et al.*, 2000), chromosomal proteins (Schwartz *et al.*, 1999), hydrogel polymers (Leong *et al.*,

1998), precipitated calcium phosphate (CaPi) (Graham and van der Eb, 1973) and calcium phosphate nanoparticles (Roy *et al.*, 2003) and natural biopolymers (Leong *et al.*, 1998; Lee *et al.*, 1998; Erbacher *et al.*, 1998; Truong-Le *et al.*, 1999; Roessler and Davidson, 1994). Polymers are increasingly being proposed as potential vectors because of their versatility in fine-tuning physico-chemical properties (Leong, 1999; Wu and Wu, 1987; Boussif *et al.*, 1995; Abdallah *et al.*, 1996; Coll *et al.*, 1999; Lemkine *et al.*, 2002).

Optimally, transgene expression in brain NS/PC should be transient. This is important because permanent transgene expression in the SVZ (i.e. by integrating viral vector with the host DNA) would either permanently stimulate NS/PC proliferation and potentially tumorigenesis or stimulate differentiation and block proliferation, thus exhausting the endogenous NS/PC population. The non-viral vectors render a transient transgene expression terminated by gradual dilution of the non-integrated episomal plasmid in continuously dividing cells.

The application of nanotechnology towards the development of non-viral vectors for safe and efficient DNA delivery is expected to help in the development of new therapeutic tools (Prasad, 2003; Luo and Saltzman, 2000). In recent studies, several groups have successfully prepared silica-based nanoparticles as potential gene transfection vectors (Csogor *et al.*, 2003; Kneuer *et al.*, 2000; Sameti *et al.*, 2003; Luo and Saltzman, 2000). In 1968, Stober *et al.* reported a stable method to produce highly monodispersed nanoparticles of silica by using a simple synthetic approach (Rossi *et al.*, 2005). This report led to a burst of research activities in the field of synthesis and applications of silica nanoparticles of various sizes. The chemically flexible surface of the silica nanoparticles can be exploited for the co-incorporation of diagnostic/therapeutic molecules for multimodal imaging and therapy. Organically modified silica nanoparticles are generally synthesized by the condensation-polymerization of their organosilane precursor molecules, where one or more of the alkoxy "arms" of the tetra-alkoxysilane precursor has been replaced by a hydrocarbon group. When this synthetic step is carried out in the confined dimensions of a microemulsion (either oil-in-water or water-in-oil), highly monodispersed and ultrafine (diameter less than 50 nm) nanoparticles are precipitated, which can be purified as an aqueous suspension through simple steps such as centrifugation or dialysis (Das *et al.*, 2002). Since these particles are mesoporous, they can serve as sustained release vehicles for encapsulated biomolecules (Slowing *et al.*, 2008). Importantly, the rich chemistry of silica can be exploited in order to synthesize particles, presenting a variety of active groups (carboxyl, thiol, or amino) on their surface, which can be further modified to incorporate imaging/diagnostic probes, genetic materials, and biorecognition molecules.

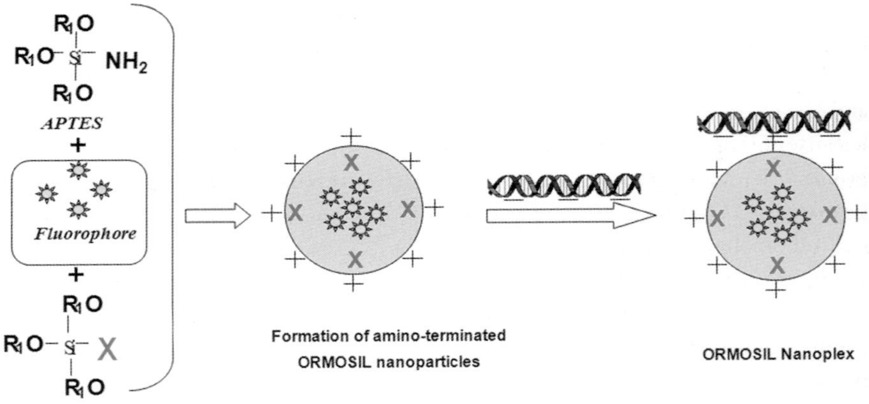

(a)

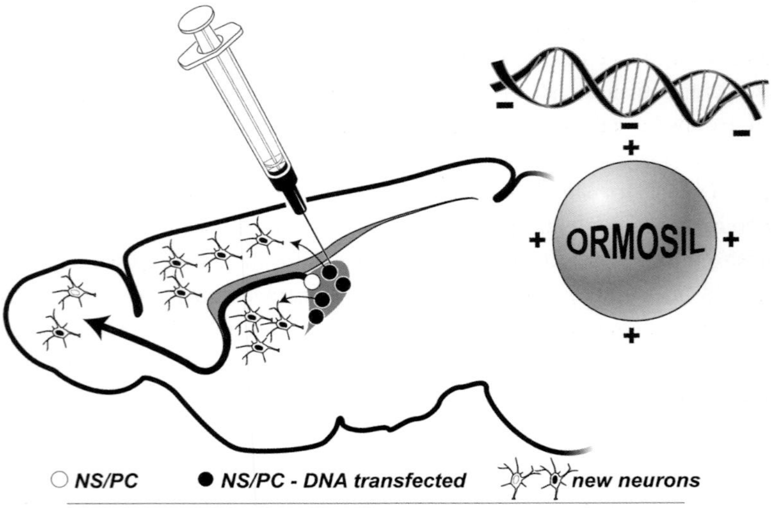

(b)

Figure 11.4. (a) Schematic formation of multimodal ORMOSIL. Various available ORMOSIL precursors can be mixed to generate nanoparticles with desired functionalities within their matrix as well as on their surface. The problem of water dispersibility has been overcome in this system by preparing nanoparticles in an aqueous system. Plasmid DNA or other genetic materials can be electrostatically incorporated on their surface. (b) By injecting ORMOSIL-DNA nanoplexes into brain ventricles, we find that targeting the INFS mechanism in NS/PC through *in vivo* gene transfer may effectively induce neuronogenesis in the adult brain.

*INFS: integrative nuclear FGFR1 signaling, NS/PC: neural stem/progenitor cells, ORMOSIL: organically surface modified silica nanoparticles.

Drs. Dhruba Bharali and Indrajit Roy (group led by Dr. Paras N. Prasad in the Department of Chemistry, SUNY, Buffalo) have synthesized organically surface-modified silica nanoparticles [referred to as ORMOSIL] based on the triethoxyvinyl-silane precursor and amino-functionalized on the surface by synchronous hydrolysis with 3-aminopropyltriethoxysilane (Bharali *et al.*, 2005; Roy *et al.*, 2005). This provides the opportunity for specifically tailored nanoparticles to be used in drug or gene delivery. Amino-terminated ORMOSIL nanoparticles can form electrostatic nanoplexes with negatively charged molecules such as plasmid DNA. This renders stabilization of the DNA against enzymatic degradation, thus allowing for efficient DNA carriage and DNA protection inside the cells. A major advantage of this system is the nanoparticles may be biodegraded through the biochemical decomposition of the Si–C bond.

The work by our interdisciplinary group has established the feasibility of using ORMOSIL as a vector for *in vivo* gene delivery into the brain with an efficiency comparable to that of a Herpes simplex virus-1 (Bharali *et al.*, 2005). Stereotaxic injections of the ORMOSIL complex with plasmid DNA encoding marker proteins such as enhanced green fluorescent protein (EGFP) into the mouse ventral midbrain, striatum and lateral ventricle have allowed us to fluorescently visualize the extensive transfection of neuronal-like cells for a few weeks. No general ORMOSIL-based toxicity was observed. With this outcome, we sought to determine if nanoplex-mediated transfection may allow us to manipulate the biology of the NS/PC cells *in vivo*, which could expand our knowledge of the mechanisms controlling their development. A compelling question is whether using ORMOSIL as a gene delivery platform could allow for effective manipulation of endogenous stem cells in the adult brain and thus the development of a new nanobiotechnology-based therapy.

11.6. INFS Targeting DNA-Nanoplexes Induce Neuronogenesis in the Adult Brain

To determine if DNA nanoplexes enable the expression of foreign genes in NS/PC *in vivo*, we performed stereotaxic injections of ORMOSIL complexed with plasmid-encoding EGFP into the brain lateral ventricles of mice. Ten days after injection, a fiber optic probe was stereotaxically inserted into the brain ventricle. The recorded images from live animals showed the presence of transfected cells in the ventricle walls expressing native EGFP fluorescence. This imaging technology revealed that there were more transfected fluorescent cells in the anterior/ventral region (Figure 11.5c) than in the posterior region of the lateral ventricle (Figure 11.5d).

Postmortem histological analysis was also performed to further identify the EGFP-expressing cells. At a different time after ORMOSIL-DNA injection, mice were anesthetized and perfused with 4% paraformaldehyde solution. The brains were additionally fixed with paraformaldehyde and sectioned. This process, although necessary for preserving the cellular structure of brain tissue, abolished EGFP fluorescence with extended fixations. Therefore, the expression of EGFP and other transfected proteins were visualized indirectly using their antibodies and immunocytochemistry (Bharali *et al.*, 2005). This approach revealed extensive EGFP transfection in cells of the SVZ (Figure 11.5a). Figure 11.5b shows a higher magnification of transfected cells close to the ventricle 7 days after the ORMOSIL-DNA injection. The presence of amino groups on the surface of ORMOSIL is critical for achieving high transfection efficiency [no significant EGFP expression was observed following injections of DNA with non-amino–terminated silica nanoparticles (Stachowiak *et al.*, 2009)].

Given the high efficacy of the ORMOSIL-mediated intraventricular transfections, we sought to determine if this approach could be used to influence the biology of the SVZ cells through targeting the INFS mechanism. To activate the INFS, we transfected DNA encoding the engineered nuclear FGFR1(SP-/NLS) (Peng *et al.*, 2001, 2002) or 23-kDa FGF2, a high molecular weight (HMW) nuclear ligand which binds to nuclear FGFR1 and elicits nuclear accumulation of endogenous FGFR1 (Dunham-Ems *et al.*, 2009). These recombinant vectors were transfected using ORMOSIL nanoplexes containing EGFP DNA to mark the transfected cells (marker proteins and INFS transgenes were expressed in the same transfected cells (Stachowiak *et al.*, 2009)). Intraventricular injections of nanoparticles complexed with DNA for FGFR1(SP-)/NLS or HMW-FGF2 markedly diminished cell proliferation in the SVZ area, demonstrating that the engineered nuclear receptor or the activation of endogenous nuclear FGFR1 allows modulation of the replication cycle of the stem/progenitor cells (Bharali *et al.*, 2005; Stachowiak *et al.*, 2009). These results conformed to *in vitro* studies of cultured human brain stem/progenitor cells in which stimulation of cell surface FGFR1 facilitated proliferation of non-differentiated cells, whereas transfection of nuclear FGFR1 caused cells to withdraw from the cell cycle and initiate neuronal differentiation (Fang *et al.*, 2005; Stachowiak *et al.*, 2003).

The *in vivo* anti-mitotic effects of nuclear FGFR1 and its ligand on SVZ cells were followed to determine whether *in vivo* activation of INFS may advance NS/PC differentiation towards neuronal or glial lineages and their migration to different brain regions. In recent years, the ORMOSIL synthesis has been further modified by increasing the density of the positive charges on the nanoparticles. This altered formulation, which has a higher content of amino groups on their surface

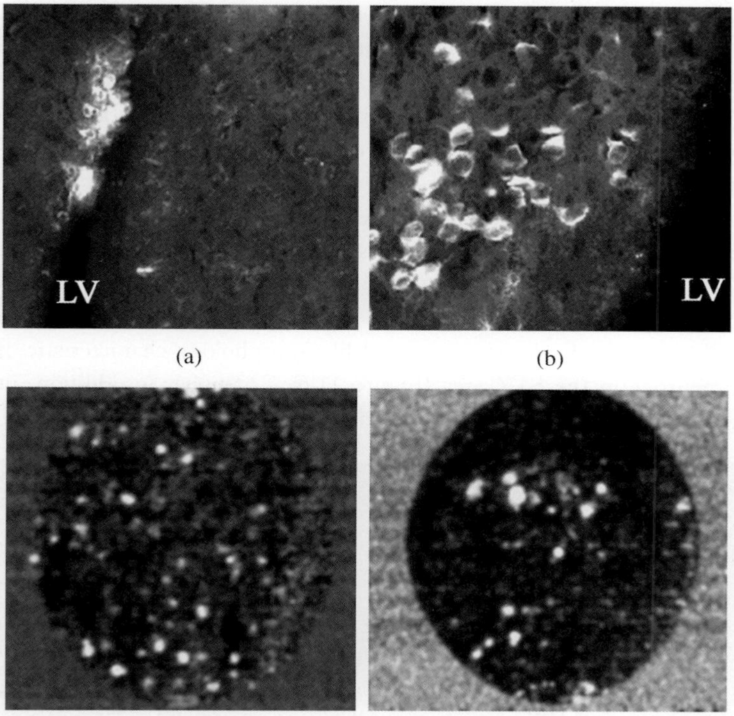

Figure 11.5. Transfection of ORMOSIL-DNA (pEGFP) nanoplex into the LV cells of the SVZ. (a–b) Seven days postmortem–brain fixation necessary to preserve the structural morphology denatures EGFP and therefore presence of expressed transfected protein is revealed by EGFP immunostaining: (a) Low magnification. (b) Higher magnification of positive region to visualize transfected cells. (c–d) *In vivo* imaging of native EGFP fluorescence in cells in the lateral ventricle. Ten days after transfection, mice were subjected to the second stereotaxic surgery and a miniature fiber optic Cell-vizio probe was inserted anterior dorsal region (c) or posterior region (d) of the lateral ventricle $>15\mu m$ from the medial ventricular wall. Dynamic sequences were recorded and selected frames are shown. Figure 11.5 was reprinted from *Proc Natl Acad USA* 102(32), pp. 11542, 2005, with permission of National Academy of Science, USA.
*LV: lateral ventrical, SVZ: subventricular zone.

(Stachowiak *et al.*, 2009), leads to the expression of transfected EGFP initially (5 days after transfection), in the immediate vicinity of the lateral ventricle (injection site), thus allowing for the study of subsequent migration and development of transfected cells.

Eighteen days after gene transfer, in mice injected with control ORMOSIL/EGFPDNA, EGFP-expressing cells were detected in the walls of the lateral ventricle (Figure 11.6; B2). In addition, EGFP was present in cells entering the rostral migratory stream (RMS) in the region connected to SVZ (Figure 11.6; B1) (Stachowiak *et al.*, 2009). The great majority of EGFP-positive cells were representative of typical round non-differentiated SVZ/RMS cells (Figure 11.6; B1, B2). Only very few transfected cells were detected in more distant brain regions like the olfactory bulb, frontal cortex and striatum.

EGFP positive cells were also found in animals transfected with FGFR1 (SP-/NLS) + EGFP or HMW-FGF2 + EGFP (Figure 11.6; C2, D2), however these cells typically displayed a neuronal-like morphology characteristic of maturating neurons in the SVZ area (Figure 11.6; C2, inset). In addition, in mice

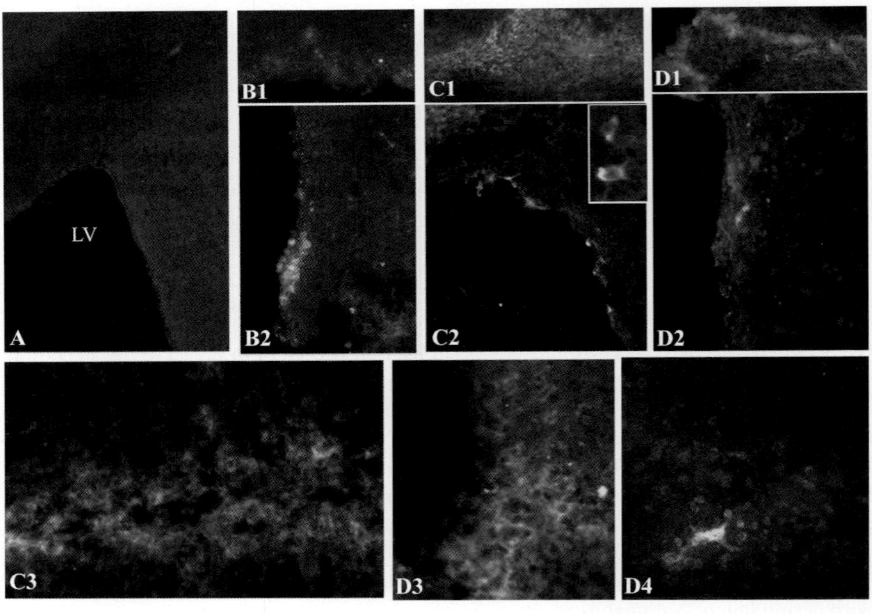

Figure 11.6. From (Stachowiak *et al.*, 2009) Expression of EGFP in mouse brain 18 days after injection of ORMOSIL-pEGFP-N2 into the brain lateral ventricle (LV). Brain sections were immunostained with EGFP antibody. (a) Nanoparticle control (no DNA) (b) ORMOSIL + control EGFP: RMS (B1), SVZ (B2); (c) ORMOSIL + EGFP + FGFR1(SP-/NLS): RMS (C1), SVZ (C2, insert — higher magnification of transfected cells), brain cortex (C3); (d) ORMOSIL + EGFP + HMW-FGF-2: RMS (D1), SVZ (D2), brain cortex (D3), Olfactory Bulb (D4). Size bar: (A) 70 um; (B1-D4) 25 um; (C2) insert 8.3 um. Figure 11.6 was reprinted from *Integrative Biology* 1, pp. 397, 2009, with permission of Royal Society of Chemistry.

receiving FGFR1(SP-/NLS) or HMW-FGF2, markedly more transfected cells entered the RMS (Figure 11.6; C1, D1) than in mice transfected with control EGFP (Figure 11.6; B1). Unlike the control transfected animals, mice receiving HMW-FGF2 or FGFR1(SP-/NLS) showed a robust presence of transfected cells in the cortex cingulate gyrus (Figure 11.6; C3, D3), olfactory bulb (Figure 11.6; D4) and in the subcortical striatum (not shown). These results indicate that the nuclear FGF receptor and its ligand stimulate morphological neuronal-like differentiation and the migration by SVZ cells to other brain regions (Stachowiak *et al.*, 2009).

To follow the developmental progression of the transfected SVZ cells, we investigated expression of the developmental stage-specific marker proteins in transfected cells identified by the expression of EGFP. In the SVZ, populations of cells that are believed to have pluripotent neural stem cell properties express glial fibrillary acidic protein (GFAP) (Seri *et al.*, 2001) and nestin (Lendahl *et al.*, 1990). Later, GFAP expression continues in the glial lineage of differentiating NS/PC, but is turned off in the neuronal lineage at the time when the cells begin to express double cortin (DCX) and become migratory neuroblasts (Couillard-Despres *et al.*, 2005). The expression of nestin also ceases abruptly with the appearance of DCX (Mellodew *et al.*, 2004).The neuroblast-specific DCX is a reliable and specific marker of neurogenesis but not neuronal regeneration/plasticity events or gliogenesis (Couillard-Despres *et al.*, 2005; Kuhn *et al.*, 1996; Brown *et al.*, 2003). Later, the expression of DCX may be diminished and a new protein β-III tubulin is induced, a cytoskeletal protein typically present in young, immature neurons. In adult differentiated neurons, neurofilament and other mature cytoskeletal proteins are expressed. Neurons that develop into the components of different neurotransmitter systems express their specific synaptic proteins. For example, dopamine neurons of the olfactory bulb express tyrosine hydroxylase (TH), a rate-limiting enzyme in dopamine synthesis.

We found that transfections with recombinant nuclear FGFR1(SP-NLS) or nuclear FGFR1 ligand HMW-FGF2 into the SVZ causes immature NS/PCs to cease proliferation, turn off the GFAP and nestin expression, and differentiate into DCX-expressing migratory neuroblasts (Stachowiak *et al.*, 2009). The FGFR1(SP-NLS) or HMW-FGF2-induced neuroblasts appeared not only in the olfactory bulb but, most notably, in the brain cortex and subcortical brain regions, including the striatum. Double immunostaining for EGFP and DCX revealed that migrating neuroblasts represent the transfected cells. These transfected cells progressed further towards neuronal differentiation by expressing neuronal markers β-III tubulin, a cytoskeletal protein typically present in young neurons, and neurofilament L, a marker of mature neurons. Cells that migrated into

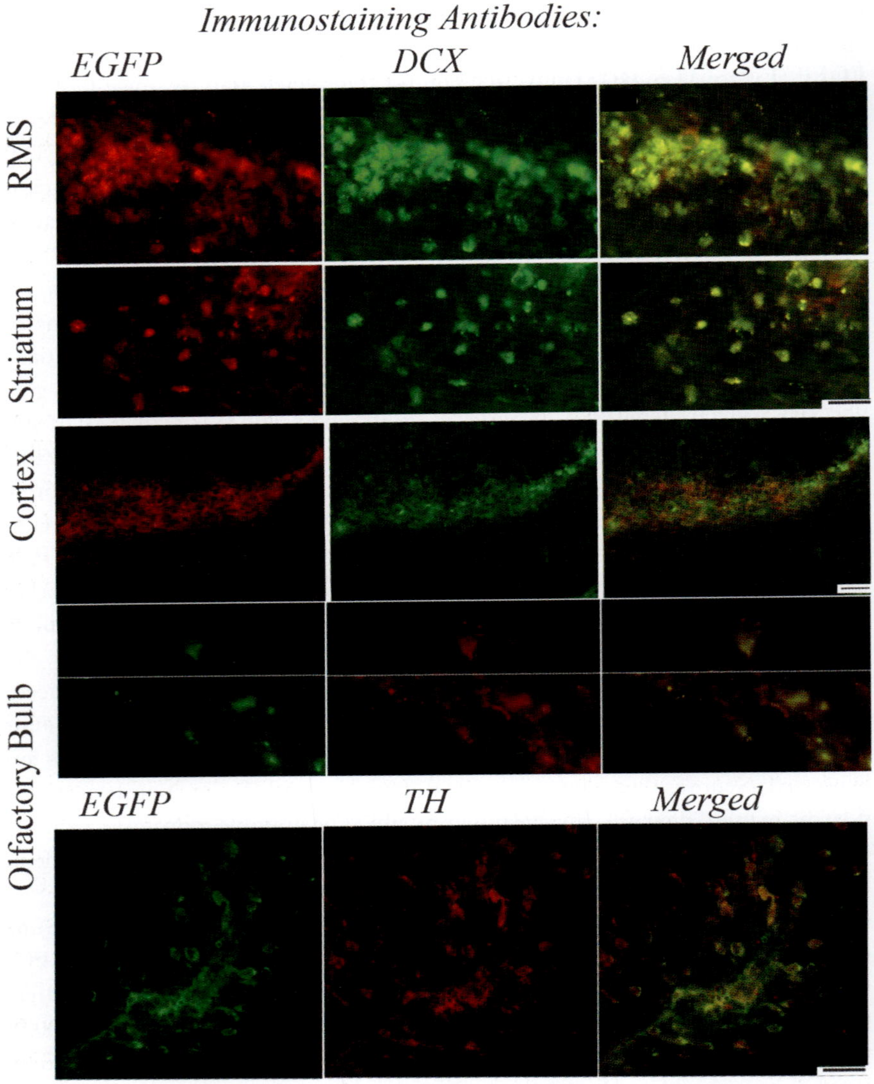

Figure 11.7. Neuronal differentiation of cells transfected with nuclear R1(SP-/NLS)/EGFP. Mice were perfused 18 days after ORMOSIL-DNA transfection into the lateral brain ventricles. Double immunostaining was performed to detect transfected cells (EGFP+) and the migratory neuroblasts (DCX+) or transfected (EGFP+) mature olfactory neurons (TH+). Single and merged images are shown (across our studies we used different secondary antibodies to ascertain that the staining can be observed with

Figure 11.7. (***Continued***). with different antibodies/fluorophores). Transfection of nuclear R1(SP-/NLS) led to a robust migration of transfected cells from SVZ to RMS, striatum, cortex and olfactory bulb (this was not observed in brains transfected with control EGFP not shownf). Majority of the R1(SP-/NLS)/EGFP transfected cells expressed DCX typical to migratory neuroblasts. In the olfactory bulb, several differentiated TH-expressing neurons were derived from the transfected (EGFP+) cells. In our report, we have also described further differentiation of transfected cells into β-III tubulin or neurofilament L-expressing neurons (not shown here). Size bar: 20 um. Figure 11.7 was reprinted from *Integrative Biology* 1, pp. 399, 2009, with permission of Royal Society of Chemistry.
*DCX: doublecortin, RMS: rostral migratory stream, SVZ: subventricular zone, TH: tyrosine hydroxylase.

the olfactory bulb expressed TH, indicating their differentiation into dopamine-producing neurons (Stachowiak *et al.*, 2009). Examples of R1(SP-/NLS)/EGFP transfected cells (EGFP+) coexpressing DCX (in RMS, striatum, cortex, olfactory bulb) or TH (olfactory bulb) are presented on Figure 11.7. The absence of transgene expression in mature astrocytes and the coexpression in DCX-positive neuroblasts (Stachowiak *et al.*, 2009) demonstrated that upon INFS stimulation, SVZ NS/PC progresses specifically towards neuronal phenotype, not towards the astrocytic lineage. The data thus strongly indicate an instructive function of INFS in the promotion of neuronal lineage.

The development of proliferating NS/PC into postmitotic neuronal lineage cells was further documented in BrdU pulse labeling-chase experiments. To label and track dividing SVZ cells, mice were injected i.p daily with BrdU for 5 days. These injections, which labeled mitotically active cells, were followed by intraventricular transfections of ORMOSIL DNA nanoplexes. Differentiation of the BrdU-labeled cells were studied 2–3 weeks later by triple-immunolabeling (Stachowiak *et al.*, 2009) for BrdU (marking cells born prior to gene transfection), EGFP (marking transfected cells) and for one of the neuronal markers (DCX), neurofilament L or TH (Stachowiak *et al.*, 2009). In mice transfected with the control nanoplex, the majority of transfected (EGFP+) NS/PC continued to proliferate, thereby diluting their BrdU label. Very few of these cells differentiated into DCX+ neuroblasts (Figure 11.8a, 1st column). In contrast, in mice transfected with R1(SP-/NLS)/EGFP or HMW-FGF-2/EGFP nanoplexes, notably more cells retained the BrdU label (indicating mitotic arrest) and coexpressed DCX, β-III tubulin, neurofilament or TH. A massive migration of EGFP+/BrDU+ cells coexpressing DCX (Figure 11.8a), β-III tubulin, neurofilament, and TH was observed, moving into the olfactory bulb, cortex and striatum. Examples of BrdU-labeled, EGFP positive cells that developed into neurofilament-L expressing

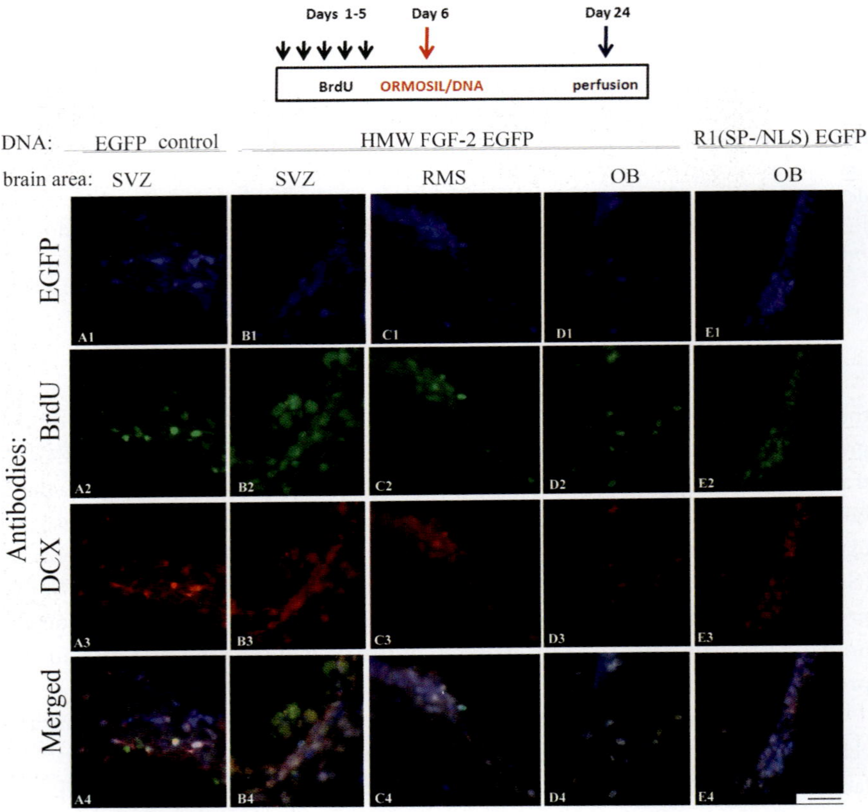

(a)

Figure 11.8. From (Stachowiak *et al.*, 2009). (a) New born cells develop into neuronal lineage neuroblasts after transfection of nuclear HMW-FGF2 or nuclear FGFR1(SP-/NLS). Mice received i.p. BrdU injections for 5 consecutive days followed by transfection of ORMOSIL with (a) control EGFP, or (b–d) HMW-FGF2 + EGFP or (e) FGFR1 (R1)(SP-/NLS)+EGFP. Mice were perfused after an additional 18 days. Brain sections were triple-immunostained for EGFP (A1-E1), BrdU (A2-E2) and DCX (A3-E3); merged images (A4-E4). Note lack of EGFP coimmunostaining with BrdU or DCX in (a) and frequent triple coimmunostaining in (b–e). Size bar = 25 μm (b) Differentiation of newly generated SVZ cells into neurofilament L (NfL) expressing neurons. Mice received i.p. BrdU injections for 5 consecutive days followed by intraventricular ORMOSIL transfections of: (a) control EGFP DNA, or (b) R1(SP-/NLS)/EGFP. Mice were perfused after an additional 18 days. Brain sections were triple-immunostained for EGFP (A1-B1), BrdU (A2-B2) and NfL (A3-B3); merged images (A4-B4). Note lack of EGFP coimmunostaining with BrdU or NfL in (a) and frequent triple coimmunostaining (b; cells expressing all three antigens) in mice transfected with R1(SP-/NLS).This outcome indicates that dividing NS/PC upon transfection of R1(SP-/NLS) leave the cell cycle and develop into mature (NfL+) neurons.

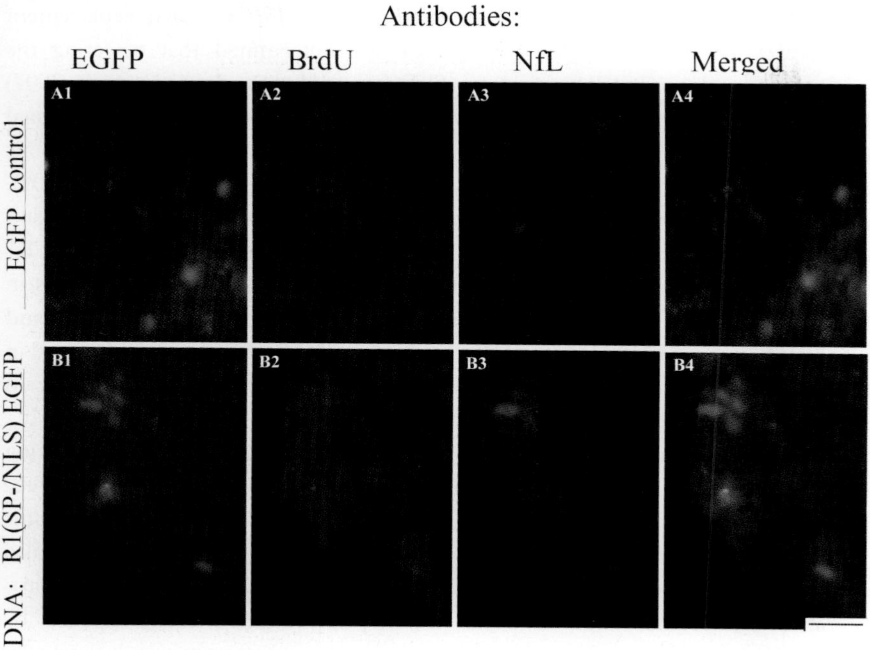

(b)

Figure 11.8. (*Continued*). Examples of such cells in striatal region region are shown. The same effects were produced by HMWFGF2 ligand (not shown here). In control DNA transfected mice the transfected (EGFP+) cells were mostly BrdU negative suggesting additional division after transfection. Thus, activation of INFS promotes differentiation of newly generated SVZ cells into neurofilament L (NfL) expressing neurons. Size bar = 25 μm. Figure 11.8a was reprinted from *Integrative Biology* 1, pp. 399–400, 2009, with permission of Royal Society of Chemistry. Figure 11.8b was reprinted from *Integrative Biology* 1, 2009, Supplemental Information, with permission of Royal Society of Chemistry.

neurons are shown in Figure 11.8b. These data thus strongly indicate an instructive function of INFS in the promotion of neuronal lineage development.

11.7. Conclusion

In the mature, intact mammalian brain, neuronogenesis is restricted primarily to the hippocampal dentate gyrus and the SVZ of the lateral ventricle/olfactory bulb axis (Altman and Das, 1965). An understanding of the mechanisms that control the transition of brain stem cells into functional neurons during development before becoming restricted in adulthood could potentially provide the opportunity for recruitment of endogenously produced NS/PC for neuronal replacement in diseased adult brain regions. Our work has determined that targeting the integrative nuclear FGFR1 signaling (INFS) module (Stachowiak *et al.*, 2007) through nanoparticle-mediated gene transfer may effectively stimulate neuronal development of SVZ NS/PC and its migration into the "non-neuronogenic" regions of the adult brain, thus reinstating neuronogenic capacity (Stachowiak *et al.*, 2009). It has been demonstrated by two independent means that increased expression of nuclear FGFR1 by transfection of either an engineered nuclear FGFR1 (SP-/NLS) or HMW-FGF2 has a distinct stimulating effect on neurogenesis in the adult SVZ, followed by exportation of new neurons into cortex, striatum and olfactory bulbs. Thus, the INFS-mediated induction of neuronogenesis observed in the present study re-emphasizes the possibility of latent neurogenesis in these brain regions.

The rate of neuronogenesis is thought to be determined by factors stimulating cell proliferation, including extracellular FGFs that interact with cell surface receptors [for review, see Guillemot (2005)]. In addition, however, mechanisms which hasten neuronal differentiation programs in progenitor cells could also stimulate neuronogenesis. Indeed, our findings demonstrate that while the proliferation and number of the GFAP/nestin-positive NS/PC are reduced by nuclear HMW-FGF2 or FGFR1(SP-/NLS), neuronal differentiation is markedly increased, thus creating a net increase in neuronogenesis (Stachowiak *et al.*, 2009). Such a mechanism is consistent with evidence suggesting that under normal conditions, the differentiation rate of the NS/PC is low due to the combination of infrequent conversions into DCX-positive neuronal precursors along with the natural developmental cell death of a proportion of the DCX-positive neuroblasts (Biebl *et al.*, 2000). Our studies indicate that the generation of neurons may be effectively regulated at the post-mitotic stage. An increase in neurogenesis could potentially be mediated by improving the rate of cell survival. However, since the INFS can stimulate neuronal differentiation in several *in vitro* neuronal models and activate expression of genes associated with differentiation (Peng *et al.*, 2002), we conclude that INFS specifically stimulates neuronal differentiation.

Stimulation of cell surface FGFR1 by the FGF peptide injected into the brain lateral ventricles effectively increases proliferation of SVZ cells (Craig *et al.*, 1996; Kawamata *et al.*, 1997). In contrast, transfected nuclear FGFR1(SP-/NLS) inhibits NS/PC proliferation but stimulates differentiation (Stachowiak *et al.*, 2007, 2009). Hence, it may be possible to sequentially stimulate proliferation and differentiation of NS/PC by co-injecting FGF2 peptide along with ORMOSIL-R1 (SP-/NLS)-EGFP into the brain ventricle. The mitogenic expansion of NS/PC by stimulation of cell surface FGF receptors with extracellular ligands (i.e. 18 kDa FGF-2) synchronized with nuclear FGFR1(SP-/NLS)-triggered differentiation could potentially maximize the generation of new neurons in adult brain neurons.

The recombinant INFS gene technology combined with our second technology, silica nanoparticles-mediated gene transfers, could allow for the INFS pathway to be directly targeted, thus stimulating endogenous neuronogenic capacity. The ability to shepherd NS/PC to neuronogenesis in the adult brain could be a ground-breaking development in the management of psychiatric and neurologic disorders where there is neuronal underdevelopment or attrition.

Further testing is needed to determine whether such nanoparticle-mediated functional gene delivery translates into actual therapy of these and other diseases. One drawback of these nanoparticles is the necessity for direct invasive injection in the brain to achieve efficient transgene expression. New strategies may be developed which would allow for similar levels of gene delivery in the brain following non-invasive, systemic injection of the nanoplexes. This could potentially be achieved by adapting the "trojan horse" approach for delivery of inorganic nanoparticles such as ORMOSIL across the blood–brain barrier, which has already shown promise for brain-specific delivery of liposomal and polymeric nanocarriers (Shi and Pardridge, 2000). Recently, such a delivery has been reported using rod-shaped nanocrystals of semiconductor materials (quantum rods) bioconjugated with the iron-transporting protein transferring (Xu *et al.*, 2008). It will be interesting to observe whether ORMOSIL nanoparticles complexed with suitable plasmid DNA can also be similarly delivered into the brain. An alternative or an additional approach could involve development of small molecules that may effectively activate the INFS module in brain stem cells when administered systemically. In collaboration with the pharmaceutical industry, our laboratories are pursuing such new therapeutic agents.

Acknowledgement

We gratefully acknowledge grant support from the New York State Stem Cell Science (NYSTEM).

Bibliography

Altman, J. and Das, G. D. 1965. Autoradiographic and histological evidence of postnatal hippocampal neurogenesis in rats, *J Comp Neurol* 124(3):319–335.

Abdallah, B., *et al.* 1996. A powerful nonviral vector for *in vivo* gene transfer into the adult mammalian brain: Polyethylenimine, *Hum Gene Ther* 7(16):1947–1954.

Boussif, O., *et al.* 1995. A versatile vector for gene and oligonucleotide transfer into cells in culture and *in vivo*: Polyethylenimine, *Proc Natl Acad Sci U S A* 92(16):7297–7301.

Brown, J. P., *et al.* 2003. Transient expression of doublecortin during adult neurogenesis, *J Comp Neurol* 467(1):1–10.

Bharali, D. J., *et al.* 2005. Organically modified silica nanoparticles: A nonviral vector for *in vivo* gene delivery and expression in the brain, *Proc Natl Acad Sci U S A* 102(32):11539–11544.

Biebl, M., *et al.* 2000. Analysis of neurogenesis and programmed cell death reveals a self-renewing capacity in the adult rat brain, *Neurosci Lett* 291(1):17–20.

Coll, J. L., *et al.* 1999. *In vivo* delivery to tumors of DNA complexed with linear polyethylenimine, *Hum Gene Ther* 10(10):1659–1666.

Check, E. 2003. Cancer risk prompts US to curb gene therapy, *Nature* 422:6927.

Csogor, Z., *et al.* 2003. Modified silica particles for gene delivery, *Materials Science & Engineering C-Biomimetic and Supramolecular Systems* 23(1–2):93–97.

Couillard-Despres, S., *et al.* 2005. Doublecortin expression levels in adult brain reflect neurogenesis, *Eur J Neurosci* 21(1):1–14.

Craig, C. G., *et al.* 1996. *In vivo* growth factor expansion of endogenous subependymal neural precursor cell populations in the adult mouse brain, *J Neurosci* 16(8): 2649–2658.

Davis, S. S. 1997. Biomedical applications of nanotechnology — implications for drug targeting and gene therapy, *Trends Biotechnol* 15(6):217–224.

Das, S., Jain, T. K. and Maitra, A. 2002. Inorganic-organic hybrid nanoparticles from n-octyl triethoxy silane, *J Colloid Interface Sci* 252(1):82–88.

Dunham-Ems, S. M., *et al.* 2009. Fibroblast growth factor receptor-1 (FGFR1) nuclear dynamics reveal a novel mechanism in transcription control, *Mol Biol Cell* 20(9): 2401–2412.

Erbacher, P., *et al.* 1998. Chitosan-based vector/DNA complexes for gene delivery: Biophysical characteristics and transfection ability, *Pharm Res* 15(9):1332–1339.

Fang, X., *et al.* 2005. Control of CREB-binding protein signaling by nuclear fibroblast growth factor receptor-1: A novel mechanism of gene regulation, *J Biol Chem* 280(31):28451–28462.

Fang, X., *et al.* Control of CBP signaling by nuclear fibroblast growth factor receptor-1 (FGFR1) — a novel mechanism of gene regulation, *J Biol Chem* in press.

Gates, M. A., *et al.* 1995. Cell and molecular analysis of the developing and adult mouse subventricular zone of the cerebral hemispheres, *J Comp Neurol* 361:249–266.

Graham, F. L. and van der Eb, A. J. 1973. A new technique for the assay of infectivity of human adenovirus 5 DNA, *Virology* 52(2):456–467.

Guillemot, F., 2005. Cellular and molecular control of neurogenesis in the mammalian telencephalon, *Curr Opin Cell Biol* 17(6):639–647.

Hu, Y., *et al.* 2004. 90-kDa ribosomal S6 kinase is a direct target for the nuclear fibroblast growth factor receptor 1 (FGFR1): Role in FGFR1 signaling, *J Biol Chem* 279(28):29325–29335.

Kawamata, T., *et al.* 1997. Intracisternal basic fibroblast growth factor enhances functional recovery and up-regulates the expression of a molecular marker of neuronal sprouting following focal cerebral infarction, *Proc Nat Acad Sci U S A*, 94(15):8179–8184.

Kuhn, H. G., *et al.* 1997. Epidermal growth factor and fibroblast growth factor-2 have different effects on neural progenitors in the adult rat brain, *J Neurosci* 17(15): 5820–5829.

Klejbor, I., *et al.* 2006. Fibroblast growth factor receptor signaling affects development and function of dopamine neurons — inhibition results in a schizophrenia-like syndrome in transgenic mice, *J Neurochem* 97(5):1243–1258.

Klejbor, I., *et al.* 2009. Serotonergic hyperinnervation and effective serotonin blockade in an FGF receptor developmental model of psychosis, *Schizophr Res* 113(2–3):308–321.

Kneuer, C., *et al.* 2000. A nonviral DNA delivery system based on surface modified silica-nanoparticles can efficiently transfect cells *in vitro*, *Bioconjug Chem* 11(6):926–932.

Kneuer, C., *et al.* 2000. Silica nanoparticles modified with aminosilanes as carriers for plasmid DNA, *Int J Pharm* 196(2):257–261.

Kuhn, H. G., Dickinson-Anson, H. and Gage, F. H. 1996. Neurogenesis in the dentate gyrus of the adult rat: Age-related decrease of neuronal progenitor proliferation, *J Neurosci* 16(6):2027–2033.

Lois, C. and Alvarez-Buylla, A. 1994. Long-distance neuronal migration in the adult mammalian brain, *Science* 264(5162):1145–1148.

Leong, K. W., *et al.* 1998. DNA-polycation nanospheres as non-viral gene delivery vehicles, *J Control Release* 53(1–3):183–193.

Lee, K. Y., *et al.* 1998. Preparation of chitosan self-aggregates as a gene delivery system, *J Control Release* 51(2–3):213–220.

Leong, K. W. 1999. Biopolymers-DNA Nanospheres, *Nonviral Vectors for Gene Therapy*, eds. H. Huang L., M-C., Wanger, E. Academic Press.

Lemkine, G. F., *et al.* 2002. Preferential transfection of adult mouse neural stem cells and their immediate progeny *in vivo* with polyethylenimine, *Mol Cell Neurosci* 19(2):165–174.

Luo, D. and Saltzman, W. M. 2000. Synthetic DNA delivery systems, *Nat Biotechnol* 18(1):33–37.

Luo, D. and Saltzman, W. M. 2000. Enhancement of transfection by physical concentration of DNA at the cell surface, *Nat Biotechnol* 18(8):893–895.

Lendahl, U., Zimmerman, L. B. and McKay, R. D. 1990. CNS stem cells express a new class of intermediate filament protein, *Cell* 60(4):585–595.

Myers, J. M., *et al.* 2003. Nuclear trafficking of FGFR1: A role for the transmembrane domain, *J Cell Biochem* 88(6):1273–1291.

Marwick, C. 2003. FDA halts gene therapy trials after leukaemia case in France, *Brit Med J* 326(7382):181.

Mellodew, K., *et al.* 2004. Nestin expression is lost in a neural stem cell line through a mechanism involving the proteasome and Notch signalling, *Brain Res Dev Brain Res* 151(1–2):13–23.

Peng, H., *et al.* 2001. Novel nuclear signaling pathway mediates activation of fibroblast growth factor-2 gene by type 1 and type 2 angiotensin II receptors, *Mol Biol Cell* 12(2):449–462.

Peng, H., *et al.* 2002. Integrative nuclear FGFR1 signaling (INFS) pathway mediates activation of the tyrosine hydroxylase gene by angiotensin II, depolarization and protein kinase C, *J Neurochem* 81(3):506–524.

Prasad, P. N., 2003. *Introduction to Biophotonics*, Hoboken, New Jersey: John Wiley and Sons, Inc.

Reilly, J. F. and Maher, P. A. 2001. Importin beta-mediated nuclear import of fibroblast growth factor receptor: Role in cell proliferation, *J Cell Biol* 152(6):1307–1312.

Ramsay, E., *et al.* 2000. Examination of the biophysical interaction between plasmid DNA and the polycations, polylysine and polyornithine, as a basis for their differential gene transfection *in-vitro*, *Int J Pharm* 210(1–2):97–107.

Roy, I., *et al.* 2003. Calcium phosphate nanoparticles as novel non-viral vectors for targeted gene delivery, *Int J Pharm* 250(1):25–33.

Roessler, B. J. and Davidson, B. L. 1994. Direct plasmid mediated transfection of adult murine brain cells *in vivo* using cationic liposomes, *Neurosci Lett* 167(1–2):5–10.

Rossi, L. M., *et al.* 2005. Stober synthesis of monodispersed luminescent silica nanoparticles for bioanalytical assays, *Langmuir* 21(10):4277–4280.

Roy, I., *et al.* 2005. Optical tracking of organically modified silica nanoparticles as DNA carriers: A nonviral, nanomedicine approach for gene delivery, *Proc Natl Acad Sci U S A* 102(2):279–284.

Stachowiak, M. K., *et al.* 2003. Integrative nuclear FGFR1 signaling (INFS) as a part of a universal "feed-forward-and-gate" signaling module that controls cell growth and differentiation, *J Cell Biochem* 90(4):662–691.

Stachowiak, M. K., Maher, P. A. and Stachowiak, E. K. 2007. Integrative nuclear signaling in cell development — a role for FGF receptor-1, *DNA Cell Biol* 26(12):811–826.

Stachowiak, M. K., *et al.* 1996a. Nuclear localization of functional FGF receptor 1 in human astrocytes suggests a novel mechanism for growth factor action, *Brain Res Mol Brain Res* 38(1):161–165.

Stachowiak, M. K., *et al.* 1996b. Nuclear accumulation of fibroblast growth factor receptors is regulated by multiple signals in adrenal medullary cells, *Mol Biol Cell* 7(8):1299–1317.

Stachowiak, E. K., *et al.* 2003. cAMP-induced differentiation of human neuronal progenitor cells is mediated by nuclear fibroblast growth factor receptor-1 (FGFR1), *J Neurochem* 84(6):1296–1312.

Stachowiak, M. K., Maher, P. A. and Stachowiak, E. K. 2007. Integrative nuclear signaling in cell development–a role for FGF receptor-1, *DNA Cell Biol* 26(12):811–826.

Stachowiak, M. K., *et al.* 1997. Growth factor regulation of cell growth and proliferation in the nervous system. A new intracrine nuclear mechanism, *Mol Neurobiol* 15(3):257–283.

Schwartz, B., *et al.* 1999. Synthetic DNA-compacting peptides derived from human sequence enhance cationic lipid-mediated gene transfer *in vitro* and *in vivo*, *Gene Therapy* 6(2): 282–292.

Sameti, M., *et al.* 2003. Stabilisation by freeze-drying of cationically modified silica nanoparticles for gene delivery, *Int J Pharm* 266(1–2):51–60.

Slowing, II, *et al.* 2008. Mesoporous silica nanoparticles as controlled release drug delivery and gene transfection carriers, *Adv Drug Deliv Rev* 60(11):1278–1288.

Stachowiak, E. K., Yu-Wei Lee, I. R., Capacchietti, M., Aletta, J. M., Prasad, P. N. and Stachowiak, M. K. 2009. Targeting novel integrative nuclear FGFR1 signaling by nanoparticle-mediated gene transfer stimulates neurogenesis in adult brain, *Integr Biol* 1:394–403.

Seri, B., *et al.* 2001. Astrocytes give rise to new neurons in the adult mammalian hippocampus, *J Neurosci* 21(18):7153–7160.

Stachowiak, E. K., Yu-Wei Lee, I. R., Capacchietti, M., Aletta, J. M., Prasad, P. N. and Stachowiak, M. K. 2009. Targeting novel integrative nuclear FGFR1 signaling by nanoparticle-mediated gene transfer stimulates neurogenesis in adult brain, *Integr Biol* in press.

Shi, N. and Pardridge, W. M. 2000. Noninvasive gene targeting to the brain, *Proc Natl Acad Sci U S A* 97(13):7567–7572.

Truong-Le, V. L., *et al.* 1999. Gene transfer by DNA-gelatin nanospheres, *Arch Biochem Biophy* 361(1):47–56.

Wu, G. Y. and Wu, C. H. 1987. Receptor-mediated invitro gene transformation by a soluble DNA carrier system, *J Biol Chem* 262(10):4429–4432.

Xu, G., *et al.* 2008. Bioconjugated quantum rods as targeted probes for efficient transmigration across an *in vitro* blood-brain barrier, *Bioconjug Chem* 19(6):1179–1185.

Zauner, W., Ogris, M. and Wagner, E. 1998. Polylysine-based transfection systems utilizing receptor-mediated delivery, *Adv Drug Deliv Rev* 30(1–3):97–113.

12 Cancer Stem Cells — Targets for Future Cancer Therapies

Andrew Puca, Valentina Malafoglia, and Antonio Giordano

Research using human and animal stem cells is an extremely dynamic area of contemporary biomedical investigation. It has contributed to new awareness about the pathways of normal and abnormal cell differentiation and tissue development. It has also shed a new profound light on cell transplantation therapies for human diseases. Although the availability of a variety of human stem cells is relatively recent, since the isolation of human embryonic stem cells in 1998, much has happened in both publicly and privately funded research centers around the world. One such seminal development has been the identification of cancer stem cells which has improved clinical challenges faced in provisions for new cancer therapies. The identification of cancer stem cells (CSCs) has provided innovative approaches by the ability to improve disease identification, selective drug targeting, and prevention of metastasis. Cancer stem cells, being the mutated counterparts of normal stem cells may also have similar functions, which would allow them to survive various cancer therapies. With the ability to selectively target cancer stem cells, it would be possible to treat patients with aggressive tumors, as well as preventing the tumors from metastasizing. Success in this area would enable advances in personalized medicine in cancer treatments by using the right combination of drugs and/or treatments. This chapter will discuss how evolving stem cells research could potentially improve the future of cancer therapy.

12.1. Stem Cells and Cancer Stem Cells

The adult human body, and all its differentiated cells, tissues, and organs arise from a small cluster of cells enclosed within the early embryo at the blastocyst stage of development. Throughout *in vivo* embryonic development, these cells will divide and differentiate in parallel with each other and with the whole of which they are a part of, eventually producing the specialized and integrated tissues and organs of the body. Embryonic stem cells are capable of becoming many different types of differentiated cells if stimulated to do so *in vitro* (Kerr *et al.*, 2003). Stem cells are a diverse class of cells, which can now be isolated from a variety of embryonic, fetal, and adult tissues. Stem cells share two characteristic properties: (1) unlimited or prolonged self-renewal, which is the capacity to maintain a pool of stem cells like themselves through self-replication, and (2) potency for differentiation, the potential to produce more differentiated cell types usually more than one and, in some cases, many. When stem cells head down the pathway toward differentiation, they usually proceed by first giving rise to a more specialized type of stem cell (sometimes called "precursor cells" or "progenitor cells"), which can, in turn, either proliferate through self-renewal or produce fully specialized or differentiated cells (Mimeault and Batra, 2008).

One major example is hematopoietic stem cell development; the hematopoietic system consists of a network of stem and progenitor cells of varying degrees of maturity interacting with other cells acting with supportive and regulatory capacities. Many cell types have been identified by their *in vivo* or *ex vivo* growth potential, i.e. colony assays, or by their physical characteristics, such as cell surface markers identified by monoclonal antibodies (Lodie *et al.*, 2002).

These studies indicate that primitive hematopoietic stem cells express a wide variety of genes, including a number of transcription factors. Presumptively, transcription factor access to DNA is a major determinant of such gene expression; therefore, several specific genes to fetal liver hematopoietic cells have been identified. In addition, when comparing fetal hematopoietic cells with adult hematopoietic cells, they found several genes that were coexpressed on fetal and adult stem cells, as well as genes specific for either fetal or adult stem cells (Mao *et al.*, 2001). The lessons from these and other stem cells may be important for our understanding and therapeutic targeting of the new class of stem cells that are associated with cancer.

12.1.1. *Cancer stem cells*

Rapid advances in stem cell research and discovery of the cancer stem cell (CSC) field have provided cause for optimism for the development of more reliable cancer therapies in the future. Strategies aimed at efficient targeting of CSCs are becoming important for monitoring the progress of cancer therapy and for evaluating new therapeutic approaches. CSCs are a subpopulation of cancer cells (found within solid tumors as well as hematological cancers) that possess characteristics normally associated with stem cells. These cells are believed to be cancer forming, in contrast to the bulk of cancer cells, which are thought to be non-cancer–forming. CSCs have stem cell properties such as self-renewal and the ability to differentiate into multiple cell types. The hypothesis suggests such cells persist in tumors as a distinct population and cause relapse and metastasis by giving rise to new tumors. Development of specific therapies targeted at cancer stem cells holds hope for the improvement of survival and quality of life of cancer patients, especially for sufferers of metastasis disease. Although this is still the subject of considerable debate, it is possible that the development of CSCs might involve at least two of the following events: (1) a change in the microenvironment of the stem cell niche within a tissue; (2) alterations in cellular metabolism, cell cycle control and/or progression and signaling pathways as a result of mutations and epigenetic changes; and (3) amplification of cell populations with an altered molecular phenotype that give rise to heterogeneous primary tumors and metastases (Klonisch *et al.*, 2008).

12.2. Identification of Cancer Stem Cells: Hematopoietic & Solid Tumors

Transmembrane markers have been used for several years to identify specific populations of cell types, as well as to deduce stem cell subtypes for differentiation assays. The accessibility of cell surface markers allowing the isolation of cells at distinct stages of hematopoietic development, including hematopoietic stem cells (HSCs) and, combined with particular and quantitative assays, greatly facilitated the identification of acute myeloid leukemia (AML) stem cells. For instance, it was found that a population of cells with the phenotype $CD34^+CD38^-$ and $CD90^{-6}$, which represented 0.1–1% of the total cells, were the only cells able to initiate AML that resembled the human disease in immunocompromised (NOD/SCID) mice. Further analysis of *in vitro* assays allowed identifying lineage panel marker negative fraction (Lin^-) in CSCs even though $CD34^+38^-90^+Lin^-$ HSCs contains the AML1/ETO translocation in non-leukemic, or pre-leukemic, HSC (Miyamoto *et al.*, 2000). The $CD34^+38^-90^-CD45RA^-Lin^-$ population has recently been identified in normal human hematopoietic tissues as a multipotent progenitor, but not a HSC (Blair *et al.*, 1997). Therefore the phenotypic expression of AML stem cells indicates that they probably originate in the hematopoietic stem/progenitor cell compartment, perhaps originating from a multipotent progenitor. Other studies have confirmed when AML progressed to blast crisis, the progression step is linked to the activation of a self-renewal program in the granulocyte-macrophage progenitor population, thus indicating cells that normally cannot self-renew have CSC properties. While CSCs may derive from normal stem cells, this may not be the case in all tumors. More differentiated cells could acquire self-renewal ability through multiple mutagenic events, leading them to become the CSCs, thus driving tumor growth (Klonisch *et al.*, 2008).

Identification of CSCs from patients is obtained by staining with labeled antibodies to various cell surface markers, either individually or in combination, and techniques such as magnetic columns or fluorescence-activated cell sorting (FACS) are used to separate labeled vs. unlabeled populations of cells. Populations are then assayed for their ability to initiate tumor growth in immunocompromised mice. Immunostaining and flow cytometric analysis of the engrafted tumor must then show recapitulation of all original populations that were in the primary tumor, and upon resorting of the CSC phenotype and retransplantation, a fully heterogeneous tumor must again begin to grow. In the case of solid tumors, the challenge of identifying and characterizing CSCs is significant. Unlike the hematopoietic system, the normal tissue developmental hierarchy has not been extensively characterized. This makes the assortment of candidate markers more intricate, and provides no basis for assessment between normal and malignant cells, or means to categorize the cell of origin. *In vivo* models of human solid tumors can also be difficult to achieve, as in some cases where orthotopic injection is not practical and is technically challenging (e.g. lung, colon, and bladder cancers).

There are recent compelling studies in which alleged CSCs have been identified based on their ability to form colonies *in vitro*. Sphere-forming ability as a measure of stem cells was first developed for central nervous system (CNS) cells, where it has been shown that a subset of cells isolated from human fetal brain, and consequently from human CNS tumors, can form spheres when cultured under the appropriate conditions (Clarke *et al.*, 2006; Uchida *et al.*, 2000; Singh *et al.*, 2003). These spheres self-renew *in vitro*, and differentiate into all of the neuronal lineages, both *in vitro* and *in vivo*. Importantly, it was subsequently demonstrated for brain tumors that the neurosphere-forming cells could be prospectively isolated from fresh tissue using the cell surface marker CD133, and that the $CD133^+$ cells did indeed initiate brain tumors *in vivo*; with no *in vitro* manipulation, this indicates that they do in fact represent CSCs (Singh *et al.*, 2003). Upon culturing of metastatic melanoma cell suspensions under suitable conditions, a subset of cells could be propagated as non-adherent spheres, which could then be induced to differentiate *in vitro* and to generate tumors *in vivo*. However, a marker has not definitively been recognized to allow the potential isolation of melanoma sphere-forming cells from fresh tumors. The ability to acutely isolate and assay subpopulations of cells from tumors behaving as CSCs is essential before performing characterizations such as gene expression profiling, to avoid artifacts introduced by culturing cells for extended periods of time. Another tumor for which CSCs have been potentially identified based on *in vitro* assays is prostate cancer. Cells that are $CD44^{+}\alpha 2\beta 1^{+}$ $CD133^{+}$ possess a significant capacity for self-renewal, differentiation, and invasiveness *in vitro* (Collins *et al.*, 2005). In this case, it also remains to be seen whether these cells can reproduce the original tumor *in vivo*.

12.3. Innovative Cancer Stem Cells Therapies

In recent years, CSCs have been recognized as important components in carcinogenesis, and they could form the basis of many (if not all) tumor types. CSCs have been isolated from cancers of the breast, brain, blood (leukemia), skin (melanoma), head and neck, thyroid, cervix, lung, organs of the gastrointestinal and reproductive tract and retina. The identification of CSC-specific marker sets and the targeted therapeutic destruction of CSCs remain a challenge. Current anticancer therapies mostly fail to eradicate CSC clones and instead favor expansion of the CSC pool and/or select for resistant CSC clones, thus leading to a fatal outcome of the disease. Despite the recent surge of interest in CSCs, the clinical significance of this population remains unclear. One prediction of the CSC model is that clinical behavior should be largely dependent on the CSC population, either in quantitative terms (such as the relative or absolute number of CSCs), or qualitative aspects related to biological features of CSCs. To date, there is little data addressing this question, although a recent study suggests that in AML, a higher percentage of blasts with the $CD34^{+}CD38^{-}$ LSC phenotype are correlated with poorer overall survival (van Rhenen *et al.*, 2005). Clearly, more studies are needed to determine the prognostic significance of CSCs, and these data will be important in demonstrating that this experimentally defined population is relevant in humans.

In addition to prognosis, the CSC hypothesis has profound implications for cancer treatment strategies. With growing evidence that cancers arise from stem cells, both researchers and clinical investigators will increasingly focus their efforts on CSCs rather than studying whole tumors. The potential for CSC-specific therapy is currently an active area of investigation, and a recent study using a mouse AML model suggests that LSCs may be selectively targeted without adverse effects on normal HSCs (Yilmaz *et al.*, 2006). Interestingly, normal HSCs express high levels of drug efflux transporters, and defining whether these transporters are present in leukemic or other CSCs will be a critical step in understanding the mechanisms of tumor drug resistance. Clearly, targeting CSCs offers the potential for improvements in therapy, and future studies are likely to include the identification of CSC-specific surface markers for antibody therapy, elucidation of CSC-specific pathways that can be pharmacologically targeted and evaluation of agents that promote the differentiation of CSCs into progenitors that do not self-renew (Figure 12.1).

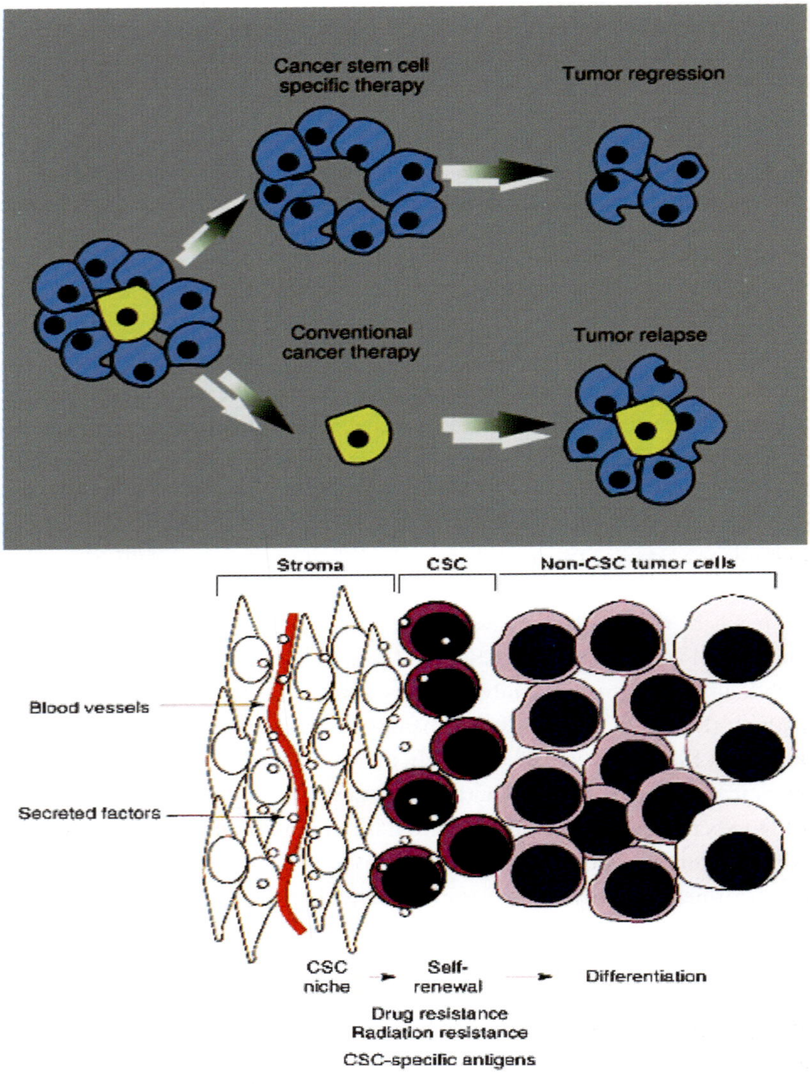

Figure 12.1. Targeting CSCs offers the potential for improvements in therapy, and future studies are likely to include the identification of CSC-specific surface markers for antibody therapy, elucidation of CSC-specific pathways that can be pharmacologically targeted and evaluation of agents that promote the differentiation of CSCs into progenitors that do not self-renew.

While the CSC hypothesis has exciting clinical implications, its widespread acceptance and application to the practice of medicine has yet to occur. It is anticipated that in the upcoming years, CSCs will be identified in additional tumor types, and knowledge of the detailed biology and clinical significance of this experimentally defined population will provide further support for the CSC hypothesis. Ultimately, focusing research efforts on the CSC may drive important advances in our understanding of cancer biology and developing potential cures for these devastating diseases.

12.4. Regulatory Pathways and Therapeutic Challenges

Regulatory pathways and proteins ensure a balance between self-renewal and apoptosis. Especially important proteins are Sonic hedgehog shh, Wnt, Notch, PTEN, Bmi-1, and Bcl- The knowledge taken from stem cells led to investigations of the same pathways in tumor stem cells. It is interesting to note that disruption of these pathways or the loss or overexpression of those proteins play a role in many tumors, and may be vital for future therapeutic implications.

12.4.1. *Wnt pathway*

The Wnt protein is a paracrine messenger starting a signaling cascade involved in cell proliferation and differentiation. Mutations leading to overexpression of the protein or a disturbance within the Wnt signaling cascade (APC/axin complex) can promote tumorigenesis. This has been shown in colorectal cancer, breast cancer, ovarian cancer, CNS and skin tumors, leukemia, and prostate cancer (Reya and Clevers, 2005). Dysregulation of the Wnt pathway-related E-cadherin can also lead to tumorigenesis and metastasis (Menke *et al.*, 2001).

12.4.2. *Notch pathway*

The Notch pathway is a juxtacrine-signaling cascade involved in embryogenesis and the regulation of differentiation, proliferation, and self-renewal in adult stem cells. Membrane-bound ligands (δ-like−1,−3,−4, and Jagged-1 and -2) induce cleavage of the intracellular domain of the Notch receptors (Notch1–4) by the γ-secretase complex. After association of the intracellular part of Notch (icN) with the transcription factor CSL and additional proteins (Mastermind-like−1,−2,−3), a multiprotein complex is formed in the nucleus leading to the transcription of several target genes. Depending on the tissue and context, activation of the Notch pathway can lead either to an accumulation of immature progenitor cells or to terminal differentiation. The role of dysregulation of the pathway in tumorigenesis must therefore always be viewed in the context of the specific tissue. In the mouse, it regulates differentiation of epidermal cells; Notch ablation leads to hyperplasia and basal cell carcinoma. A deregulation of the Notch signaling cascade by overexpression of Jagged-1 has been found in advanced prostate cancers (Santagata *et al.*, 2004). In clear cell renal cell carcinomas, overexpression of Notch1 and Jagged-1 could be shown *in vitro* and *in vivo*. By blocking the Notch signaling cascade, a significant inhibition of tumor growth could be achieved in a mouse model (Sjolund *et al.*, 2008).

12.4.3. *PTEN pathway*

The PTEN tumor suppressor is involved in cell cycle regulation in almost all tissues. By inhibiting uncontrolled growth and cell division, it can inhibit tumorigenesis. It is also involved in cell migration, cell adhesion, and angiogenesis. Growth factors activate phosphatidylinositol 3-kinase (PI 3-kinase), leading to synthesis of the second messenger phosphatidylinositol (3,4,5)-triphosphate (PIP3). The phosphatase PTEN catalyzes the opposite reaction, dephosphorylating PIP3 to yield PIP2. PIP3 can activate Akt via the protein kinase PDK1, the Akt activation in turn lead to the inactivation of pro-apoptotic proteins and to increased cell proliferation. Thus, the loss of PTEN function results in an increased concentration of PIP3 and in Akt hyperactivation, leads to the protection from various apoptotic stimuli (Di Cristofano and Pandolfi, 2000). PTEN is also connected with various other tumor-relevant signaling pathways. PTEN-inactivating mutations have been found in various tumors, including endometrial carcinoma, glioblastoma, astrocytoma, melanoma, and in bladder and kidney cancer (Cantley and Neel, 1999). In bladder cancer, a correlation between the decrease of PTEN protein expression and tumor stage and grade was shown (Tsuruta *et al.*, 2006). In advanced renal cell carcinoma, PTEN expression is frequently reduced (Shin Lee *et al.*, 2003). PTEN seems to be the most frequently mutated gene in prostate cancer and is predominantly associated with aggressive and advanced tumors. Additionally, an association between androgen-independent tumor growth and a PTEN mutation has been shown. While PTEN mutations lead to a predisposition for prostate cancer in mouse models, such a connection could not yet be shown for humans (Dong, 2001; Chow *et al.*, 2006). PTEN loss-of-function mutations are also often associated with resistance to conventional chemotherapy by activating MDM2-mediated anti-apoptotic mechanisms (Zhou *et al.*, 2003).

12.4.4. *Polycomb family*

Bmi-1 is a member of the Polycomb family of transcriptional repressors involved in accurate cell differentiation, proliferation, and self-renewal. Overexpression of Bmi-1 leads to repression of INK4a (p16), resulting in an inactivation of Rb. This prompts the cell to enter the cell cycle and proliferate. At the same time, Arf is repressed, leading to a reduction of p53-mediated apoptosis (Park *et al.*, 2004). Deregulation of the Polycomb protein complex by mutation of Bmi-1 leads to increased tumor cell proliferation. Leukemic stem cells from Bmi-$1^{+/+}$ mice could, in contrast to leukemic stem cells of Bmi-1 knockout animals, induce leukemia in host mice (Lessard and Sauvageau, 2003).

12.4.5. *Pluripotency pathway*

The self-renewal of stem cells is maintained through the promotion of proliferation, the prevention of differentiation and is enabled by a self-organizing network of three transcription factors (Oct3/4, Sox2 and Nanog), which form positive feedback loops (Chapter 5). The overexpression of Oct3/4 and Sox2, plus two additional factors KlF4 and c-Myc, were found sufficient to establish pluripotency in fibroblast nuclei and offers a promising technological advance now widely replicated in many laboratories. Whether disabling this pathway in cancer cells could block cancer growth remains to be established. If so, the development of drugs that can block the self-organizing pluripotency network may hold promise for future cancer therapy.

12.4.6. *INFS pathway*

One potential target for the anti-cancer therapy could be the novel developmental gene-controlling mechanism: integrative nuclear fibroblast growth factor receptor (FGFR)-1 signaling (INFS) described in Chapter 4. Enhancement of the INFS may be used to re-establish coordinated gene regulation in order to interrupt cancer stem cell proliferation and promote differentiation. By activating the INFS mechanism using DNA transfection, a blockade of cancer cell proliferation and stimulation of differentiation have been achieved *in vitro*. Development of small molecules that may effectively activate the INFS module could lead to a new class of anti-cancer drugs.

12.5. Summary

Despite the recent surge of interest in CSCs, the clinical significance of this population remains unclear. One prediction of the CSC model is that clinical behavior should be largely dependent on the CSC population, either in quantitative terms such as the relative or absolute number of CSCs, or qualitative aspects related to biologic features of CSCs. There is little data addressing this question, although studies suggests that in AML, a higher percentage of blasts with the $CD34^{+}CD38^{-}$ LSC phenotype is correlated with inferior overall survival (van Rhenen *et al.*, 2005). Clearly, more studies are needed to determine the prognostic implications of CSCs; this data will be important in representing that this experimentally distinct population is relevant in humans.

In addition to prognosis, the CSC hypothesis has profound implications for cancer treatment strategies. With growing evidence that cancers arise from stem cells, both researchers and clinical investigators will increasingly focus their efforts on CSCs rather than studying the tumor as a whole. The potential for CSC-specific therapy is currently a very active area of research. AML model using mouse studies suggests that LSCs may be selectively targeted without adverse effects on normal HSCs (Yilmaz *et al.*, 2006). Interestingly, normal HSCs express high levels of drug efflux transporters, and defining whether these transporters are present in leukemic or other CSCs is a vital and critical step in understanding the mechanisms of tumor drug resistance. Targeting CSCs offers the potential for improvements in therapy, and future studies are likely to include the identification of CSC-specific surface markers for antibody therapy, elucidation of CSC-specific pathways that can be pharmacologically targeted and evaluate the factors that promote differentiation of CSCs into progenitors which do not self-renew.

While the CSC hypothesis has stimulating clinical implications, its widespread recognition as well as its application to the practice of medicine has not occurred. The future of CSC therapies is to identify populations in various tumor types, and understanding of the detailed biology and clinical significance of this experimentally defined population will support the CSC hypothesis. Ultimately, further progress in stem cells research will drive important advances in our understanding of cancer biology and developing potential cures.

Bibliography

Blair, A., Hogge, D. E., Ailles, L. E., Lansdorp, P. M., Sutherland, H. J. 1997. Lack of expression of Thy-1 (CD90) on acute myeloid leukemia cells with long-term proliferative ability *in vitro* and *in vivo*. *Blood* 89:3104–3112.

Cantley, L. C., Neel, B. G. 1999. New insights into tumor suppression: PTEN suppresses tumor formation by restraining the phosphoinositide 3-kinase/AKT pathway. *Proc Natl Acad Sci U S A* 96:4240–4245.

Chow, L. M., Baker, S. J. 2006. PTEN function in normal and neoplastic growth. *Cancer Lett* 241:184–196.

Clarke, M. F., Dick, J. E., Dirks, P. B., *et al.* 2006. Cancer stem cells — perspectives on current status and future directions: AACR Workshop on cancer stem cells. *Cancer Res* 66:9339–9344.

Collins, A. T., Berry, P. A., Hyde, C., Stower, M. J. and Maitland, N. J. 2005. Prospective identification of tumorigenic prostate cancer stem cells. *Cancer Res* 65:10946–10951.

Di Cristofano, A. and Pandolfi, P. P. 2000. The multiple roles of PTEN in tumor suppression. *Cell* 100:387–390.

Dong, J. T. 2001. Chromosomal deletions and tumor suppressor genes in prostate cancer. *Cancer Metastasis Rev* 20:173–193.

Kerr D. A., J. Llado, M. J. Shamblott, *et al.* 2003. Human embryonic germ cell derivatives facilitate motor recovery of rats with diffuse motor neuron injury. *J Neurosci* 23: 5131–5140.

Klonisch, T., Wiechec, E., Hombach-Klonisch, S., *et al.* 2008. Cancer stem cell markers in common cancers — therapeutic implications. *Trends Mol Med.*

Lessard, J. and Sauvageau, G. 2003. Bmi-1 determines the proliferative capacity of normal and leukaemic stem cells. *Nature* 423:255–260.

Lodie, T. A., Blickarz, C. E., Devarakonda, T. J., *et al.* 2002. Systematic analysis of reportedly distinct populations of multipotent bone marrow-derived stem cells reveals a lack of distinction. *Tissue Eng* 8:739–751.

Mao, J., Wang, J., Liu, B., *et al.* 2001. Low-density lipoprotein receptor-related protein-5 binds to Axin and regulates the canonical Wnt signaling pathway. *Mol Cell* 7:801–809.

Menke, A., Philippi, C., Vogelmann, R., *et al.* 2001. Down-regulation of E-cadherin gene expression by collagen type I and type III in pancreatic cancer cell lines. *Cancer Res* 61:3508–3517.

Mimeault, M and Batra S. K. 2004. Recent progress on tissue-resident adult stem cell biology and their therapeutic implications. *Stem Cell Rev* 4:27–49.

Miyamoto, T., Weissman, I. L., Akashi, K. 2000. AML1/ETO-expressing nonleukemic stem cells in acute myelogenous leukemia with 8;21 chromosomal translocation. *Proc Natl Acad Sci U S A* 97:7521–7526.

Park, I. K., Morrison, S. J., Clarke, M. F. Bmi1, stem cells, and senescence regulation. *J Clin Invest* 113:175–179.

Reya T and Clevers, H. 2005. Wnt signalling in stem cells and cancer. *Nature* 434:843–850.

Santagata, S., Demichelis, F., Riva, A., *et al.* 2004. JAGGED1 expression is associated with prostate cancer metastasis and recurrence. *Cancer Res* 64:6854–6857.

Shin Lee, J., Seok Kim, H., Bok Kim, Y., Cheol Lee, M. and Soo Park, C. 2003. Expression of PTEN in renal cell carcinoma and its relation to tumor behavior and growth. *J Surg Oncol* 84:166–172.

Singh, S. K., Clarke, I. D., Terasaki, M., *et al.* 2003. Identification of a cancer stem cell in human brain tumors. *Cancer Res* 63:5821–5828.

Sjolund, J., Johansson, M., Manna, S., *et al.* 2008. Suppression of renal cell carcinoma growth by inhibition of Notch signaling *in vitro* and *in vivo. J Clin Invest* 118:217–228.

Tsuruta, H., Kishimoto, H., Sasaki, T., *et al.* 2006. Hyperplasia and carcinomas in Pten-deficient mice and reduced PTEN protein in human bladder cancer patients. *Cancer Res* 66:8389–8396.

Uchida, N., Buck, D. W., He, D., *et al.* 2000. Direct isolation of human central nervous system stem cells. *Proc Natl Acad Sci USA* 97:14720–14725.

van Rhenen, A., Feller, N., Kelder, A., *et al.* 2005. High stem cell frequency in acute myeloid leukemia at diagnosis predicts high minimal residual disease and poor survival. *Clin Cancer Res* 11:6520–6527.

Yilmaz, O. H., Valdez, R., Theisen, B. K., *et al.* 2006. Pten dependence distinguishes haematopoietic stem cells from leukaemia-initiating cells. *Nature* 441:475–482.

Zhou, M., Gu, L., Findley, H. W., Jiang, R., Woods, W. G. 2003. PTEN reverses MDM2-mediated chemotherapy resistance by interacting with p53 in acute lymphoblastic leukemia cells. *Cancer Res* 63:6357–6362.

Index